MW00852510

Complete Building Construction

revised by Eugene Leger

An Audel® Book

Macmillan Publishing Company
New York

Maxwell Macmillan Canada
Toronto

Maxwell Macmillan International
New York   Oxford   Singapore   Sydney

FOURTH EDITION

Copyright © 1978 by Howard W. Sams & Co., Inc.
Copyright © 1983 by The Bobbs-Merrill Co., Inc.
Copyright © 1986, 1993 by Macmillan Publishing Company, a division of Macmillan, Inc.

Macmillan Publishing Company
866 Third Avenue
New York, NY 10022

Maxwell Macmillan Canada, Inc.
1200 Eglinton Avenue East, Suite 200
Don Mills, Ontario M3C 3N1

Macmillan Publishing Company is part of the Maxwell Communication Group of Companies.

Production services by the Walsh Group, Yarmouth, ME.

Library of Congress Cataloging-in-Publication Data

Complete building construction.—4th ed. / rev. by Eugene Leger
    p.   cm.
  Updated ed. of: Complete building construction / edited by John Phelps. 2nd ed. c 1986.
  "An Audel book."
  Includes index.
  ISBN 0–02–517882–2
  1. Building—Amateurs' manuals. I. Leger, Eugene (Eugene H.) II. Phelps, John, 1932–   Complete building construction.
TH148.C648 1993
690—dc20   92–5338
                                      CIP

10 9 8 7 6 5 4 3 2 1
Printed in the United States of America

Contents

Preface

The fourth edition of *Complete Building Construction* may well be considered a new book. It contains seven new chapters. Eleven of the old chapters have been substantially revised, expanded, or both, to treat new developments. Some earlier material has been retained with slight modifications. Many of these changes are a result of environmental concerns, energy conservation, and increasing building and operating costs.

The book's objective is to provide in a single volume a compendium of the best of current building design and construction practices, and information that is most useful to those who must decide which building materials and what construction methods to use. The emphasis is on the why of construction.

Chapter 1, *Location of Structure on Site,* deals with subjects rarely treated in books on residential construction: corner lots, nonconforming (grandfathered) lots, covenants, and how these affect the location of a house on a lot. Zoning and setbacks are also covered.

Chapter 3, *Foundations,* covers in detail different types, methods of construction, and design of footings. The Air Freezing Index, Degree Days and capillary break are topics new to this edition. The controversial issue of crawl space ventilation is discussed in detail, and current research is introduced. Waterproofing, dampproofing, as well as ground water management, are thoroughly covered. Coverage of Permanent Wood Foundations and Frost-Protected Shallow Foundations, common in Scandinavia, but rare in the United States, is included.

The new chapter on concrete, Chapter 4, provides the builder with a good working knowledge of the properties and types of concrete and cements, and their proper handling and placement. The controversial issue of the right and wrong use of welded wire fabric is discussed.

There is comprehensive coverage of plywoods, strandboards, hardboard, engineered lumber, and I-section joists in Chapter 7. Adhesives, nails, and floor gluing are discussed in depth, as is exterior wall framing in Chapters 10 and 11.

Chapter 16, **Windows,** has been rewritten to include low-E glass, heat mirror, switchable glazings and Aerogels. A new chapter on insulation, Chapter 17, is comprehensive, includes reflective insulations, and the latest findings on reflective barriers for use in attics. The new insulation, cotton, is also covered.

Chapter 18, **Interior Walls and Ceilings,** covers two products new to the United States—Gypsonite and FiberBond. **Attic ventilation**—the arguments, for and against, the many types—is examined thoroughly in Chapter 23. A new chapter has been added on **Radon**—Chapter 24.

It is hoped that you will find this new edition even more useful than the previous ones, and that the information will help you to build more cost-effective, energy-efficient structures.

EUGENE LEGER

CHAPTER I

Location of Structure on Site

Basic Conditions

There are a number of conditions that determine what kind of building may be erected, and where on the lot it may be located:

1. Covenants,
2. Zoning Ordinances,
3. Well location,
4. Septic system location,
5. Corner lots,
6. Nonconforming lots,
7. Natural grades and contours.

Covenants

Covenants are legally binding regulations that may, for example, limit the minimum size of a house, prohibit utility buildings, or ban rooftop TV antennas. Because covenants are *private agreements,* they are not enforceable by local government. A lot may be zoned for duplexes but the covenants may allow *only* single family residences. When buying lots, check the deed or with the city building department to see if there are covenants.

1

Zoning Ordinances

Zoning regulates how much of the site may be occupied by a building, restricts the minimum size of a dwelling, limits its height, and establishes *setbacks*. These are the minimum distances permitted between a building and the property lines around it. Because setbacks can vary according to soil conditions, confirm setback requirements with local Zoning Administrator.

Wells and Septic Systems

Building lots requiring a septic system and well can make locating the house difficult. An Approved Septic System Design shows the location of proposed house, well, septic system, and the required safety zone distances. A typical safety zone may require that a house with footing drains be located 25 feet from the septic tank, 35 feet from the leach field, and the well 75 feet from the septic tank. If footing drains are not required, the house can be 5 feet and 10 feet respectively from the tank and leach field. Well distance is constant. A buyer of a lot with Approved Septic Designs may not like the location of the house, and want it changed. Lot size, shape, natural grades, contours, and safety zone requirements may not allow moving the house. If safety zone distances can be maintained, house relocation may be approved. Otherwise, another soil percolation (perk test) must be performed, and a new design submitted to the state for approval. The well can be relocated, but some zoning ordinances do not allow it in the front yard setback.

Corner Lots

Corner lots front two streets. They have two front yard setbacks, a rear yard setback, but no side yard setback. Which street the house faces is the builder's or buyer's decision, but local government Subdivision Regulations may prohibit two driveways. Because of the two front yard setbacks, the lot area within the setbacks is somewhat reduced. If a septic system and well are required, fitting all of this on a smaller lot is tricky, and will be more difficult if wells and swimming pools are prohibited in the front yard setback. Many states limit how close the leach field can be to the property line. Local ordinances also may prevent locating the leach field between the side yard setbacks and the property line.

Designing the septic system requires digging deep hole test pits to examine the soil at various depths to locate the Seasonable High Water Table (SHWT), the presence or absence of water, ledge, stumps, or debris, and to obtain a soil profile. This information is recorded on the design plan. These data tell how expensive excavation may be, and how far down the bottom of the basement should be. On lots with town sewers, dig test holes 8 feet to 10 feet deep where the house will set, to find the depth of water table, and if ledge is present.

Nonconforming Lots

Nonconforming (grandfathered) lots are those whose area, frontage, depth, or setbacks do not conform to present zoning ordinances. Getting a building permit may be difficult. As with smaller corner lots, trying to fit house, well, septic system, and safety zone within the setbacks can be very demanding, if not impossible.

Natural Grades and Contours

Natural grades and contours also affect location of septic system, house, well, driveways. Is the lot on a hill, on flat land or in a valley? What are the soil types and how do they affect site use? Is the soil-bearing capacity adequate for the proposed construction? Heavily treed lots are a mixed blessing. Trees provide shade on the south and west, and act as a buffer on the north. After the site is cleared of trees, where does one dispose of the stumps? If the local dump will not accept them, who does? Will local conservation commission allow them to be buried on the lot? If the house is built on raised fill, what effect will this have on drainage of water toward abutter's property? Are there stagnant ponds, marshes, or other breeding sources of mosquitoes? If wetlands exist, is enough land left for building after subtracting the wetlands area from the total lot area?

Staking Out House Location

With site analysis completed and a specific location chosen, the next step is to locate each corner, and lay out the building lines. Staking a building on a level rectangular lot is simple. On a sloping, odd-shaped lot it is more difficult. In both cases accuracy is important. There are two methods of staking out the house location:

1. Measuring from a known reference line,
2. Using a transit-level.

Staking Out from a Known Reference Line

When a building is to be erected parallel to the property line, the property line is a known, identifiable line. The property line becomes the *reference* point and makes a builder's level unnecessary. First, make certain that corner markers or monuments, usually granite in the front, and iron pipes or pins in the rear, are in place. If markers are missing, call the surveyor. From the plot plan (Fig. 1–1) find the set-back distances. To stake out proceed as follows:

Caution—Taping is more difficult than it seems to be. The distances to be measured are *horizontal*, not sloped distances. If the lot is sloped and you are downhill from the reference marker, use plumb bobs and hand levels to keep the tape level. If the ground is fairly level, lay the tape on the ground rather than supporting each end. On sloping lots, pull hard on the tape to remove most of the sags. In this instance a steel tape is best.

1. Prepare ten or more 3-foot long stakes by drawing diagonals on the flat head to locate the center, and drive a nail where the lines cross.
2. Locate the right rear property marker D. Measure 45'-0" from D toward the front granite marker B. This is the rear yard setback distance. Drive a stake. This stake is marked *E1* in Fig. 1–2.
3. Locate the left rear property marker . Measure 45'-0" from C toward the front granite marker A. This stake is marked *E2*.
4. Stretch a line tightly across the lot between stakes E1 and E2 to locate the rear yard setback line (Fig. 1–2). Next, the two rear cor-

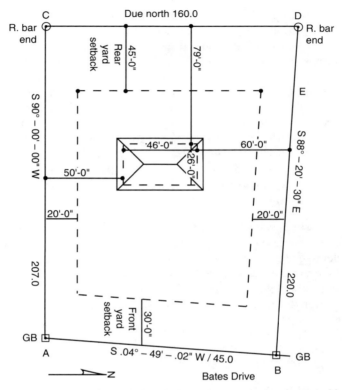

Fig. 1–1. Plot plan showing property lines and corner markers, located and identified, house location, and setback lines.

ners of the house must be located. The plan shows the house is 34 feet from the rear yard set back line. From the left stake *E2* measure 34′-0″ toward the front the front and drive a stake, *F2*. From the right stake *E1* measure 34′-0″ toward the front and drive a stake, *F1*. Consult the plot plan to see how far in the house corners will be from the left and right property lines.

5. From the left stake *F2* measure in 50′-0″ to the right, and drive a stake. This is the *left rear corner* of the house. From the right stake *F1* measure in 60′-0″ to the left, and drive a stake. This is the *right rear corner* of the house. The distance between these two stakes is the length of the rear of the building. Confirm that this distance, 46'-0", agrees with the length given on the plot plan (Fig. 1–1).

6. Get the depth of the house from the plot plan. From the left rear corner stake measure 26'-0" toward the front yard, and drive a stake. This is the *left front corner* of the house. From the right rear corner stake measure 26'-0" toward the front yard, and drive a stake. This is the *right front corner* of the house. The distance between these two stake is the length of the front of the building. Confirm that it agrees with the length shown on the plot plan. If the property lines form a 90-degree angle at the corners, the left and right sides of the building should be parallel with the left and right property lines. The front and rear lengths should be parallel with the front and rear property lines.

On a nonrectangular lot where the corners do not form a 90-degree angle, this method will not work because the building lines will not be parallel to the property lines. The setback lines should be staked out, and the corner of the building closest to the property line, but within the setback, should be located. The building should be staked out from this point, with a dumpy level or transit level, using the method described under *Offset Stakes.*

Laying Out with a Transit-Level

There are two types of surveyor's levels in common use: the automatic optical level, also known as a *dumpy level* or *builder's level* (Fig. 1–3), and the *transit-level* (Fig. 1–4). The optical level is fixed horizontally and cannot be used to measure angles. The transit-level can be moved horizontally or vertically, and can be used to measure vertical angles, run straight lines, and determine whether a column, building corner, or any vertical structure is plumb. The *laser level* (Fig. 1–5), common in commercial construction, is slowly replacing the transit-level in residential construction.

To lay out the building (Fig. 1–1) using transit-level (Fig. 1–2) a reference point, or *bench mark,* is needed, and the rear right corner marker serves this purpose.

Caution—When setting up the transit over a marker on a slope, put two of the tripod legs on the downhill side, and the other leg on the uphill side. Locate the top of the tripod as close as possible to the marker.

1. Level and plumb the transit over marker *D*. Sight down to the opposite corner marker *B*.
2. The rear yard setback is 45′-0″. Measure 45′-0″ from marker *D*. Take one of the previously prepared stakes, align the 45′-0″ mark on the tape measure with the center of the stake. Release the tran-

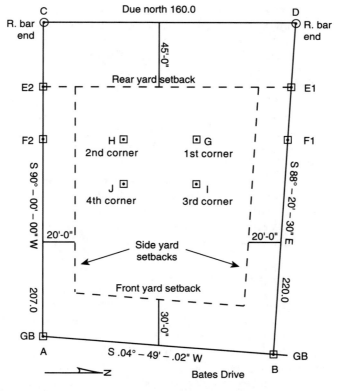

Fig. 1–2. Steps 1 to 8. Laying out with a transit-level.

sit telescope, and lower it until the cross hairs, the nail in the center of the stake, and the 45'-0" mark agree. This is point *E*.

3. The house is 34 feet from the rear setback line. From point *E1* measure 34'-0". While holding the tape 34'-0" mark at the nail in the center of the stake, raise the telescope until the cross hairs are exactly on the 34'-0" tape mark, and drive the stake. This is point *F1*.

4. Move the transit to mark *F1*, level and plumb it, and sight back on marker *B*. Now turn the telescope 90 degrees to the right.

5. The distance from the property line (Fig. 1–1) to the right side of the building is 60'-0". From mark *F1* measure 60'-0", drive a stake. Lower the telescope until the horizontal cross hair is on the 60'-0" mark on the tape. This is the *first* corner of the building, and is point *G*.

6. Move the transit to point *G*, level and plumb it. Measure 46'-0" from point *G*. This is the length of the building. Now raise the telescope until the horizontal cross hair coincides with the 46'-0" mark on the tape. Align the center of the stake with the 46'-0" mark on the tape, and drive the stake. Point *H* has been located, and is the *second* corner of the building.

7. With the transit still over point *G*, turn it 90 degrees to the left. Measure 26'-0" from point *G*. Lower the telescope until the horizontal cross hair is on the 26'-0" mark on the tape. Align the nail with the 26'-0" tape mark, and drive the stake. Point *I* is established and is the *third* corner of the building.

8. Level and plumb the transit over point *I*, and sight back to point *G*. Rotate the telescope 90 degrees to the left. From point *I* measure 46'-0". Lower the telescope until the horizontal cross hair in on the 46'-0" mark on the tape. Align the center of the stake with the 46'-0" mark on the tape, and drive the stake. This, the *fourth* and final corner of the building, is point *J*.

Batter Boards and Offset Stakes

Now that the building corners have been established, building lines must be set up to mark the boundaries of the building. Batter boards are used to mark permanently the excavation and foundation lines. The forms for the foundation walls will be set to these building lines. The batter boards should be installed 4 to 6 feet back from the building corner stakes. Suspend a plumb bob over the building corner stakes to locate exactly the lines over the corner stakes. When all the

Fig. 1–3. Automatic level. *(Courtesy The Lietz Company)*

Fig. 1–4. Transit-level. *(Courtesy The Lietz Company)*

Fig. 1–5. Laser level. *(Courtesy The Lietz Company)*

building lines are in place, be sure that the measurements between the lines agree with the measurements shown on the blueprints. Measure the two diagonals of the batter board lines to be sure that the building lines are square.

Offset stakes, an alternative to batter boards, are stakes that are offset several feet away from the corner markers. Set up and level the transit over one of the corner stakes which we will call A. Site down the telescope to establish a reference point called B, and drive a stake. Set the 360 scale at 0. Now rotate the telescope until the scale indicates a 90-degree turn. Set up the leveling rod the required distance from the transit, sight down the telescope to establish point C, and drive a stake. Line AC is perpendicular to line AB, forming a right angle where the lines intersect at point A. Lines stretched between the pairs of stakes intersect at point A, one of the house corners.

Pythagorean Theorem Method

The squareness of the corner can be checked by using the Pythagorean Theorem, to determine the length of the *hypotenuse* in a right angle triangle. The theorem says that the square of the hypotenuse of any right angle triangle is equal to the sum of the squares of the other two sides: $C^2 = A^2 + B^2$. Imagine a triangle with one 9-foot side (A), a 12-foot side (B), and a hypotenuse, 15 feet, C.

If we square 9 (A), that is, multiply 9 by itself, we get 81. The square of 12 (B) is 144. The sum of the squares is: 81 plus 144 = 225 (C). $C^2 = A^2 + B^2$. But 225 is the square of the hypotenuse; we need the square root of the hypotenuse, that is, the number which when multiplied by itself equals 225. Most pocket calculators have a square root function key. Enter 225, press the square root key, and the number that appears is 15. If the corner is square, if side A is perpendicular, that is, at 90 degrees to side B, the diagonal should measure exactly 15 feet. Any multiple of 3 can be used. In the example, we used a 9–12–15 triangle: $3 \times 3 = 9$. $3 \times 4 = 12$. $3 \times 5 = 15$. Nine, twelve, and fifteen, are multiples of 3.

Other Important Documents

Certified Plot Plan

A *Certified Plot Plan* (Fig. 1–6) is an as-built, showing how the property was actually built, as opposed to how it was proposed to be built. It is very important to check state regulations as to who may legally certify a plot plan. A professional engineer (PE) may be qualified to survey the property, but in some states, New Hampshire, for example, unless one is a *Licensed Land Surveyor*, one cannot certify the plot plan.

Certificate of Occupancy

One of the most importance pieces of paper in the life of a builder is the *Certificate of Occupancy*, or CO. The CO is the final piece of

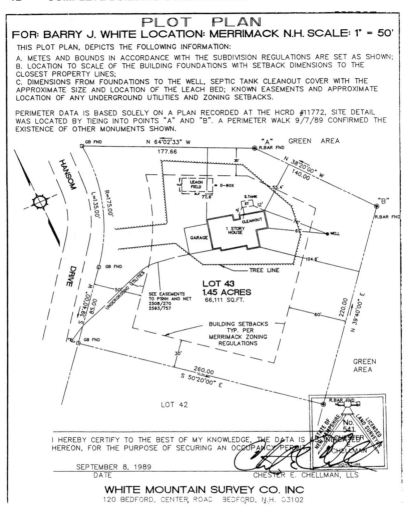

Fig 1–6. Certified plot plan.

paper, the sign-off, that says the construction of the building is complete and it is ready to be occupied. Not all municipalities require a CO before the property can be legally lived in. Any town that has adopted the *BOCA* or the *UBC* building codes will require a CO. In addition, many banks require a CO before the passing of papers can take place.

Very often, as a condition for getting a CO, municipalities require the builder to submit a Certified Plot Plan. If a well is the source of water, a certificate of the water test may also be required. The structure does not necessarily have to be 100 percent completed, but this requirement varies within a state, and from state to state. Check with the building code official as to the requirements.

CHAPTER 2

Concrete

Concrete is an ancient and universal building material. Excavations in Jericho unearthed concrete floors that date back to 7000 BC. The nearly 2½ million limestone blocks of the Great Pyramid of Giza were mortared with cement. The Roman Pantheon, constructed in 27 BC, was the largest concrete structure in the world until the end of the 19th century. The materials necessary to manufacture concrete are cheap and plentiful, and are found in every part of the world.

Until about the latter part of the nineteenth century, concrete was made with natural cement, and was unreinforced. Natural cement is made from naturally occurring calcium, lime, and clays. In 1824 an English builder, Joseph Aspdin, patented an artificial cement he called *portland cement,* after the gray limestone on the isle of Portland, whose color it resembled. Ninety-five percent of the cement used in the world today is portland cement. Concrete has been called a noble material, because it does not burn or rot, is relatively inexpensive, has high compressive strength, is easy to work, and oddly enough, is relatively light. But it could be stronger. In order to carry heavy loads, it must be reinforced with steel bars. Over time it shrinks, and its low tensile strength leads to cracking.

Cements

When portland cement, sand, crushed stone (aggregates), and water are mixed together, they combine chemically to form crystals that bind the aggregates together. The result is a rock-like material called concrete. As the concrete hardens, it gives off considerable heat, called the heat of hydration. Hydration (hardening), continues for years if the concrete does not dry out. The strength of concrete is dependent on the amount of water per pound of cement, or gallons of water per bag of cement.

The ingredients used to make portland cement vary, but basically consist of lime, iron, chalk, silica, sand, alumina, and other minerals. These materials are separately ground, blended, and heated to about 2700 degrees Fahrenheit in a rotating kiln to produce pellets, called *clinker*. The clinker is ground together with a small amount of gypsum into a very fine powder called portland cement. It is sold in bulk or in bags. A standard bag contains 94 pounds of cement, and has a volume of one cubic foot (ft^3).

Types of Portland Cement

The American Society for Testing & Materials (ASTM) Specification for Portland Cement (C150–78) establishes the quality of cement and identifies eight different types:

Type I is most commonly used for general construction, and is called normal cement.

Type IA is a normal air-entraining cement.

Type II is a modified cement for use with concrete in contact with soils or water containing sulfates, which are salts of sulfuric acid. Sulfates attack concrete and can cause the concrete to crack and break up.

Type IIA is a moderate sulfate-resistant, air-entraining cement.

Type III is a high early strength cement that is as strong in 3 days as is Type I or Type II cement in 28 days. Because Type III generates lots of heat, which could damage the concrete, it should not be used in massive structures. It also has poor resistance to sulfates.

Type IIIA is a high early strength air-entraining cement.

Type IV is a low heat of hydration cement, developed for use in massive structures such as dams. If the concrete cannot get rid of the heat as it dries out, its temperature can increase by 50 or 60 degrees Fahrenheit. The temperature increase causes the soft concrete to increase in size. As it hardens and cools off, shrinkage causes cracks to develop; the cracking may be delayed and not show up until much later. These cracks weaken the concrete and allow harmful substances to enter and attack the interior of the concrete.

Type V is a special high-sulfate-resistant cement for use in structures exposed to fluids containing sulfates, such as sea water, or other natural waters.

Normal Concrete

Normal concrete is concrete made with fine aggregates (sand), and regular aggregates (crushed stone or gravel), and water. No air-entraining admixtures have been added. Air-entrained concrete, lightweight concrete, heavyweight concrete, polymer concrete, and fiber-reinforced concrete (FRC) are *not* normal concrete.

Admixtures

Anything, other than cement, aggregates, and water, added to concrete to change its properties is called an *admixture*. More than 70 percent of all ready-mixed concrete contains water-reducing admixtures. Romans added lard, blood, milk, and other material to make concrete more workable. There are many types of admixtures, the most common of which are:

1. Air-entraining agents,
2. Accelerators,
3. Retarders,
4. Water reducers,
5. High range water reducers,
6. Pozzolans.

Air-Entraining

Normal concrete contains a small amount of air. By adding an air-entrainment admixture, the amount of air in the concrete can be increased by 10 percent or more by volume. The air, in the form of tiny bubbles, makes a more workable and longer-lasting concrete. The bubbles are so small that a cubic inch (in^3) of air-entrained concrete may contain 7 million bubbles. The stiffness of concrete, called its slump, can be changed by air-entrainment. Adding air only to a normally stiff 2-inch slump concrete will result in a 5- or 8-inch slump mixture.

Air-entrainment greatly increases the resistance of the finished concrete to repeated cycles of freezing and thawing. All concrete exposed to weathering and attack by strong chemicals should be air-entrained. The American Concrete Institute (ACI) Building Code requires air-entrainment for all concrete exposed to freezing temperatures while wet.

Because adding air to concrete reduces its strength, the water/cement ratio, the water content, or both must be reduced. These adjustments alone will not return the concrete to the required strength if more than 6 percent air is present. The air content must be limited to about 4 percent, with 6 percent as an upper limit. However, large amounts of air are added to create lightweight nonstructural concrete with thermal insulating properties.

Accelerators

Accelerating admixtures speed up the setting and hardening of concrete. The objective of an accelerator is to reduce curing time by developing a 28-day strength in 7 days. The most common accelerator, calcium chloride, is used primarily in cold weather. Concrete hardens very slowly at temperatures below 50 degrees Fahrenheit. Because too much calium chloride can cause corrosion of reinforcing steel, it should be limited to 2 percent of the weight of the concrete. Accelerators are sometimes mistakenly called antifreezes, or hardeners, but they have little effect on lowering the temperature of the concrete. Concrete made with warm water and heated aggregates may make the use of accelerators unnecessary. For more information see American Concrete Institute (ACI), "Guide for Use of Admixtures in Concrete," 212.2R.

Retarders

Pouring concrete during hot, dry, windy weather (hot-weather concreting), can result in concrete that: (1) dries too fast; (2) sets too fast; (3) requires more water to make concrete workable; (4) has rapid loss of slump; (5) is more likely to crack from plastic shrinkage because the surface may dry before curing starts; (6) loses strength.

Retarders are admixtures that slow down the initial setting and curing of the concrete before it can be placed and finished. Most retarders are also water reducers.

Water Reducers

Water reducing admixtures reduce the amount of water needed to produce a cubic yard (yd^3) of concrete of a given slump. Because the workability is not reduced, a higher-strength concrete results.

Superplasticizers

High-range water reducers are called superplasticizers, or supers. They can turn a 3-inch slump concrete into a 9-inch slump concrete without lowering the strength. The amount of water can be reduced by as much as 30 percent to produce high-strength concrete—6000 psi or more. Supers allow concrete to flow around corners, through the tangle of rebar, and into hard-to-reach cavities. Shrinkage cracking is reduced, no water is added, and strength is not reduced. But superplasticizers are more expensive than normal reducers. The action of many plasticizers lasts only 30 minutes at normal temperatures and must be added at the job site. Extended-life supers are available that maintain the increased slump for one to three hours and can be added at the batch plant.

Pozzolans

Pozzolans are natural volcanic ash or artificial materials that react with lime in wet concrete to form cementing compounds. They help to reduce the temperature of curing concrete and improve its workability. However, because pozzolans react slowly, longer curing time is necessary. The most commonly used pozzolan is fly ash, a waste by-product of coal-burning power generating plants. Pozzolan, named after a town in Italy, is where the Romans first mixed ash from Mt. Vesuvius with lime, water, and stone to make a form of concrete. Pozzolans were used in the Roman Aqueduct.

Handling and Placing Concrete

Concrete is a forgiving material. It is often abused by careless construction practices, yet it performs, but not always as it should. How long concrete lasts depends on how it is manufactured, and on the aggregates, admixtures, transporting, handling, placing, curing, and finishing.

Concrete is not a liquid, but a slurry mixture of solids and liquid. Because it is not a stable mixture, excessive vibrating, moving it long distances horizontally, or dropping from excessive heights, can result in the mixture segregating, that is, coming apart. The coarse aggregates work their way to the bottom of the form, while the cement paste and water rise to the top.

Segregation of the aggregate causes loss of strength and watertightness. Segregation must be avoided during all phases of concrete placement: from the mixer (truck) to the point of placement, to consolidation, to finishing. It must be thoroughly consolidated, should fill all corners and angles, and be carefully worked around rebar or other embedded items. The temperature of fresh concrete must be controlled during all operations: from mixing through final placement, and protected after finishing.

Methods of Placing

Caution—Fully-loaded ready-mix trucks may weigh as much as 80,000 pounds. When slabs are being poured in an existing foundation, the truck must be at a 45-degree angle to the foundation; never parallel to it. The truck pressure on the soil increases the lateral pressure against the foundation, and could crack or cave in the foundation wall. The rule-of-thumb is to keep the truck wheels as far away from the foundation as the foundation is deep: if the excavation is 8 feet deep, the truck wheels should be 8 feet or more away from the wall. Ramps may have to be built to protect curbs and sidewalks from cracking under the weight of the truck.

To prevent segregation, the mixer should discharge the concrete as close as possible to its final location. If the mixer cannot get close enough, use (a) chutes, (b) a mixer-mounted conveyor, (c) a portable conveyor, (d) a motorized buggy, (e) hand buggies (Georgia buggies), (f) small-line concrete pump.

Mixer-mounted conveyors are belt conveyors mounted on a ready-mix truck. They can move as much as 100 yds^3 of concrete per hour, reach 40 feet horizontally, over 25 feet vertically, and 10 feet below grade.

If the mixer can get no closer than 30 feet, gas-powered portable conveyors up to 30 feet long can be used. The conveyor will place 50 yds^3 of concrete an hour for flat work, and 30 yds^3 an hour for 12-foot high walls. The conveyor will fit through 30-inch wide openings (Fig. 2–1).

Riding, or walk-behind motorized buggies can handle 10 to 21 ft^3 of concrete, fit through 36-inch and 48-inch openings, and travel at 5 to 10 mph (Fig. 2–2).

Fig. 2–1. Gas-powered portable conveyor *(Courtesy Morgen Manufacturing Co.)*

Fig. 2–2. Walk-behind motorized buggy. *(Courtesy Morrison Division of Amida Industries, Inc.)*

Small-line concrete pumps (Fig. 2–3) can move 25 to 40 yds^3 per hour. With the exception of hand buggies, all these methods offer speed. Ready-mix suppliers allow only so many minutes per yd^3 to discharge the load, after which there is a charge per minute. The most economical method depends on the job conditions, the equipment, and the contractor's experience.

Do not place concrete in one end of the form and move it across or through the forms. Do not use rakes or vibrators to move the concrete.

Minimize the distance concrete drops to help prevent segregation of aggregates. Maximum drop with rebar should be no more than 5

Fig. 2–3. Truck-mounted small line concrete pump. *(Courtesy Morgen Manufacturing Co.)*

feet. Keep the maximum drop to 8 feet, and use *dropchutes* to break the fall of the concrete (Fig. 2–4).

Consolidation

Consolidation keeps the separate ingredients in the concrete together. In order to eliminate trapped air, fill completely around rebar

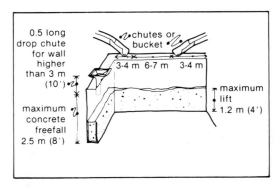

Fig. 2–4. Concrete drop chutes. *(Courtesy Canada Mortgage and Housing Corporation)*

and corners, and to ensure that the concrete is in contact with the surface of the forms, it must be consolidated, or compacted. Do not use vibration because it is difficult to control overvibration, which can cause segregation. Hand spade the concrete to remove entrapped air at the face of the forms. Consult ACI 309, "Recommended Practice for Consolidation of Concrete."

Reinforcing

The marriage of steel and concrete overcame the major weakness of concrete: lack of tensile—stretching—strength. Concrete and steel expand and contract at nearly the same rate as temperatures change. The alkalinity of the concrete protects the steel from corrosion. The concrete bonds very tightly to the steel. This combination allows concrete to be used for every type of construction.

The term *rebar* is more commonly used than the phrase *deformed steel bar*. Rebars are hot-rolled with surface ribs or deformations for better bonding of the concrete to the steel (Fig. 2–5). Rebars are available in 11 standard sizes, and are identified by a size number which is equal to the number of eighths of an inch (⅛ inch) of bar diameter. A number 3 rebar is ⅜ inches in diameter, and a number 8 rebar is ⅝ inches or 1 inch in diameter. Both standard billet-steel and axle-steel rebars are produced in two strength grades: 40 and 60. The standard grade for building construction is Grade 60. The grade number indi-

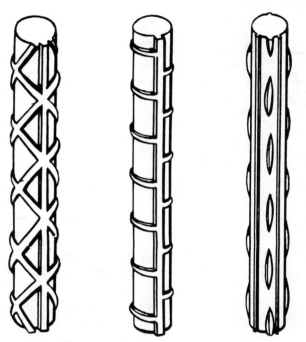

Fig. 2–5. Deformed bars are used for better bonding between concrete and steel bars.

cates the strength in 1000 pounds per square inch (ksi). For more information, see *Manual of Standard Practice,* published by the Concrete Reinforcing Steel Institute (CRSI).

Bar Supports

Three types of bar support material are available: wire, precast concrete, and molded plastic. The bars must be free of mud, oil, rust, and form coatings. The placement of rebar must be according to the plans. Bottom bars must be placed so that at least two inches of concrete *cover* is left below and to the sides of the rods. The cover protects the rebar against fire and corrosion. The bar spacing must be wide enough to allow aggregates to move between them. To guarantee that

Table 2–1. Reinforcing Bar Numbers and Dimensions

Bar sizes*			
Old (inches)	New numbers	Weight (lbs. per. ft.)	Cross-sectional area (sq. in.)
¼	2	0.166	0.05
⅜	3	0.376	0.1105
½	4	0.668	0.1963
⅝	5	1.043	0.3068
¾	6	1.502	0.4418
⅞	7	2.044	0.6013
1	8	2.670	0.7854
1/square	9	3.400	1.0000
1⅛ square	10	4.303	1.2656
1¼ square	11	5.313	1.5625

the rebars will be at the correct cover height, steel *chairs* or long *bolsters* must be used (Fig. 2–6). Do not use uncapped chair feet on bare ground. They can rust and the rust can spread upward to the rebar. Use capped chairs, bolsters, concrete blocks, or molded plastic bar supports. Plastic supports may expand (coefficient of temperature expansion) at a different or higher rate than the concrete, especially in areas

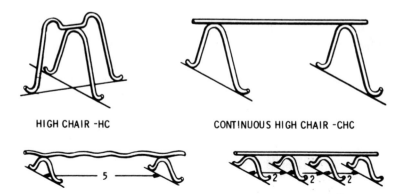

HIGH CHAIR -HC   CONTINUOUS HIGH CHAIR -CHC

Fig. 2–6. Chairs or bolsters used to support bars in concrete beams.

of wide temperature variations. Check with the manufacturer to be certain that plastic bar support coefficient of temperature expansion is similar to concrete's.

Welded-Wire Fabric

Welded-wire fabric (WWF) looks like fencing, and is manufactured with plain or deformed cold-drawn wire, in a grid pattern of squares or rectangles (Fig. 2–7). The fabric is available in 150- to 200-foot rolls, 5 to 7 feet wide. Spacings range from 2 to 12 inches, as well as custom spacings. It is also available in sheets, 5 to 10 feet wide and 10 to 20 feet long. Fabric is usually designated WWF on drawings. The

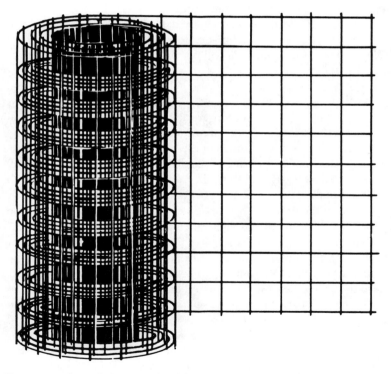

Fig. 2–7. Concrete slab reinforcement is usually in the form of screen mesh.

sizes of WWF are given by a spacing followed by the wire size: WWF 6 × 6-W1.4 × W1.4. This indicates smooth wire, size W1.4 spaced at 6 inches in each direction, and a wire diameter of 0.135 inches. The 6 × 6-W1.4 × W1.4 WWF is the type most commonly used in residential slabs.

Probably no two building materials are more often misused, their purposes more widely misunderstood, and called things they are not, than welded-wire fabric and vapor diffusion retarders (so-called vapor barriers). The term most commonly applied to WWF in residential construction is *reinforcing*. Although WWF may be installed in a slab, the slab is considered to be a *plain* slab (no steel) because the *amount* of steel is less than the minimum necessary to reinforce the slab.

Therefore, not only is WWF misused, but misnamed, as well. The small percentage of steel used in light residential slabs has little effect in increasing the load-carrying ability of the slab. It does not help to distribute the loads to the subgrade, does not permit a reduction in the slabs thickness for equal load, will not stop curling, and does not reinforce the slab.

The basic purpose of WWF is to hold tightly together any random cracks that may happen between the joints. Welded-wire fabric does not stop cracking, will not stop cracking, cannot stop cracking. The widespread, but incorrect, common practice of placing WWF on the subgrade/subbase and pouring concrete over it, makes the fabric useless. The steel must be correctly sized and located within the upper 2 inches of the top surface of the slab. Placing it at the bottom of the slab and then pulling it up through the concrete with a hook is as bad as leaving it on the ground. Its proper location within the slab cannot be guaranteed. This practice is therefore a waste of time and money.

To make stronger floors requires much more steel than is used for crack control. There must be a top layer and a bottom layer to resist the stresses which alternate from top to bottom of the slab as moving loads cause the slab to deflect over soft and hard spots in the subgrade. The small amount of steel normally used will be about 0.1 percent of the slab's cross-sectional area. If this 0.1 percent were placed at the top and bottom of a 6-inch slab—twice as much as normally used—the floor would be stronger by a mere 3 percent. Clearly then, doubling the amount of steel used for crack control does not reinforce, or make for a stronger floor.

Stronger floors can be achieved for considerably less money by simply increasing the slab thickness. The bending strength of a 4-inch concrete slab is a function of the square of its thickness. Four squared (4×4 or 4^2)= 16. How much stronger is a 5-inch slab than a 4-inch slab? Five squared = 25. The ratio of 25:16 is 25 divided by 16 = 1.56×100 = 156 percent. The 5-inch slab is 56 percent stronger than 4-inch slab, with only a 25 percent increase in concrete cost. A 6-inch slab is 125 percent stronger than the 4-inch slab, with only a 50 percent increase in material costs. There is no increase in subgrade preparation cost, or floor finishing costs.

We have seen that WWF usually used in residential slabs does not stop cracking and does not reinforce or strengthen the floor. Is WWF necessary? If there is a properly compacted, uniform, subgrade/sub-base, without hard or soft spots, and short joint spacing (joint spacing in feet a maximum of 2 slab thicknesses in inches), *WWF is unnecessary.*

However, when for appearance, or in order to reduce the number of joints, longer joint spacing is necessary, *WWF is necessary.*

Chairs

The success of WWF in controlling cracks depends on its location. Because shrinkage cracks are almost always widest at the surface, and to keep crack width as small as possible, WWF must be installed 1½ to 2 inches below the slab's surface. When slabs are exposed to salt water, or deicing salts, the WWF should be protected by lowering it to the middle of the slab. Because the WWF must be laid flat and be free of curling, sheets are the best choice. Sheets are available in 25-foot lengths in the western part of the country.

Capped metal chairs, concrete brick, or plastic Mesh-ups™, manufactured by Lotel, Inc. (Fig. 2–8) can be used. As noted previously, check with the manufacturer regarding the plastic's coefficient of temperature expansion. The spacing of chairs depends on the wire size and wire spacing. Common practice is to place one for every 2 or 3 feet of mesh, at the point where the wires intersect. The bottom wire of the intersection must be placed in the lower slot of the Mesh-up. On sand subbases, use a plastic sand disk to prevent the chairs' sinking into the subbase.

RE-RINGS are manufactured by Structural Components, E. Longmeadow, Massachusetts. Made of PVC, they are suitable for use with 5×10 and 8×20, and rolled steel re-mesh and re-rods. It is an

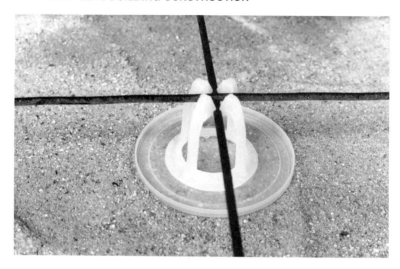

Fig. 2–8. **2-inch plastic chair for use with 3- to 5-inch slabs shown with sand disk.** *(Courtesy Lotel, Inc.)*

Fig. 2–9. RE-RINGS PVC chair. *(Courtesy Structural Components, Inc.)*

excellent choice for use with multiple layers of WWF (Fig. 2–9). The coefficient of thermal expansion is very close to that of concrete.

Concrete Slump

There are many tests performed to determine if concrete meets job specifications. Most of these acceptance tests have been standardized by ASTM, and include tests of strength, slump, air content, unit weight, temperature, and impact rebound. Little, if any, testing is performed in residential construction; if any were done, it would be limited to slump testing.

Slump is a measure of how consistent, fluid, and workable a batch of freshly-mixed concrete is. Any change in the slump may mean that the amount of water, the temperature, hydration or setting is changing. Basically, slump measures the amount of water in the mix.

To perform a slump test, a slump cone, a ruler, a scoop, and a standard tamping rod are required. The slump cone is made of sheet metal and is 12 inches high. The opening in the top of the cone is 4 inches in

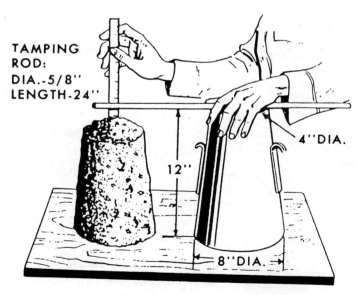

TAMPING ROD:
DIA.-5/8"
LENGTH-24"

4"DIA.

12"

8"DIA.

Fig. 2–10. Measuring concrete slump.

diameter, and 8 inches in diameter at the bottom. The tamping rod is 24 inches long, and 5/8 inch in diameter, with the end rounded like a half circle. If a non-standard tamping rod is used, the results are not considered valid for accepting or rejecting the concrete. The metal cone is first wetted, the concrete layered in, and each layer rodded 25 times. The cone is carefully lifted off, and the wet concrete allowed to slump under its own weight. The slump is measured to the nearest 1/4 inch (Fig. 2–10).

CHAPTER 3

Foundations

Though the foundation supports a building, it is the earth which is the ultimate support. The foundation is a *system* comprised of foundation wall, footing, and soil. The prime purpose of an efficient structural foundation system is to transmit the building loads directly to the soil without *exceeding* the bearing capacity of the soil. A properly designed and *constructed* foundation system transfers the loads uniformly, minimizes settlement, and anchors the structure against racking forces and uplift. Because soil type and bearing capacity is the crucial factor in the foundation system, the foundation must be designed and built as a *system*. Too many residential foundations are designed and built *without* any concern for the soil.

Types of Foundations

The many types of foundations can be separated into two broad groups: shallow foundations and deep foundations. Shallow foundations consist of four types: deep basements (8-foot walls), crawl spaces, slabs-on-grade, and frost-protected shallow foundations. They include spread footings, mat or raft footings, long footings, strap footings (Fig. 3–1). Deep foundations extend considerably deeper into the earth, and

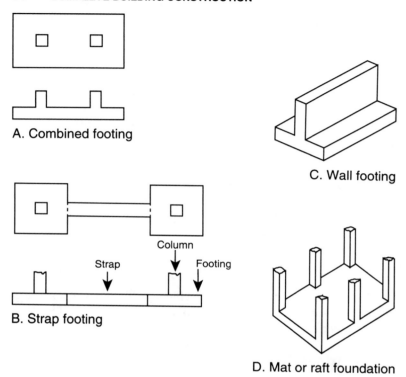

A. Combined footing

C. Wall footing

B. Strap footing

D. Mat or raft foundation

Fig. 3–1. Shallow foundation types.

include drilled caissons or piers, groups of piles (Fig. 3–2), driven and cast-in-place concrete piles, and floating foundations.

There are a number of different construction systems that can be used. Cast-in-place concrete is the most widely used material for residential foundations, followed by concrete block. Other methods include precast foundation walls, cast-in-place concrete sandwich panels, masonry or concrete piers, All Weather Wood Foundations (AWWF)—now called Permanent Wood Foundations (PWF), or Preserved Wood Foundations in Canada—and expanded polystyrene (EPS) blocks, polyurethane blocks, and other similar systems using EPS blocks filled with concrete.

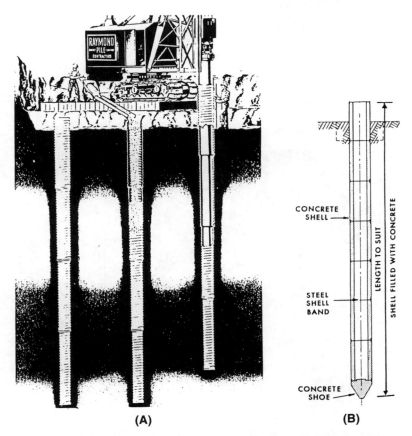

(A)   (B)

Fig. 3–2. (A) A step-taper pile. On the right, the pile is being driven with a steel core, and in the center, it is being filled with concrete. At the left is a finished pile. (B) Concrete-shell pile. The steel bands are used to seal the joints between the sections. *(Courtesy Raymond International, Inc.)*

Footings

Footings, which may be square, rectangular, or circular, are strips of concrete or filled concrete blocks placed under the foundation wall. Gravel or crushed stone footings are used with Permanent Wood

Foundations. The purpose of the footings is to transfer the loads from walls, piers, or columns, to the soil. The spread footing is the most common type used to support walls, piers, or columns. The National Concrete Masonry Association has developed a system of solid interlocking concrete blocks called IDR Footer Blocks™ [Fig. 3–3 (A) and (B)]. The minimum width of the footing is based on the foundation wall thickness. An 8-inch thick foundation wall would have an 8-inch wide footing. However, footings are made wider than the foundation wall,

Fig. 3–3. (A) IDR Footer Blocks™. *(Courtesy National Concrete Masonry Association)*

Fig. 3–3. (B) IDR Footer Blocks in place and first course of blocks mortared in place. *(Courtesy National Concrete Masonry Association)*

and the extra width projects or cantilevers equally beyond each side of the wall (Fig. 3–1). Contrary to widespread belief, the purpose of footings is *not* for spreading out and distributing the loads to the soil. The extra width is used to support the wall forms while the concrete is poured, or as a base for concrete masonry blocks or brick.

Foundation Design Details

> *Foundations ought to be twice as thick as the wall to be built on them; and regards in this should be had to the quality of the ground, and the largeness of the edifice; making them greater in soft soils, and very solid where they are to sustain a considerable weight. The bottom of the trench must be level, that the weight may press equally, and not sink more on one side than on the other . . .*
> Andrea Palladio, **The Four Books of Architecture,** 1570

Footing size is determined by the size of the imposed loads, and the bearing capacity of the soil. Table 3–1 lists soil-bearing capacities.

Table 3–1. General Characteristics and Typical Bearing Capacities of Soils

Group Symbols	Typical Names	Drainage Characteristics	Frost Heave Potential	Volume Change	Backfill Potential	Typical Bearing Capacity	Range (psi)	General Suitability
GW	well-graded gravels and gravel-sand mixtures, little or no fines	excellent	low	low	best	8000 psf	1500 psf to 20 tons ft^2	good
GP	poorly-graded gravels and gravel-sand mixtures, little or no fines	excellent	low	low	excellent	6000 psf	1500 psf to 20 tons ft^2	good
GM	silty gravels, gravel-sand silt mixtures	good	medium	low	good	4000 psf	1500 psf to 20 tons ft^2	good
GC	clayey gravels, gravel-sand-clay mixtures	fair	medium	low	good	3500 psf	1500 psf to 10 tons ft^2	good
SW	well-graded sands and gravelly sands, little or no fines	good	low	low	good	5000 psf	1500 psf to 15 tons ft^2	good
SP	poorly-graded sand and gravelly sands, little or no fines	good	low	low	good	4000 psf	1500 psf to 10 tons ft^2	good
SM	silty sands, sand-silt mixtures	good	medium	low	fair	3500 psf	1500 psf to 5 tons ft^2	good
SC	clayey sands, sand-clay mixtures	fair	medium	low	fair	3000 psf	1000 psf to 8000 psf	good

Table 3-1. General Characteristics and Typical Bearing Capacities of Soils (continued)

Group Symbols	Typical Names	Drainage Characteristics	Frost Heave Potential	Volume Change	Backfill Potential	Typical Bearing Capacity	Range (psi)	General Suitability
ML	inorganic silts, very fine sands, rock flour, silty or clayey fine sands	fair	high	low	fair	2000 psf	1000 psf to 8000 psf	fair
CL	inorganic clays of low to medium plasticity, gravelly clays, sandy clays, silty clays, lean clays	fair	medium	medium	fair	2000 psf	500 psf to 5000 psf	fair
MH	inorganic silts, micaceous or diatomaceous fine sands or silts, elastic silts	poor	high	high	poor	1500 psf	500 psf to 4000 psf	poor
CH	inorganic clays of medium to high plasticity	poor	medium	high	bad	1500 psf	500 psf to 4000 psf	poor
OL	organic silts and organic silty clays of low plasticity	poor	medium	medium	poor	400 psf or remove	generally remove soil	poor
OH	organic clays of medium to high plasticity	no good	medium	high	no good	remove	—	poor
PT	peat, muck and other highly organic soils	no good	—	high	no good	remove	—	poor

39

In simplest terms, bearing capacity is the soil's ability to hold up a structure. More technically, *bearing capacity* is the pressure that a structure imposes onto the mass of the earth without overstressing it. The *ultimate bearing capacity* is the loading per square foot (ft^2) that will cause shear failure in the soil: the foundation settles or punches a hole in the soil. For one- and two-story residential dwellings, our concern is with the *allowable*, or *design* bearing capacity, designated by the symbol q_a. This is the load the soil will support without unsafe movement or collapsing.

Sizing the Footings

There are two methods of determining the size of footings:

1. By Rule-of-Thumb,
2. By Calculation.

Rule-of-Thumb—Because residential loads are comparatively light, for average bearing soils [2000 pounds per square foot (psf) or greater] the size of the footing may be found from the rule-of-thumb which states:

> *The nonreinforced width of the footing should not exceed twice the width of the foundation wall, and should be at least as high as the wall is wide; but in residential construction never less than 6 inches high.*

An 8-inch foundation wall would have a footing 16 inches wide by 8 inches high. With a 10-inch wall, the footing would be 20 inches wide by 10 inches high. The 2:1 thickness-to-width ratio of the footing should be maintained. If the footings are too wide for a given thickness, they could fail in shear. Key the footings to help the foundation wall to resist lateral earth pressure when backfilling is done *before* the basement floor is poured.

Calculation—In order to calculate the required footing size, the total live and dead loads acting on the footing must be determined. *Live load* is the weight of people, furniture, wind, snow; *Dead load* is the weight of the building materials such as foundation exterior walls, roofing, siding, and other materials. Table 3–2, Typical Weights of Some Building

Table 3–2. Typical Weights of Some Building Materials

Roof	Lb per sq ft
Wood shingles	3
Asphalt shingles	3
Copper	2
Built-up roofing, 3 ply & gravel	5.5
Built-up roofing, 5 ply & gravel	6.5
Slate, ¼″ thick	10
Mission tile	13
1″ Wood decking, paper	2.5
2″ × 4″ Rafters, 16″ o.c.	2
2″ × 6″ Rafters, 16″ o.c.	2.5
2″ × 8″ Rafters, 16″ o.c.	3.5
½″ Plywood	1.5
Walls	
4″ Stud partition, plastered both sides	22
Window glass	5
2″ × 4″ Studs, 1″ sheathing	4.5
Brick veneer, 4″	42
Stone veneer, 4″	50
Wood siding, 1″ thickness	3
½″ Gypsum wallboard	2.5
Floors, Ceilings	
2″ × 10″ wood joists, 16″ o.c.	4.5
2″ × 12″ wood joists, 16″ o.c.	5
Oak flooring, 25/32″ thick	4
Clay tile on 1″ mortar base	23
4″ Concrete slab	48
Gypsum plaster, metal lath	10
Foundation Walls	
8″ Poured concrete, at 150 lb per cu ft	100
8″ Concrete block	55
12″ Concrete block	80
8″ Clay tile, structural	35
8″ Brick, at 120 lb per cu ft	80

Materials, and Fig. 3–4, Calculations for exterior wall footings, illustrate how to find the total weight on one lineal, or running, foot of footing.

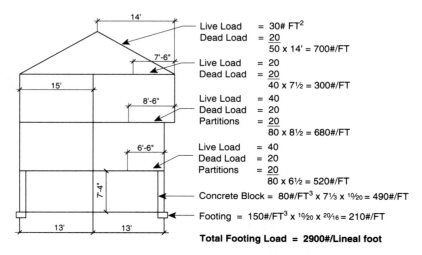

Fig. 3–4. Calculations for exterior wall footings.

Example—Using the data given in Fig. 3–4, calculate the size of the footing required to support a load of 2900 pounds per foot (lb/ft) bearing on a soil with a design capacity (q_a) of 2000 pounds per square foot (psf).

$$\text{Footing Area (ft}^2) = \frac{\text{footing load (lb/ft)}}{q_a \ (\text{lb/ft}^2)}$$

$$\text{Area} = \frac{2900 \ \text{lb/ft}}{2000 \ \text{lb/ft}^2} = 1.45 \ \text{ft}^2$$

The required footing area per lineal foot of wall is 1.45 ft². The footing width = 12 inches × 1.45 = 17.4 inches. Round to 18 inches.

Problem—Given a footing width of 18 inches and a soil-bearing capacity (q_a) of 2000 psi, what is the load on the soil?

$$L = \frac{\text{footing load (lb/ft)}}{\text{footing width (ft)}} = \frac{2900 \ \text{lb/ft}}{1.5 \ \text{ft}} = 1933 \ \text{psf}$$

The footing thickness (8 inches) will have to be increased to 10 inches to maintain the approximate 2:1 ratio.

Reinforcing

There is little consistency in the use of rebar in residential foundations: all or none. Yet, few residential foundations fail because of a lack of rebar. Frost heaving and backfilling damage are more common. Reinforcing of residential foundations is not required by the CABO *One and Two Family Dwelling Code* as long as the height of unbalanced fill for an 8-inch thick wall does not exceed 7 feet maximum. However, when soil pressures exceed 30 pounds per cubic foot (pcf), groundwater or unstable soils are present, or when in seismic zones 2 or 3 (Fig. 3–5), foundations must be reinforced. The small amount and size of

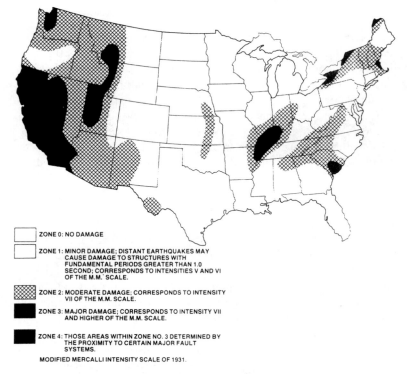

ZONE 0: NO DAMAGE

ZONE 1: MINOR DAMAGE; DISTANT EARTHQUAKES MAY CAUSE DAMAGE TO STRUCTURES WITH FUNDAMENTAL PERIODS GREATER THAN 1.0 SECOND; CORRESPONDS TO INTENSITIES V AND VI OF THE M.M. SCALE.

ZONE 2: MODERATE DAMAGE; CORRESPONDS TO INTENSITY VII OF THE M.M. SCALE.

ZONE 3: MAJOR DAMAGE; CORRESPONDS TO INTENSITY VII AND HIGHER OF THE M.M. SCALE.

ZONE 4: THOSE AREAS WITHIN ZONE NO. 3 DETERMINED BY THE PROXIMITY TO CERTAIN MAJOR FAULT SYSTEMS.

MODIFIED MERCALLI INTENSITY SCALE OF 1931.

Fig. 3–5. Seismic zone map of the United States.

rebar used in residential foundations has no effect on the strength of concrete. A 40′ × 24′ two-story house with full basement weighs about 149,000 pounds. An 8-inch wall, with a perimeter of 128 lineal feet, has a surface area of 85.3 ft^2. The 149,000 pounds exerts 12.1 pounds per square inch (psi) on this area, which is less than ½ of 1 percent of the compressive strength of 2500 psi concrete. What then is the purpose of the rebar?

The settlement of different parts of the foundation at different times, and the shrinkage of concrete as it dries, causes cracks in the wall. The extent of the shrinkage can be controlled by using a number of techniques: (a) Low water/cement ratio concrete, properly cured by keeping it moist; (b) Low slump concrete (3 inches). Because of its stiffness it is difficult to work, and vibrators may be necessary. A slump of 5 to 7 inches can be had by using water reducers and plasticizers.

Caution—Vibrators should be used only by experienced concrete workers.

(c) Rebar, used primarily to help control cracking when the foundation moves, will not stop the cracking—a force sufficient to crack the foundation will bend or snap the rebar. Rebar, will however, help hold the pieces together. Another way of controlling foundation cracking is by placing control joints where the cracks are most likely to happen (Fig. 3–6). Footings more than 3 feet wide should be thickened, or reinforced to resist the bending stresses, and to limit cracking and settlement at the base of the wall. Install two No. 4 rebars, 2 inches below the top of the footing, and in the long direction, parallel to the wall.

Stepped Footings

Stepped wall footings (Fig. 3–7) are used on sloping sites in order to maintain the required footing depth below grade. Care must be used to guarantee that (a) the vertical step is not higher than three-fourths of the length of the horizontal step, (b) concrete is poured monolithically (at the same time), (c) all footings and steps are level, (d) all vertical steps are perpendicular (at 90 degrees) to the horizontal step, (e) the vertical and horizontal steps are the same width.

Foundations must rest on undisturbed soil, and should never be set on wet clay, poorly drained soils, compressive soils, organic soils, or frost-susceptible soils. Compressive soils are weak, easily compressed

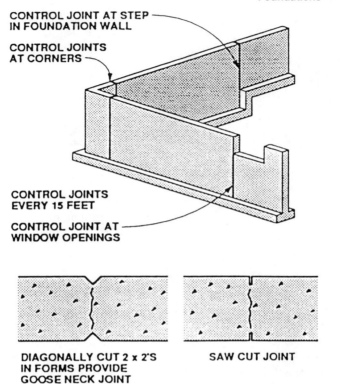

CONTROL JOINT AT STEP
IN FOUNDATION WALL

CONTROL JOINTS
AT CORNERS

CONTROL JOINTS
EVERY 15 FEET

CONTROL JOINT AT
WINDOW OPENINGS

DIAGONALLY CUT 2 x 2'S
IN FORMS PROVIDE
GOOSE NECK JOINT

SAW CUT JOINT

Fig. 3–6. Control joints in concrete walls. *(Courtesy Canada Mortgage and Housing Corporation)*

(crushed) soils. Organic soils are often black or dark grey, contain decaying/decayed vegetation, and may give off an unpleasant odor. They are easily compressed, spongy, and crumble readily (friable). If the soil is disturbed, it must be mechanically compacted, or removed and replaced with concrete. Footings set on compacted soil must be designed by an engineer.

Frost Protection

Never set foundations during freezing weather unless the underlying soil has been kept frost-free. Foundations must not be placed on

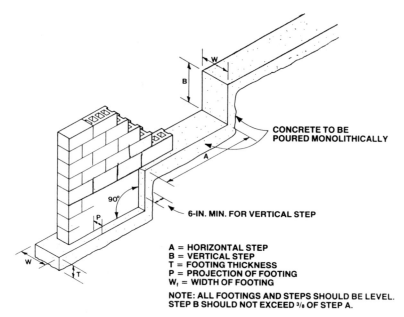

CONCRETE TO BE
POURED MONOLITHICALLY

6-IN. MIN. FOR VERTICAL STEP

90°

A = HORIZONTAL STEP
B = VERTICAL STEP
T = FOOTING THICKNESS
P = PROJECTION OF FOOTING
W₁ = WIDTH OF FOOTING

NOTE: ALL FOOTINGS AND STEPS SHOULD BE LEVEL.
STEP B SHOULD NOT EXCEED ³/₈ OF STEP A.

Fig. 3–7. Stepped wall footings.

frozen ground unless it is permanently frozen ground (permafrost). Frozen soil usually has a high water content which weakens the soil (it loses its crushing strength) as warm weather thawing turns the frost into water. The soil is easily compressed by the weight of the building, and settlement of the building takes place. Because all parts of the building may not settle uniformly, structural damage can result. For frost action to occur three conditions are necessary: frost-susceptible soil, enough water, and temperatures low enough to freeze the water/soil combination. If any one of these conditions is missing, soil freezing and frost heaving will not happen.

One of several methods used to prevent frost damage is to place the footings at or below the frost line. The three model building codes, *National Building Code* (BOCA) in the Northeast, the *Standard Building Code* (SBCCI) in the Southeast, and the *Uniform Building Code* (UBC) in the West, as well as the *CABO One and Two Family Dwelling Code,* all require that footings extend to or below the frost line.

Only the BOCA code says that the footings do not have to extend to the frost line when ". . . otherwise protected from frost . . .". Because BOCA is a performance code, it does not specify what "otherwise protected" means. The methods and materials used to frost-protect the footings are left to the designer or builder.

Frost lines in the USA are based on the maximum frost depth ever recorded in a particular locality. This information comes from grave diggers and utility workers repairing frozen water lines. There are two other methods of determining frost depths: air temperatures, and an *air-freezing index* (F), which is based on the number of hours each year that the average hourly temperature falls below 32 degrees Fahrenheit. The average of the three coldest winters in 30 years are assumed to be the normal frost depth. Because the grave diggers'/utility workers' records rarely include presence of snow, soil conditions, wind and other important data, the results are often questionable. Minneapolis, with 8200 heating degree days (HDD), quite cold, requires footings to be placed 42 inches below grade. New York City with 5200 HDD, cold, requires a 48-inch footing depth.

Seasonal *heating degree days* (HDD or DD) are a measure of the number of days the average temperature is below the base of 65 degrees Fahrenheit. If during a 24-hour day, the maximum temperature was 50 degrees, and the minimum temperature was 20 degrees, the average temperature was 1/2 (50 + 20) = 35. 65 − 35 = 30. There were 30 DD for that day. By adding up the daily degree days, we get the number of degree days in a year, which tell us how cold it was. Miami, Florida has 199 DD, while Bismarck, North Dakota has 9075 DD.

Adfreezing

Placing footings at or below the frost line is no guarantee they will not encounter freezing. Frost depth can vary widely in any one geographic area. Frost depth is affected by many factors including air temperature, soil moisture content, soil type, vegetation, snow cover, solar radiation, and wind velocity. If the foundation is backfilled with frost-susceptible soil—such as silty sands or silty clays—the foundation could be damaged by *adfreezing*. This happens when the soil freezes to the surface of the foundation. As the frozen soil heaves, it lifts the foundation up out of the ground. In heated basements these forces are minimal and intermittent. In unheated basements, garages, crawl spaces, the adfreeze bond and tangential forces may be quite high. If the foundation cannot resist the heaving forces, the yearly heaving adds up and could result in the foundation rising up many inches to several feet.

Crawl Spaces

Nearly 90 percent of crawls spaces are located in 15 states, and are concentrated in hot humid areas of the USA. Traditionally, crawl spaces were designed to be unheated, ventilated, with the overhead floor insulated. In the colder northern regions, crawl space temperature is close to the outdoor air temperature. Even with insulated floors (wiring, plumbing, and floor bridging can make this difficult) winter time ventilation results in substantial heat loss through the floor. Duct work and pipes located there must be insulated against heat loss, gain or both, and to prevent freezing.

Crawl space ventilation was recommended until further research positively established its need in the early 1950s as a means of eliminating crawl space moisture. In 40 years, a recommendation became a requirement—crawl spaces *must* be ventilated, except those used as underfloor plenums, year around. The size and location of the vents is also specified. Although both the BOCA and the *CABO One And Two Family Dwelling Code* permit a reduction in the required net venting area if ". . . an approved vapor barrier is installed over the ground surface . . .", ventilation remains the primary required means of eliminating moisture. Floor failures (mostly in the South, but also in California), high moisture levels in crawl spaces throughout the country and in Canada, and the dominance of crawl space basements in humid climates, has led to a questioning of the accepted guidelines for the venting and insulating of these spaces.

A 15-month California study by the USDA Forest Products Laboratory, research by Oak Ridge National Laboratory and other scientists throughout the country concluded that ventilation of crawl spaces is not only unnecessary, but in hot, humid, regions or areas with hot, humid summers, it can cause rotting of the wood framing. Ventilation air, rather than removing moisture, brings hot, humid, air into the cool crawl space, where it condenses. The moisture can accumulate faster than the ventilation can remove it. These studies confirm that with proper drainage and good ground cover, crawl space ventilation has little effect on wood moisture content.

Caution—Although ventilation is not necessary, *groundwater and moisture control is.*

(1) Slope the ground 6 inches in 10 feet away from the crawl space walls. (2) Install gutters and downspouts to direct rain and groundwater away from the crawl space. (3) Lay a heavy duty reinforced polyethylene such as TU-TUF™, or Ethylene Propylene Diene Monomer (EPDM) moisture retarder over the crawl space earth, extend it up the sides of the foundation walls, and tape it there. (4) Cover the plastic with a concrete slab (preferred), or with a layer of sand. (5) Insulate the foundation walls, not the floor above. In hot humid climates, especially, but also in areas with hot humid summers, insulating the floor decouples the house from the cooling effect of the earth, and increases the air conditioning load (Fig. 3–8).

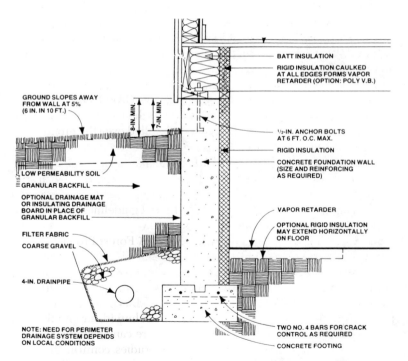

Fig. 3–8. Concrete crawl space wall with interior insulation. *(Courtesy ORNL)*

Slabs-on-Grade

The term slabs-on-grade actually includes industrial, commercial, apartment, residential, single-family dwelling slabs, parking lot slabs and pavements. Slabs-on-grade may be of 3 types: grade beam with soil supported slab; a notched grade beam supporting the slab, or monolithically cast slab and grade beam (thickened edge) (Fig. 3–9). The American Concrete Institute (ACI) classifies slabs into six types (ABCDEF), and recognizes five methods of design.

Type A slabs. Contain no reinforcing of any type. They are designed to remain uncracked from surface loads, and are usually constructed with type I or II portland cement. Joints may be strengthened with dowels, or thickened edges. Construction and contraction joint spacing should be minimal to reduce cracking.

Type B slabs. Similar to type A, but contain small amounts of shrinkage and temperature reinforcement in the upper half of the slab. Contrary to widespread belief, the reinforcing, usually 6 × 6, W1.4 × W1.4 WWF, will not prevent the slab from cracking. Its primary purpose is to hold tightly closed any cracks that may form between the joints. WWF should be placed on chairs to ensure its location in the top half of the slab. Greater joint spacing is allowed than in type A slabs.

Type C slabs. Similar to type B but contain more reinforcing, and are constructed using shrinkage-compensating concrete, or type K cement. Type C slab joints can be spaced further apart than in types A or B. Although the concrete shrinks, it first expands by an amount slightly more than the expected shrinkage. Reinforcement in the upper half of the slab limits the expansion of this part of the slab, and helps control the drying shrinkage. The subbase helps to control the expansion of the lower part of the slab. These slabs must be separated from fixed structures by a compressible material that allows slab expansion.

Type D slabs. Use post-tensioning to control cracks, and are usually constructed with type I or II portland cement. The prestressing permits greater joint spacing than in types A, B, and C.

Type E slabs. Designed to be uncracked, and structurally reinforced by post-tensioning, steel, or both. They can be designed to resist the forces produced by unstable soils.

Type F slabs. Expected to crack from surface loads. They are structurally reinforced with one or two layers of deformed bars or heavy wire mesh. Correct placement of the steel is important. Because of the expected cracking, joint spacing is not critical.

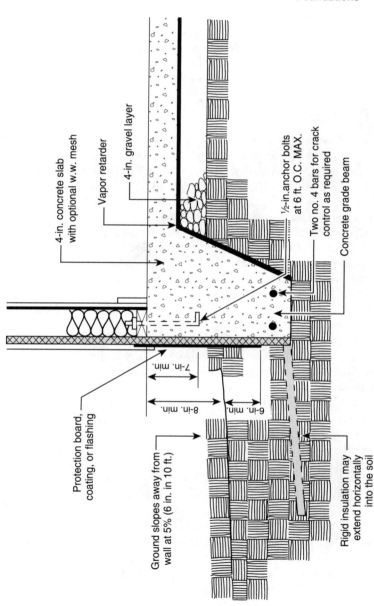

Fig. 3-9. Slab-on-grade and integral grade beam with exterior insulation. *(Courtesy ORNL)*

Fiber Reinforced Concrete (FRC) is an alternative to the use of wire-welded fabric for the control of shrinkage and temperature cracking. Fibers can be low carbon or stainless steel, mineral such as glass, synthetic organic substances such as carbon, cellulose, or polymeric (polyethylene, polypropylene), or natural organic substances such as sisal.

Site Preparation

One of the most important factors in the design, construction, or both of slabs-on-grade and deep foundations, is the condition of the site. If an approved Septic System Design exists, consult it for information on depth of water table, Seasonal High Water Table (SHWT), presence of debris, stumps, logs, and soil types. Ask the building inspector about previous structures on this lot. Request a copy of soils map, or consult with Zoning Administrator as to whether or not the lot's setbacks are based on soil conditions. If soil information is not available from local officials, call USDA Soil Conservation Service for help. On lots with no known history, or on virgin lots, dig test holes 8 to 10 feet deep where house will sit to locate water table, hidden streams, ledge, debris, or buried foundations that could have a negative effect on the slab. Vegetation, contours, drainage, and their possible negative impact on the slab should be noted. A thorough site investigation is the best insurance against legal action.

Excavate at or below the frost line, and remove all expansive (clays), compressive, and frost-susceptible soils. Be careful when excavating or backfilling to avoid creating hard and soft spots (Fig. 3–10). Use only stable, fully compactible material for replacement fill. A subbase is not required, but it can even out subgrade irregularities. Crushed stone, ¾ inch preferred, gravel free of fines, and coarse sand (¹⁄₁₆-inch grains minimum) are suitable, and provide a capillary break. Compact the subbase material to a high density, and do not exceed 5 inches in thickness. Thicker subbases do not significantly increase subgrade support, nor do they allow a reduction in slab thickness.

Capillarity and Capillary Breaks

Capillarity is the upward movement of liquids in small diameter, or capillary (hair-like), tubes. It causes moisture to move into porous materials. A sponge sitting on a wet surface will draw water up into

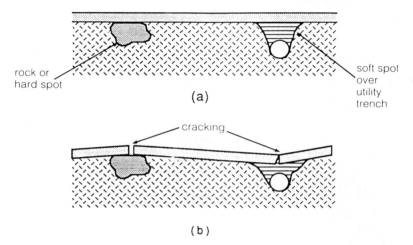

rock or
hard spot

soft spot
over
utility
trench

(a)

cracking

(b)

Fig. 3–10. Hard and soft spots.

itself against the force of gravity, by capillary suction. If a liquid can wet
(stick to) the sides of a capillary tube, and the diameter is small enough,
the liquid will rise in the tube. If it cannot wet the pores, and the pore
diameter is too large, upward movement of liquid will not happen.
Capillarity does not exist in nonporous substances such as glass, steel,
and many plastics. But, the space between two pieces of nonporous
material can become a capillary, if they are close enough together.
Rain-deposited water on clapboard siding can be sucked up between
the laps and held there in spite of gravity.

Proper grading, backfilling with good drainage material such as
gravel—a capillary break—and an effective drainage system to drain
water away from the foundation are necessary for the control of rainwater and melting snow. Unless the water table is high, no water will enter
through cracks in the walls and slab. But this does not guarantee a dry,
warm basement. Because the foundation is a network of capillary
pores, it will draw up moisture from the damp earth, wetting the walls
and slab unless the capillary suction is broken. The result is a damp,
musty cellar, the development of mold and mildew, high interior humidity, salt deposits (efflorescence) on the walls, and an increase in
heat loss as the water evaporates from walls and slab. The capillary action in concrete can draw up water to a height of 6 miles.

Dampproofing

There are two methods of breaking the capillary suction. The first and most common method is dampproofing with bituminous compounds, parging, and waterproofing. The second method utilizes a drainage screen such as rigid fiberglass board, or an open-weave material with air gaps too large to allow capillarity.

Caution—*Dampproofing is not waterproofing.* Dampproofing compounds are designed to retard the transmission (capillarity) of water vapor through concrete. It will not stop water under hydrostatic pressure from penetrating the foundation.

Bitumen is a generic term for asphalt or coal-tar pitch, of which there are two types: cutbacks and emulsions. *Cutbacks* are solvent-thinned; *emulsions* are water-based, solvent-free. Raw asphalt must be heated in order to be applied, and cannot be sprayed on. Bitumens may be applied hot or cold, and brushed, sprayed or trowelled on. If extruded expanded polystyrene (XPS) insulation is to be used on the exterior, dampproofing is unnecessary. XPS is a capillary break. If XPS and dampproofing are used, do not use cutback dampproofing. The solvents in cutbacks are Volatile Organic Compounds (VOCs)—they attack XPS, and are harmful to the environment. *Volatile* means the solvents evaporate into the air. *Organic* refers to hydrocarbons that combine with other compounds and sunlight to form ozone, the primary component in smog. *Emulsion* dampproofing manufactured by Monsey, W.R. Meadows, and Sonneborn, contains no VOCs and does not attack XPS.

Bitumens suffer from a number of disadvantages:

(a) they will not bridge normal-sized structural cracks;
(b) they are subject to to brittleness, cracking, and splitting;
(c) they are ultraviolet sensitive and can bubble from the sun's heat;
(d) they dissolve in the presence of water, especially hard water because of the salts, and eventually vanish;
(e) they can be ruptured or damaged by backfilling, and may require the use of a protection board.

One manufacturer claims his asphalt emulsion will withstand sustained contact with or immersion in soft or hard water. Unfortunately, the other disadvantages remain.

Parging is a ¾-inch or thicker coating of mortar trowelled onto the exterior of the foundation to increase its resistance to moisture penetration. Unequal foundation settlement, drying shrinkage, and temperature drops can crack both the foundation and the parging, and allow moisture to leak in.

Bitumens and other liquid seals are not effective when applied directly to concrete masonry block. The blocks should be parged to seal the large pores, and the parging dampproofed to seal the capillaries. Unfortunately, if the blocks crack, the parge cracks and moisture enters.

Polyethylene sheeting is the lowest cost, best long-term dampproofing solution available. After the form ties have been broken, wrap the entire foundation with a cross-laminated polyethylene such as TU-TUF, SUPER SAMPSON™, GRIFFOLYN™, CROSS-TUFF™, or RUFFCO™. Do NOT use RUFFCO WRAP™. Secure the plastic to the top of the foundation, and let it drape down to the base of the footings. The footing drain system's crushed stone will keep the poly in place. These plastics can stand up to abuse, but for best results leave plenty of slack, tape the joints, and be careful when backfilling. The plastic remains on the foundation during construction. It will bridge any cracks, provide waterproof concrete, eliminate dampproofing and expensive parging. The top of the footings must also be dampproofed to break the capillary path between them and the foundation wall. Use polyethylene plastic sheeting because it allows the form chalk lines to be seen.

Waterproofing

Caution— Waterproofing should not be attempted by any but experienced workers under professional supervision.

Because of a 1000 percent increase in the number and types of waterproofing products, the complexity of the subject, and the skill and experience required to apply them, what follows is a brief coverage of the main points. Waterproofing is the surrounding of a structure with a true highly impermeable membrane. Such a membrane totally prevents water, passive or under pressure, from moving through building materials. A waterproof membrane is only one element in a system which includes:

1. A properly-designed, cured and finished foundation,
2. Grading and landscaping to control rain, melting snow, and flooding,
3. Proper backfilling to prevent percolation and control capillary suction,
4. A subsurface drainage system to draw down the groundwater table and drain away percolated water.

Bituminous Waterproofing—Built-up membranes may be asphalt or coal tar, and may be applied hot or cold. Asphalt is the main ingredient in cold-applied waterproofing. Although coal-tar pitch is more permanent than asphalt, in the early 1970s the Occupational Safety and Health Association (OSHA) declared that many of the compounds in coal tars were toxic and had to be removed. Their removal has resulted in coal tars becoming brittle when used underground.

Rubberized Asphalt contains about 5 percent rubber, and comes in sheets or rolls. It may have a polyethylene sheet inside or on the outside. The polyethylene is ultraviolet sensitive and must be protected. Excessive vibration of the poured concrete can cause dusting of the surface. If the foundation surface is not clean and smooth, the membrane will not adhere properly. Bituthene is a polyethylene-coated rubberized asphalt commonly used as flashing on roofs to protect them from ice dams.

Polymeric Asphalt is a combination of asphalt and certain materials which react chemically to produce a mixture of asphalt and rubber. Polymeric asphalts have good crack bridging, are little affected by water, and remain flexible down to 0 degrees F. Koch Materials Company (formerly marketed by Owens/Corning) markets a polymeric asphalt called TUFF-N-DRI™, through Koch Certified Independent Waterproofing Contractors. It is hot-spray applied. Koch Materials Company guarantees the product for 10 years when properly applied with *positive drainage and grading*. This acknowledges that proper grading and drainage is part of the waterproofing system. The TUFF-N-DRI may have to be protected if backfill contains sharp pieces of rock.

Ethylene Propylene Diene Monomer (EPDM) is a single-ply synthetic rubber. Its weight causes it to stretch, making vertical application difficult. It is one of the common roofing materials, and below ground is used basically for flashing.

Neoprene is a synthetic rubber, resistant to chemical attack, and used underground as flashing. It is more pliable and easier to splice together than other synthetic rubbers. Although tough and durable, it is subject to cracking. The cracking results in considerable stress on the membrane and it begins to tear along the crack.

Bentonite clays are of volcanic origin. Millions of years ago, volcanic ash falling into the oceans mixed with the saltwater to form sodium bentonite. Bentonite is a type of rock made up principally of Montmorillonite, and is named after the town of Montmorillon, France, where it was first discovered. Bentonite is so fine that one cubic inch has trillions of individual particles whose total surface area is nearly an acre—43,000 ft^2. The chemistry of bentonite is too complex to explain. Basically, when mixed with water, a molecular change takes place that creates a highly expansive, impermeable seal against water. Raw bentonite may be sprayed or trowel-applied, or dry bentonite may be sprayed with polymer binder. It dries quickly, should be protected and backfilled as soon as possible. Bentonite can also be obtained in cardboard panels, which are installed shingle-fashion with 1½-inch headlaps. The cardboard biodegrades leaving a bentonite slurry as the waterproofing material. The panels can be installed using unskilled labor, but the joints and seams in any waterproofing membrane are the weak spots and can be the undoing of the waterproofing. Care in installing must be used. Cold weather application is possible because the panels are applied dry, but they must be protected against wetting.

Drain Screens

The classic drain screen system is composed of free draining gravel placed directly against the foundation wall, and footing drains. The gravel allows water to readily move downward, rather than horizontally, toward the subsurface drainage system. The footing drains lower the groundwater table, and collect and move percolated water away from the foundation (Fig. 3–11). This system worked well, and kept basements dry even if holes or cracks existed in the wall. Over the years competitive pressures reduced the use of freely draining materials. Clayey, silty, and organic soils were substituted. Backfilling with dirty soils, and trying to compensate with footing drains, does not stop perched water from entering holes in the foundation. Water draining down through permeable layers is stopped by impermeable layers. It

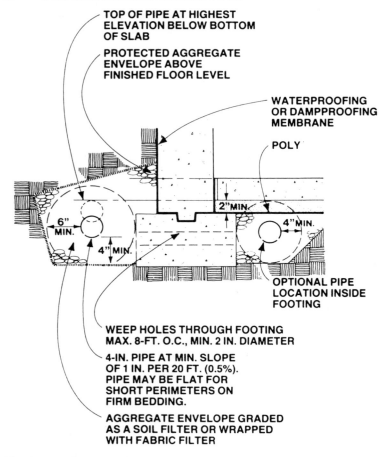

TOP OF PIPE AT HIGHEST
ELEVATION BELOW BOTTOM
OF SLAB

PROTECTED AGGREGATE
ENVELOPE ABOVE
FINISHED FLOOR LEVEL

WATERPROOFING
OR DAMPPROOFING
MEMBRANE

POLY

2" MIN.

6" MIN.

4" MIN.

4" MIN.

4" MIN.

OPTIONAL PIPE
LOCATION INSIDE
FOOTING

WEEP HOLES THROUGH FOOTING
MAX. 8-FT. O.C., MIN. 2 IN. DIAMETER

4-IN. PIPE AT MIN. SLOPE
OF 1 IN. PER 20 FT. (0.5%).
PIPE MAY BE FLAT FOR
SHORT PERIMETERS ON
FIRM BEDDING.

AGGREGATE ENVELOPE GRADED
AS A SOIL FILTER OR WRAPPED
WITH FABRIC FILTER

Fig. 3–11. Footing with perimeter drainage. *(Courtesy ORNL)*

sits, or perches, and builds up to become a perched water table. This buildup is stopped by horizontal flow through the soil, and unfortunately, through holes in the foundation as well.

A technique developed by the Scandinavians and used in Canada since the 1970s uses rigid fiberglass board as an exterior insulation, and as a drainage screen. The high-density fiberglass (6.5 lb/ft^3) has an R value of R-4.5 per inch. Its R value is not affected by moisture because it drains moisture and does not absorb it. Canadian builders report little or not compression of the fiberglass from backfilling.

The glass fibers are layered, and oriented parallel to each other. This orientation makes it easier for water to run down the fibers, under the influence of gravity, rather than across fiber layers. Even in wet soil the water does not penetrate more than ⅛ inch into the insulation. The bulk of the remaining fibers stay dry. The fiberglass boards must be installed vertically full length down to the footings, where they must connect directly to the perimeter subsurface drainage system. The fiber orientation serves as drainage pathways that conduct the water down through the outer surface of the insulation to the footing drains. This eliminates water pressure against the foundation, and even with holes in the wall, water will not enter. Free-draining membranes can replace the gravel/coarse sand drain screen, but to do so could result in serious damage to the foundation, as noted in the sections, *Backfilling*, and *Correct Backfilling Practice*. But the membranes must be connected to a perimeter subsurface drainage system to prevent ponding of the water at the bottom. The above-grade portion of the membrane must be capped with flashing to prevent water from entering the top edge. In the USA, Koch Materials Company markets through Koch Certified Waterproofing Contractors, a fiberglass drainage membrane called WARM-N-DRI™ (Fig. 3–12).

Other free-draining materials are manufactured from molded expanded polystyrene, and extruded expanded polystyrene, and other plastics. Only a few will be discussed. Dow Chemical, manufacturer of STYROFOAM™, produces a $2' \times 8'$ Styrofoam board called Thermadry (Fig. 3–13). It is available in two thicknesses: 1.5 and 2.25 inches. The panel has ¼-inch vertical and horizontal channels cut into one side. A filtration fabric covers the channels to prevent the build-up of silt in the channel. It is about five times as expensive as regular Styrofoam. Plastic mats such as MIRADRAIN™, and TERRADRAIN™, can also be used, but they too are quite expensive. Both come in panels made up of rows and columns of raised dimples which provide the drainage channels. The panel is covered with a filtration fabric to prevent blocking of the drainage channels.

Subsurface Drainage System

Underground drainage dates back to ancient Rome, where roofing tile was used as drain tile. Over the centuries the short, half-round tiles

Fig. 3–12. Fiberglass drain screen. *(Courtesy of Koch Materials Co.)*

were replaced with longer circular tubes, and eventually holes were added to one side. Today, corrugated plastic tubing, with slots located in the valleys of the corrugations along the full length of the tubing, is replacing the tile with holes on one side. Tubes with slots everywhere eliminate the guesswork—Where do the holes go? Up? Down? Or sideways? (Fig. 3–14).

The simplest form of subsurface drainage is a French drain, which consists of a drain envelope only. The drain envelope may be laid in a trench or on the subgrade. It may be coarse angular sand, pea gravel, crushed stone, crushed slag, coarse soils, without a drain pipe, to collect and move water. The particles should not be larger than 1 inch, and

Fig. 3–13. Styrofoam™ Thermadry drain screen. *(Courtesy Dow Chemical)*

must be free from stones, frozen earth, large clods, and debris. A protective filter must be installed completely around the drain envelope. Unless there is a high seasonal water table or underground springs, the gravel/crushed stone drain envelope is adequate. Otherwise use a drain pipe enclosed in a drain envelope, which is surrounded completely with a filter. The subsurface drainage system is more commonly called footing drains.

The purpose of a gravel/crushed stone drain envelope, often incorrectly called a filter, is to stop all soil—except for the very fine particles suspended in the drainage water—from entering the drain pipe, and to permit water to move freely into the drain. Gravel has long been used

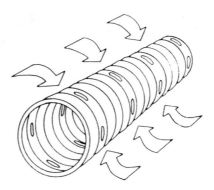

Fig. 3–14. Corrugated plastic tubing. Water entries through slots along full length of drain. *(Courtesy American Drainage Systems)*

for this purpose. Many contractors mistakenly believe that the gravel envelope is a filter, and do not install one around the envelope. Any drain envelope used as a filter is doomed to failure. Initially, it will keep sediment from entering the drain pipe, but as with any filter, it begins to cake up, the flow of water is reduced, and eventually the drainage system fails. Because the performance of the subsurface drainage system is crucial to the performance of the drain screen, it fails when the footing drains fail.

Few footing drains have filters. Typically, #15 felt, or rosin paper, is placed over, but not around, the drainage envelope. Wet rosin paper turns to papier-mâché, leaches into the pipe and eventually clogs it. Water and silt run off the felt, into the laps or joints, or between the edge of the felt and the foundation wall. Never use either of these materials here or on the crushed stone in the leachfield bed. Install a hay, straw, pine needle, or synthetic fabric filter completely around the drainage envelope or drainpipe. Unfortunately, these systems are material- and labor-intensive. Hay and straw can be difficult to install on windy days. A more efficient, less costly solution is slotted polyethylene pipe with an external synthetic fabric drainage envelope/filter sleeve, such as manufactured by American Drainage Systems (ADS).

ADS tubing is a corrugated polyethylene tubing with slots in the valleys of the corrugation running the full length of the tube. Diameters from 3 inches to 24 inches, in coil lengths to 300 feet, are available. A 4-inch diameter pipe is usually used in residential foundations.

Fig. 3–15. (A) Drain Guard. *(Courtesy American Drainage Systems)*

DRAIN GUARD[R] and ADS SOCK™ come with the drain enve-lope/filter installed at the factory. They are self-contained systems that make a gravel or crushed stone drain envelope and filter unnecessary. Slots around the diameter of the pipe assure unrestricted water intake [Fig. 3–15 (A) and (B)].

Caution—In many areas of the country when gravel is either unavail-able or too expensive, sand is used for the drain screen. In others areas clay may be a problem. In these instances, the ADS Sock should be enclosed in a gravel or crushed stone drain envelope.

Place the footing drain at the base of the footing, not on the foot-ing. Do not allow the drain pipe to be higher at any point than the underside of the floor slab. Some authorities call for sloping the drain, even as little as 0.2 percent. The slope should be constant, and transi-tions between sections should be smooth to avoid creating sediment

Fig. 3–15. (B) Drain Sock. *(Courtesy American Drainage Systems)*

traps. However, if the drain line is laid on relatively flat ground, water will not accumulate at the base of the footings, as long as the exit point to daylight is sloped downward. The end of the exit to daylight should be covered with heavy wire mesh animal guard screen. If exit to daylight is not possible, do not lead it to a dry well. The dry well will eventually silt-up, and fail. Route the drainpipe under the footing to a sump inside the basement.

Backfilling

Backfilling is part of several systems: foundation, drain screen, dampproofing/waterproofing, and subgrade drainage. When, for reasons of economy, a contractor: (a) backfills as soon as possible against green concrete to eliminate ramps and speed up the framing; (b) uses compressive, frost-susceptible, organic soils, which may also contain sharp rocks, boulders, and building debris; (c) does not properly place and distribute the soil in layers; (d) ignores the possible collapse of the unbraced foundation; (d) denies the need for a drain screen, he can expect callbacks and warranty claims for cracked, leaky basements, or lawsuits.

Basement cracks and leaks are some of the most common sources of complaints and warranty claims. Some cracking is inevitable, and unavoidable. However, uncontrolled cracking can be avoided using any of the techniques discussed in the section on *Reinforcing*. Our concern here is the cracks and leaks caused by ignoring that backfilling is part of many systems, the use of unstable fill, and improper backfilling procedures.

Protecting the Wall During Backfilling

The walls are best protected by not backfilling until the basement floor slab and the first floor platform are in place. At a bare minimum, key the footings and pour the slab. If the floor is to be poured after the forms are stripped, a keyed footing is unnecessary. Otherwise, key the footing, and use diagonal bracing until the floor is poured and the platform (deck) is installed.

Correct Backfilling Practice

Do not backfill until the foundation has reached at least two-thirds of its rated strength in seven days. If necessary, use high-early strength concrete. The earth pressure against "green" concrete is one of the major causes of cracking.

Use only freely draining soils, free of rocks 6 inches or larger, or building debris, through which water will move down faster than it can move horizontally. Dirty soils can clog the drainpipe filter, cause perched water tables, and adfreezing of the foundation. Begin backfilling diagonally, at the foundation corners, and then fill in the sides. This helps distribute the soil pressures (Fig. 3–16).

Place backfill in 6-inch-thick lifts, and tamp or compact. Although coarse soils tend to compact, a thick, uncompacted layer of gravel can settle an inch or more. Soil settlement next to the foundation, and the resulting water leakage into the basement through cracks, is a common problem. Compacting can prevent this, but care must be used with vibrating plate compactors next to the foundation. Compacting wedges

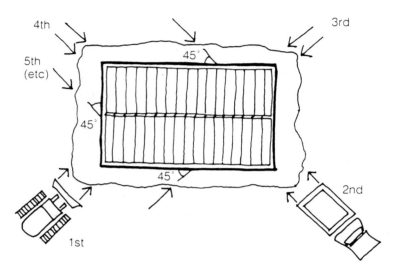

Fig. 3–16. Backfilling diagonally. *(Courtesy Canada Mortgage and Housing Corp.)*

the fill against the wall, greatly increasing the lateral earth pressure on the foundation wall. Overcompaction can create pressures great enough to crack the wall.

Caution—Keep hand-operated vibration plate compactors 6 inches (minimum) away from the foundation. Heavy vibrating roller compactors should be kept 12 to 18 inches away from the wall. Do not allow any heavy equipment to operate parallel to or at right angles to the wall (Fig. 3–17).

The gravel backfill should be 24 inches below the foundation sill. Next to the foundation wall, place 12 inches or more of native soil on the gravel, and taper it down to zero inches, 10 feet away from the foundation. This sloping semi-impervious layer protects the drain screen below, and drains water away from the foundation. Finish off with loam, grading to slope it away from the foundation. Use grass next to the foundation, and avoid crushed stone, marble chips, shrubs and flower beds. If buyer wants shrubs and flower beds, keep them as far

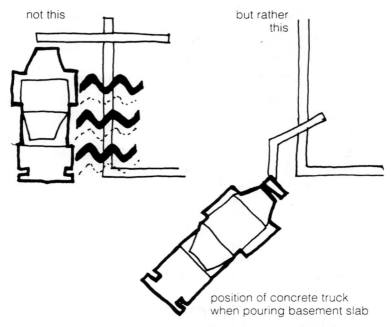

Fig. 3–17. Avoid heavy equipment parallel to the foundation. *(Courtesy Canada Mortgage and Housing Corporation)*

away from the foundation as possible. Drain leaders must terminate on splash block, never below ground.

Permanent Wood Foundations

The Permanent Wood Foundation (PWF) has been in use since the early 1940s, and proven successful in more than 300,000 homes and other structures in the USA and Canada. The PWF is recognized by BOCA, UBC, SBCCI, *CABO One And Two Family Dwelling Code,* FmHA, HUD/FHA and VA.

The PWF has a number of advantages:

(a) Freedom from form contractor schedules.
(b) Can be erected in almost any weather including freezing.

(c) No special framing or framing crew required; contractor uses his regular framers.

(d) Basement is more readily and less expensively turned into habitable space.

(e) Less expensive to insulate than poured concrete or masonry block.

Site Preparation

Site preparation is the same as for conventional foundations. After excavation is completed, lay a minimum of 4 inches of gravel, crushed stone, or coarse sand as a base for the footings, the concrete slab, and as a drainage layer. The base must extend several inches beyond the footing plate. These materials must be clean, and free of clay, silt, or organic matter. Size limitations are: gravel, ¾ inch maximum; coarse sand, ⅟₁₆ inch minimum; crushed stone, ½ inch maximum. Fine sand or pea stone could lead to capillary action, and must be avoided.

The gravel/crushed stone base must be twice the width of the footer plate, and the depth three-fourths the width of the footer plate. A 2″ × 10″ footer plate would require a base depth of 8 inches. After the gravel/crushed stone base is levelled, the footing plate is laid. Use a 2″ × 8″ plate with a 2″ × 6″ wall, and a 2″ × 10″ plate with a 2″ × 8″ wall. Continuous poured concrete footings may be used as an alternative to the plates. However, they must be placed on the stone/sand/gravel base to preserve the continuity of the drainage base. Or, the concrete footings can be set on trenches of crushed stone. Trench depth should be 12 inches minimum. Trench width is determined by footing width. Use a 12″ × 7″ concrete footing for a one story house, and a 15″ × 7″ footing with a two story house. To prevent clogging of the crushed stone drainage base, cover the trench with 12 inches of hay, or use a synthetic filter fabric on the outside of the foundation (Fig. 3–18).

Lumber Treatment

All lumber used in the foundation must be pressure treated with Chromated Copper Arsenate (CCA) in accordance with the requirements of the American Wood Preservers Bureau AWPB-FDN Standard. Each piece must bear the stamp of an approved agency certified to inspect preservative-treated lumber. The mark AWPB-FDN indicates that the lumber is suitable for foundations (FDN), has the re-

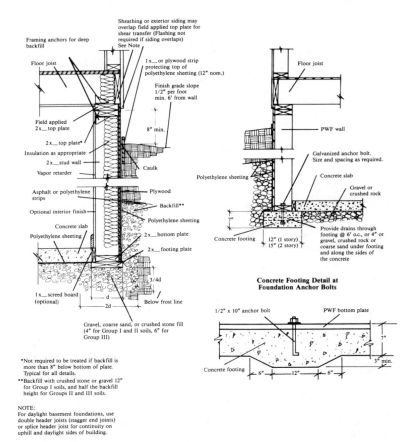

Fig. 3–18. Basement foundation on concrete footing.

quired 0.60 pounds of preservative per cubic foot of wood, and has been kiln-dried to 19 percent moisture content. All cut or drilled lumber must be field-treated with preservative, by repeated brushing until the wood absorbs no more preservative. Any cut end 8 inches or higher above grade, or top plates, headers, or the upper course of plywood, need not be treated. Extend footing plates beyond the corners to avoid cutting them.

Fasteners

All fasteners used in PWF must be corrosion-resistant. The polyethylene or fiberglass moisture retarders cannot be relied on to keep the moisture content low. Aluminum nails, steel nails coated with cadmium, zinc, and cadmium tin, are not suitable for use below ground. Use only type 304 or 316 stainless steel nails. Above grade, hot-dipped galvanized, aluminum, silicon bronze or stainless steel nails may be used.

Wall Framing

The slab may be poured before or after the walls are erected. A concrete slab is an easier work surface than gravel/crushed stone. Many framers find it easier to level the walls on a concrete footing than on a gravel base. Foundation wall framing is no different than above-grade wall framing. Stud crowns should face the plywood, as earth pressure tends to straighten them. Studs are set 16 inches on center and nailed to bottom plates with stainless steel nails. The footing plate can be pre-attached to the bottom wall plate, and the joints staggered one stud. Caulk the external ⅝-inch plywood sheathing at all joints before nailing it to the studs [Fig. 3–19 (A) and (B)]. SilPruf is one of many caulks that may be used. A list of caulks and sealants may be found in American Plywood Association's publication H405, *Caulks and Adhesives for Permanent Wood Foundation System.*

After all walls are raised and braced, wrap the entire exterior of the foundation with one of the cross-laminated polyethylene products such as Tu-Tuf or others mentioned in the section on *Dampproofing.* Attach the poly sheathing to the foundation at the gradeline with a 12-inch pressure-treated grade strip, and caulk it. Drape the poly all the way down to the footings. Tape all laps, and leave plenty of slack (Fig. 3–20). With concrete footings, allow the poly to drape over and beyond the footing. An alternative to the polyethylene is the fiberglass drain screen, discussed under *Drain Screens,* which also provides some extra insulation value. Verify that the fiberglass drain screen is connected to the subsurface drainage system, which in this instance is either the crushed stone in the trench or the gravel/crushed stone base. The subsurface drainage system must be protected with 12 inches of hay, or a synthetic filter fabric. Completed PWF is shown in Fig. 3–21.

Fig. 3–19. (A) Prefab wall sections being erected. *(Courtesy Wood Products Promotion Council)*

Fig. 3–19. (B) Wall sections are caulked before they are nailed together. *(Courtesy Wood Products Promotion Council)*

Fig. 3–20. Polyethylene secured to exterior of PWF with 12″ pressure treated grade strip. *(Courtesy Wood Products Promotion Council)*

Fig. 3–21. Completed PWF ready for concrete slab and floor platform, before backfilling. *(Courtesy Wood Products Promotion Council)*

Backfilling

Backfilling cannot begin until the concrete slab is poured, and the floor joists and plywood subfloor are installed. Each floor joist must be secured to the sill plates with Simpson H1 hurricane braces (Fig. 3–22). Where the joists are parallel to the foundation, the first bay must be braced with solid blocking at 24 inches on center. The second bay is blocked 48 inches on center. Follow the procedures outlined in the previous sections titled *Backfilling* and *Correct Backfilling Practice.*

Frost-Protected Shallow Foundations

All buildings codes in the USA require that foundation footings be placed at or below the maximum expected frost depth. The possible exception to this is the 1990 BOCA code which provides that footings on solid rock ". . . or otherwise protected from frost. . . ." do not have to extend to the frost line. Stone-Age Laplanders built their stone houses directly on the ground, and piled snow around the exterior walls as in-

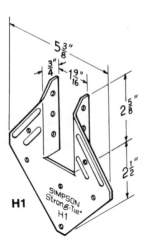

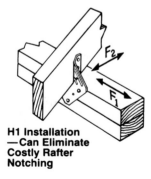

H1 Installation
—Can Eliminate
Costly Rafter
Notching

Fig 3–22. Simpson hurricane braces and installation method. *(Courtesy Simpson Strong-Tie Company)*

sulation. Heat from fires inside the house used for cooking and heating kept the exterior wall safe from frost heaving.

In the early 1930s, Frank Lloyd Wright developed his Usonian House concept. The first Usonian house built in the USA was the 1935 Hoult house in Wichita, Kansas. Usonian houses were a radical departure from the standard American compartmented box with a basement. Wright did away with the basement, and massive concrete foundations as well.

The Hoult house foundation did not go to the frost line. Instead, the foundation was a drained gravel bed on top of which was poured a monolithic slab with a slightly-thickened edge extending six inches below grade (Fig. 3–23). The 1936 Jacobs house in Wisconsin and the 1951 Richardson house in New Jersey were also built on monolithic slabs extending no more than six inches below grade. These houses incorporated underfloor heating—radiant heat, which protected the foundation from frost. Wright may have been the first individual in modern times to use frost-protected shallow foundations, and to understand that footings extending to the frost line are not necessary.

The basis of frost-protected shallow-foundation technology is that heat lost through the foundation keeps the ground under the foundation from freezing and heaving. The thawing effect on the earth sur-

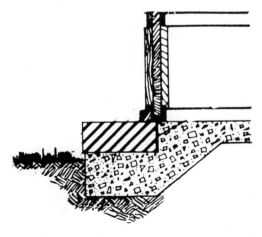

Fig. 3–23. Shallow foundation used by Frank Lloyd Wright in the 1935 Hoult house.

rounding the building tends to move the frost depth closer to the surface; that is, the frost does not penetrate as deeply as it would with an unheated building. Therefore, it is unnecessary to extend the footings to the frost line or deeper. If the foundation is insulated, frost will not penetrate beneath the footings even if the footings are only a 12 inches below grade. The primary heat loss path is beneath the footings (Fig. 3–24).

Norwegian and Swedish engineers, drawing on American research, started in 1946 to perfect the frost-protected shallow foundation. Their research led them to conclude that heat loss through the slabs would protect the ground under the foundation from freezing, and that it was not necessary to extend the footings to the frost line. The Swedish building code allows frost-protected shallow foundations 10 inches below grade anywhere in the country. The Canadian National Building Code now allows shallow foundations. Over one million houses have been built on shallow foundations in Scandinavia. Most of the shallow foundation research has been conducted in Scandinavia,

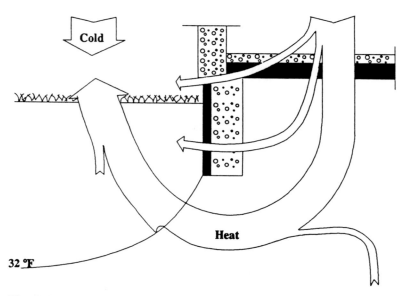

Fig. 3–24. Heat loss path under footings. (*Courtesy NAHB National Research Center*)

Canada, China, and Russia. There has been little frost-protected shallow foundation research in this country. Eich Construction Company of Spirit Lake, Iowa has built about 50 homes in northwest Iowa (8000 DD, 48-inch frost depth) on shallow foundations. The author has designed and built two houses on FPSFs in southern New Hampshire (7100 DD, 42-inch frost depth). These houses are entering their fourth winter, free of problems from frost, or frost heaving.

Shallow Foundation Design Details

There are a number of steps involved in the design/construction of frost protected shallow foundations (FPSF): (1) site preparation; (2) determining freezing degree index; (3) slab insulation; (4) foundation wall insulation; (5) optional horizontal ground insulation, exterior to the foundation.

Site Preparation

Remove all frost-susceptible clays, silts, organic, and compressible soils. A high water table or wet ground could clog the drainage layer with fine soil particles. If these conditions are present, place a synthetic filter fabric down first before laying the drainage layer. Lay down a subbase of six inches of clean gravel (free of fines and dust), coarse sand ($\frac{1}{16}$ inch grains minimum), or screened crushed stone $\frac{3}{4}$ inch to $1\frac{1}{2}$ inches in size. Ideally, the subbase should be deep enough to keep the subgrade free of frost. The gravel base serves as the footings, a drain screen, a capillary break, and is not frost-susceptible. It must be well compacted.

Freezing Degree Days

The air-freezing index, discussed in the section on *Frost Protection,* is a more accurate method of determining frost depth. The higher the freezing index number, the deeper the frost penetration. Fig. 3–25 is based on the *average* values for the three coldest years in the past 30 years. The Norwegian guidelines used here are based on the maximum values for the coldest winter that will occur in 100 years. The values in Fig. 3–25 should be multiplied by 1.4 when designing frost-protected shallow foundations.

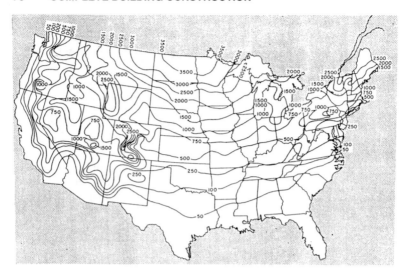

Fig. 3–25. Freezing degree days for the continental United States. *(Courtesy U.S. Army CRREL)*

Subslab Insulation

Subslab insulation keeps the slab warm and reduces heat loss from the house to the ground. R values can range from R-5 to R-10 (1 inch to 2 inches of XPS at R-5 per inch). When used under the load-bearing part of monolithic slabs (thickened edge), or under footings, the loading of the XPS must be calculated. When a load or compressive force is applied to the XPS—or any plastic insulation—it starts deforming or sagging. If the load is constant over a long period of time, the sagging will increase. This gradual increase in the compression of the foam is known as *compressive creep*. The Scandinavians have not experienced any problems with the foam creeping or settling. However, Dow Chemical recommends limiting the compressive creep to 2 percent of the Styrofoam's thickness over a 20-year period. By keeping the load on the Styrofoam to one-third of its design compressive strength, the creep of the insulation will be less than the 2 percent over 20 years.

Example—Square Edge Styrofoam has a rated compressive strength of 25 lb/in^2 (psi). The footing load is 8.5 psi. Using a safety factor of 3, is the 25 psi square edge Styrofoam adequate to support the footing dead load?

1. One-third of 25 psi = 25/3= 8.33 psi. Footing load = 8.5 psi.
2. 25 psi × 144 in^2 = 3600 psf. 3600/3 = 1200 psf. Footing load = 8.5 psi × 144 in^2 = 1224 psf. The answer is yes.

With heavier loads, Dow HI-40, HI-60, or HI-115, rated at 40, 60, and 100 psi respectively, can be used. However, these commercial foams are difficult to obtain in small quantities. Pittsburgh Corning Corporation manufactures a readily available product called FOAMGLAS™. Made of pure glass, it has a compressive strength of 100 psi, and an R value of R-2.78 per inch. A higher density Foamglas is available with a compressive strength of 175 psi. The manufacturer also recommends a safety factor of 3.

Foundation Wall Insulation Thickness

Exterior foundation insulation is most effective for frost protection and slab temperature. Heat from the building is directed down to the wall's bottom edge. The above-grade portion of the XPS must be protected. Acrylic paint can be used, or a stucco mixture of 4 parts MAXCRETE™ to one part THOROSEAL™ Acryl 60 may be troweled on. Before coating, tape the joints in the XPS with self-adhering nylon mesh. Of course, insulation placed on the inside needs no protection. When interior insulation is used, it must also be placed under the wall or the footing as well. As noted above, XPS with adequate compressive strength must be used under walls or footings.

The height above grade of the slab, and the freezing index determines the foundation wall insulation thickness. Table 3–3 lists the thickness for slabs-on-grade structures at various heights above grade. The higher the slab (floor) is above grade the thicker the insulation, in order to ensure that sufficient heat is retained under the foundation.

Table 3–3. Insulation Thickness for Various Floor Heights Above Grade

Maximum Frost Index h °C (see Fig. 3–25) (Freezing Degree Day Equivalents)	Floor's Height Above Grade, mm		
	300 (12″) or less	301 to 450 (12″+) (18″)	451 to 600 (18″+) (24″)
30,000 (2,250) or less	40 (1⅝″)	50 (2″)	60 (2⅜″)
40,000 (3,000)	50 (2″)	60 (2⅜″)	70 (2¾″)
50,000 (3,750)	60 (2⅜″)	70 (2¾″)	80 (3⅛″)
60,000 (4,500)	80 (3⅛″)	90 (3½″)	100 (4″)

(Courtesy NAHB NRC)

Ground Insulation

The need for ground insulation depends on the maximum frost index and the depth of the foundation. In soils susceptible to frost heaving, or in colder climates, additional insulation must be placed horizontally at the base of the foundation at the corners, where heat losses are

Table 3–4. Minimum Foundation Depth with Insulation Only at Corners

Maximum Frost Index h °C (Freezing Degree Day Equivalents) (See Fig. 3–25)	Foundation Wall		
	Vertical Insulation on Outside of Wall	Vertical Insulation on Inside of Wall Insulated concrete block, thickness 250 mm (10″)	Necessary Ground Insulation (polystyrene 30 kg/m³ [1.87 lb/ft³]) at corners and (t × B × L) in mm (See Fig. 5.3)
30,000 (2,250) or less	0.40 (16″)	0.40 (16″)	Ground insulation not necessary
35,000 (2,625)	0.40 (16″)	0.50 (20″)	50 × 500 × 1,000
40,000 (3,000)	0.50 (20″)	0.60 (24″)	(2″ × 20″ × 40″)
45,000 (3,375)	0.60 (24″)	0.70 (28″)	(polystyrene)
50,000 (3,750)	0.70 (28″)	0.85 (33″)	50 × 500 × 1,500
55,000 (4,125)	0.85 (33″)	1.05 (41″)	(2″ × 20″ × 60″)
60,000 (4,500)	1.00 (39″)	1.20 (47″)	(polystyrene)

Minimum foundation depth in meters with ground insulation only at the corners and outside unheated rooms is given. If the foundation walls have interior insulation, optional ground insulation must also be laid under the foundation wall. *(Courtesy NAHB NRC)*

Table 3–5. Thickness, Width, and Length of Foundation Insulation Required When Foundation 16″ Below Grade

Maximum Frost Index h °C (Freezing Degree Day Equivalents) (See Fig. 3–25)	Ground Insulation (polystyrene 30 kg/m³ [1.87 lb/ft³]	
	At Corners: Thickness, width, and length (t × B × L) in mm	Along Walls: Thickness and width (t × B) in mm
30,000 (2,250) or less	Not necessary	Not necessary
35,000 (2,625)	50 × 500 × 1,000 (2″) (20″) (40″)	50 × 250 (2″) (10″)
40,000 (3,000)	50 × 750 × 1,000 (2″) (30″) (40″)	50 × 250 (2″) (10″)
45,000 (3,375)	50 × 750 × 1,500 (2″) (30″) (60″)	50 × 250 (2″) (10″)
50,000 (3,750)	80 × 750 × 1,500 (3⅛″) (30″) (60″)	50 × 500 (2″) (20″)
55,000 (4,125)	80 × 1,000 × 1,500 (3⅛″) (40″) (60″)	80 × 500, 50 × 750 (3⅛″) (20″) (2″) (30″)
60,000 (4,500)	80 × 1,000 × 1,500 (3⅛″) (40″) (80″)	80 × 750 (3⅛″) (30″)

Necessary ground insulation with a foundation depth of 0.4 m (16″). If the foundation wall has inside insulation, the ground insulation must also be laid under the foundation wall. *(Courtesy NAHB NRC)*

higher. In very cold climates, horizontal ground insulation is placed completely around the perimeter of the foundation. Table 3–4 lists the minimum foundation depth with ground insulation only at the corners, and outside unheated rooms. Table 3–5 lists the thickness, width and length of foundation insulation needed when foundation is 16 inches below grade.

CHAPTER 4

Finishing and Curing Concrete

Screeding

To screed is to strike-off or level slab concrete after pouring. Generally, all the dry materials used in making quality concrete are heavier than water. Thus, shortly after placement, these materials will have a tendency to settle to the bottom and force any excess water to the surface. This reaction is commonly called "bleeding." This bleeding usually occurs with non-air-entrained concrete. It is of utmost important that the first operations of placing, screeding, and darbying be performed before any bleeding takes place. The concrete should not be allowed to remain in wheelbarrows, buggies, or buckets any longer than is absolutely necessary. It should be dumped and spread as soon as possible and struck-off to the proper grade, then immediately struck-off, followed at once by darbying. These last two operations should be performed before any free water is bled to the surface. The concrete should not be spread over a large area before screeding—nor should a large area be screeded and allowed to remain before darbying. If any operation is performed on the surface while the bleed water is present, serious scaling, dusting, or crazing can result. This point cannot be overemphasized and is the basic rule for successful finishing of concrete surfaces.

The surface is struck off or rodded by moving a straightedge back and forth with a sawlike motion across the top of the forms or screeds. A small amount of concrete should always be kept ahead of the straightedge to fill in all the low spots and maintain a plane surface. For most slab work, screeding is usually a two-man job because of the size of the slab.

Tamping or Jitterbugging

The hand tamper or jitterbug is used to force the large particles of coarse aggregate *slightly* below the surface in order to enable the cement mason to pass his darby over the surface without dislodging any large aggregate. After the concrete has been struck-off or rodded, and in some cases tamped, it is smoothed with a darby to level any raised spots and fill depressions. Long-handled floats of either wood or metal, called *bull floats,* are sometimes used instead of darbies to smooth and level the surface.

The hand tamper should be used sparingly and in most cases not at all. If used, it should be used only on concrete having a low slump (1 inch or less) to compact the concrete into a dense mass. Jitterbugs are sometimes used on industrial floor construction because the concrete for this type of work usually has a very low slump, with the mix being quite stiff and perhaps difficult to work.

Finishing

When the bleed water and water sheen have left the surface of the concrete, finishing may begin. Finishing may take one or more of several forms, depending on the type of surface desired.

Finishing operations must not be overdone, or water under the surface will be brought to the top. When this happens, a thin layer of cement is also brought up and later, after curing, it becomes a scale that will powder off with usage. Finishing can be done by hand or by rotating power-driven trowels or floats. The size of the job determines the choice, based on economy.

The type of tool used for finishing affects the smoothness of the concrete. A wood float puts a slightly rough surface on the concrete. A steel (or other metal) trowel or float produces a smooth finish. Extra-rough surfaces are given to the concrete by running a stiff-bristled broom across the top.

Floating

Most sidewalks and driveways are given a slightly roughened surface by finishing with a float. Floats may be small, handheld tools (Fig. 4–1), with the work done while kneeling on a board (Fig. 4–2), or they may be on long handles for working from the edge. Fig. 4–3 shows a workman using a long-handled float, and Fig. 4–4 shows the construction details for making a float.

When working from a kneeling board, the concrete must be stiff enough to support the board and the workman's weight without deforming. This will be within two to five hours from the time the surface water has left the concrete, depending on the type of concrete, any admixtures included, plus weather conditions. Experience and testing the condition of the concrete determines this.

Floating has other advantages. It also embeds large aggregate beneath the surface, removes slight imperfections such as bumps and voids, and consolidates mortar at the surface in preparation for smoother finishes, if desired.

Floating may be done before or after edging and grooving. If the line left by the edger and groover is to be removed, floating should follow the edging and grooving operation. If the lines are to be left for decorative purposes, edging and grooving will follow floating.

Fig. 4–1.   Typical wood float.

Fig. 4–2. Using the hand wood float from the edge of a slab. Often the worker will work out the slab on a kneeling board.

Troweling

Troweling, when used, follows floating. The purpose of troweling is to produce a smooth, hard surface. For the first troweling, whether by power or by hand, the trowel blade must be kept as flat against the surface as possible. If the trowel blade is tilted or pitched at too great an angle, an objectionable "washboard" or "chatter" surface will result. For first troweling, a new trowel is not recommended. An older trowel that has been "broken in" can be worked quite flat without the edges digging into the concrete. The smoothness of the surface could be improved by timely additional trowelings. There should necessarily be a

Fig. 4–3. The long-handled float.

lapse of time between successive trowelings in order to permit the concrete to increase its set. As the surface stiffens, each successive troweling should be made by a smaller-sized trowel to enable the cement mason to use sufficient pressure for proper finishing.

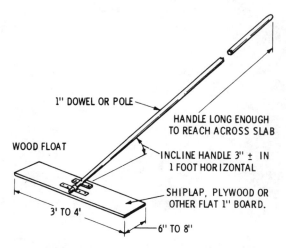

1" DOWEL OR POLE

HANDLE LONG ENOUGH
TO REACH ACROSS SLAB

WOOD FLOAT

INCLINE HANDLE 3" ± IN
1 FOOT HORIZONTAL

SHIPLAP, PLYWOOD OR
OTHER FLAT 1" BOARD.

3' TO 4'

6" TO 8"

Fig. 4–4. Construction details for making a long-handled float.

Brooming

For a rough-textured surface, especially on driveways, brooming provides fine scored lines for a better grip for car tires. Brooming lines should always be at right angles to the direction of travel.

For severe scoring, use a wire brush or stiff-bristled push broom. This operation is done after floating. For a finer texture, such as might be used on a factory floor, use a finer-bristled broom. This operation is best done after trowelling to a smooth finish.

Brooming must be done in straight lines (Fig. 4–5), never in a circular motion. Draw the broom toward you, one stroke at a time, with a slight overlap at the edge of each stroke.

Fig. 4–5. A stiff-bristled broom puts parallel lines in the concrete for a better grip.

Grooving and Edging

In any cold climate there is a certain amount of freezing and thawing of the moist earth under the concrete. When water freezes, it expands. This causes heaving of the ground under the concrete, and this heaving can cause cracking of the concrete in random places.

Sometimes the soil base will settle because all air pockets were not tamped out, or because a leaky water pipe under the soil washed some of it away. A root of a nearby tree under that part of the soil can cause it to lift as the root grows. For all these reasons, the concrete can be subjected to stresses that can cause random cracking, even years later.

To avoid random cracking due to heaving, grooves are cut into the concrete at intervals. These grooves will become the weakest part of the concrete, and any cracking will occur in the grooves. Since, in many cases, heaving or settling cannot be avoided, it is better to have cracks occur in the least conspicuous place possible.

Run a groover across the walk, using a board as a guide to keep the line straight, as shown in Fig. 4–6. About a 1 inch deep groove will be cut, and at the same time, a narrow edge of smoothed concrete will be made by the flat part of the groover.

A rounded edge should be cut along all edges of concrete where it meets the forms, with an edging tool. Running it along the edge of the concrete, between the concrete and the forms, puts a slight round to the edge of the concrete, which helps prevent the edges from cracking off and also gives a smooth-surfaced border. See Fig. 4–7 for details.

In Fig. 4–8, masons are putting the finishing touches on a concrete sidewalk. One man is using an edger, while two men are floating the surface. In Fig. 4–9, a floor slab for a home has just been finished. It has a rough texture since the floor will be covered with carpeting when the house is up and ready for occupancy. Water and sewer lines were laid in position before the concrete was poured.

Finishing Air-Entrained Concrete

Air entrainment gives concrete a somewhat altered consistency that requires a little change in finishing operations from those used with non-air-entrained concrete.

Air-entrained concrete contains microscopic air bubbles that tend

Fig. 4–6. Cutting a groove in a walk. If any cracking occurs, it will be in the groove, where it is less conspicuous.

to hold all the materials in the concrete (including water) in suspension. This type of concrete requires less mixing water than non-air-entrained concrete, and still has good workability with the same slump. Since there is less water, and it is held in suspension, little or no bleeding occurs. This is the reason for slightly different finishing procedures. With no bleeding, there is no waiting for the evaporation of free water from the surface before starting the floating and troweling operation. This means that general floating and troweling should be started sooner—before the surface becomes too dry or tacky. If floating is done by hand, the use of an aluminum or magnesium float is essential.

Fig. 4–7. An edger being used to round off the edge of a driveway.

A wood float drags and greatly increases the amount of work necessary to accomplish the same result. If floating is done by power, there is practically no difference between finishing procedures for air-entrained and non-air-entrained concrete, except that floating can start sooner on the air-entrained concrete.

Practically all horizontal surface defects and failures are caused by finishing operations performed while bleed water or excess surface moisture is present. Better results are generally accomplished, therefore, with air-entrained concrete.

Curing

Two important factors affect the eventual strength of concrete.

1. The water/cement ratio must be held constant. This was discussed in detail in previous chapters.

Fig. 4–8.   Finishing a concrete sidewalk.

2. Proper curing is important to eventual strength. Improperly cured concrete can have a final strength of only 50% of that of fully cured concrete.

It is hydration between the water and the cement that produces strong concrete. If hydration is stopped due to evaporation of the water, the concrete will become porous and never develop the compressive strength it is capable of producing.

The following relates various curing methods and times compared to the 28-day strength of concrete when moist-cured continuously at 70°F.

1. Completely moist-cured concrete will build to an eventual strength of over 130% of its 28-day strength.
2. Concrete moist-cured for 7 days, and allowed to air dry the re-

Fig. 4–9. The finished slab of a home under construction. Water and sewer pipes were placed before concrete was poured.

mainder of the time, will have about 90% of the strength of example 1 at 28 days and only about 75% of eventual strength.

3. Concrete moist-cured for only 3 days will have about 80% of the 28-day strength of example 1 and remain that way throughout its life.

4. Concrete given no protection against evaporation will have about 52% of 28-day strength, and remain that way.

Curing, therefore, means applying some means of preventing evaporation of the moisture from the concrete. It may take the form of adding water, applying a covering to prevent evaporation, or both.

Curing Time

Hydration in concrete begins to take place immediately after the water and cement are mixed. It is rapid at first, then tapers off as time goes on. Theoretically, if no water ever evaporates, hydration goes on continuously. Practically, however, all water is lost through evaporation, and after about 28 days, hydration nearly ceases, although some continues for about a year.

Actually, curing time depends on the application, the temperature, and the humidity conditions. Lean mixtures and large massive structures, such as dams, may call for a curing period of a month or more. For slabs laid on the earth, with a temperature around 70°F and humid conditions with little wind, effective curing may be done in as little as 3 days. In most applications, curing is carried for 5 to 7 days.

Table 4–1 shows the relative strength of concrete between an ideal curing time and a practical time. The solid line is the relative strength of concrete kept from any evaporation. Note that its strength continues to increase, but at a rather slow rate with increase in time. The dotted lines is the relative strength of concrete that has been cured for 7 days, then allowed to be exposed to free air after that. Strength continues to build until about 28 days, then levels off to a constant value after that.

Table 4–1.   Relative Concrete Strength Versus Curing Method

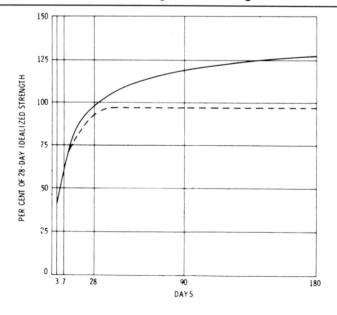

Curing methods should be applied immediately on concrete in forms, and immediately after finishing of flat slabs.

Curing Methods

On flat surfaces such as pavements, sidewalks, and floors, concrete can be cured by *ponding*. Earth or sand dikes around the perimeter of the concrete surface retain a pond of water within the enclosed area. Although ponding is an efficient method for preventing loss of moisture from the concrete, it is also effective for maintaining a uniform temperature in the concrete. Since ponding generally requires considerable labor and supervision, the method is often impractical except for small jobs. Ponding is undesirable if fresh concrete will be exposed to early freezing.

Continuous *sprinkling* with water is an excellent method of curing. If sprinkling is done at intervals, care must be taken to prevent the concrete from drying between applications of water. A fine spray of water applied continuously through a system of nozzles provides a constant supply of moisture. This prevents the possibility of "crazing" or cracking caused by alternate cycles of wetting and drying. A disadvantage of sprinkling may be its cost. The method requires an adequate supply of water and careful supervision.

Wet coverings such as burlap, cotton mats, or other moisture-retaining fabrics are extensively used for curing. Treated burlaps that reflect light and are resistant to rot and fire are available.

Forms left in place provide satisfactory protection against loss of moisture if the top exposed concrete surfaces are kept wet. A soil-soaker hose is an excellent means of keeping concrete wet. Forms should be left on the concrete as long as practicable.

Wood forms left in place should be kept moist by sprinkling, especially during hot, dry weather. Unless wood forms are kept moist, they should be removed as soon as practicable and other methods of curing started without delay.

The application of *plastic sheets* or *waterproof paper* over slab concrete is one of the most popular methods of curing. To do this, sprinkle a layer of water over the slab and lay the sheets on top. Tack the edges of the sheets to the edge forms or screeds to keep the water from evaporating. If the sheets are not wide enough to cover the entire area with one piece, use a 12-inch overlap between sheets. Use white-

Fig. 4–10. Plastic sheeting is a popular covering for curing concrete slabs.

pigmented plastic to reflect the rays of the run, except in cold weather when you want to maintain a warm temperature on the concrete. Waterproof paper is available for the same application. Keep the sheets in place during the entire curing period. Fig. 4–10 shows plastic sheeting being laid on a newly finished walk.

The use of a liquid *curing compound* that may be sprayed on the

Fig. 4–11. A concrete slab with curing compound sprayed on the surface.

Table 4–2. Curing Methods

Method	Advantage	Disadvantage
Sprinkling with water or covering with wet burlap	Excellent results if constantly kept wet.	Likelihood of drying between sprinklings. Difficult on vertical walls.
Straw	Insulator in winter.	Can dry out, blow away, or burn.
Curing compounds	Easy to apply. Inexpensive.	Sprayer needed; inadequate coverage allows drying out; film can be broken or tracked off before curing is completed; unless pigmented, can allow concrete to get too hot.
Moist earth	Cheap, but messy.	Stains concrete, can dry out, removal problem.
Waterproof paper	Excellent protection, prevents drying	Heavy cost can be excessive. Must be kept in rolls; storage and handling problems.
Plastic film	Absolutely watertight, excellent protection. Light and easy to handle.	Should be pigmented for heat protection. Requires reasonable care and tears must be patched; must be weighed down to prevent blowing away.

concrete is increasing in popularity. It is sprayed on from a hand spray or power spray. It forms a waterproof film on the concrete that prevents evaporation. Its disadvantage is that the film may be broken if the concrete bears the weight of a man or vehicle before the curing period is completed. An advantage is that it may be sprayed on the vertical portions of cast-in-place concrete after the forms are removed.

Fig. 4–11 shows a slab after applying a curing compound. The black appearance is the color of the curing compound which is sprayed on top of the concrete. A white-pigmented compound is better when the concrete is exposed to the hot sun. Table 4–2 lists various curing methods, their advantages and disadvantages.

CHAPTER 5

Concrete Block

Many people think of concrete blocks as those gray, unattractive blocks used for foundations and warehouse walls, but this is not true today. Modern concrete blocks come in a variety of shapes and colors and are used for many purposes, including partition walls (Fig. 5–1).

Block Sizes

Concrete blocks are available in many sizes and shapes. Fig. 5–2 shows some of the sizes in common use. They are all sized on the basis of multiples of 4 inches. The fractional dimensions shown allow for the mortar (Fig. 5–3). Some concrete blocks are poured concrete made of standard cement, sand, and aggregate. An 8″ × 8″ × 16″ block weighs about 40 to 50 lbs. Some use lighter natural aggregates, such as volcanic cinders or pumice; and some are manufactured aggregates such as slag, clay, or shale. These blocks weigh 25 to 35 lbs.

In addition to the hollow-core types shown, concrete blocks are available in solid forms. In some areas they are available in sizes other than those shown. Many of the same type have half the height, normally 4 inches, although actually 3⅝ inches to allow for mortar. The 8″ × 8″ × 16″ stretcher (center top illustration of Fig. 5–2) is most frequently used. It is the main block in building a yard wall or a building

(A) Using grille block in a partition wall.

(B) Using standard block and raked joints in a partition wall.

Fig. 5–1.   Examples of concrete block used in home construction.

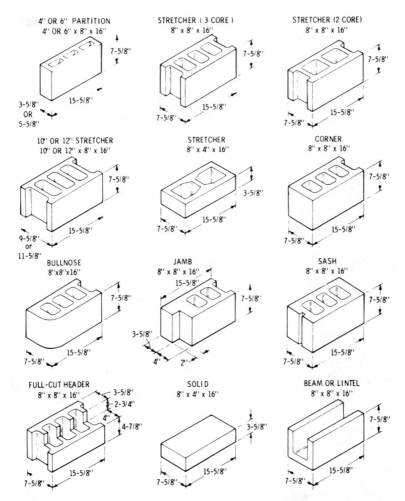

Fig. 5–2. Standard sizes and shapes of concrete blocks.

wall. Corner or bullnose blocks with flat finished ends are used at the corners of walls. Others have special detents for window sills, lintels, and door jambs.

Compressive strength is a function of the face thickness. Concrete blocks vary in thickness of the face, depending on whether they are to

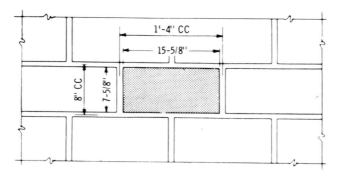

Fig. 5–3. Block size to allow for mortar joints.

be used for non-load-bearing walls, such as yard walls, or load-bearing walls, such as for buildings.

Decorative Block

In addition to standard rectangular forms, concrete masonry blocks are made in unusual designs and with special cast-in colors and finishes to make them suitable architectural designs for both indoor and outdoor construction. A few decorative blocks are described and illustrated in the next few pages.

Split Block

Resembling natural stone, split block is made from standard 8-inch-thick pieces split by the processor into 4-inch-thick facing blocks. The rough side faces out. Split block is usually gray, but some have red, yellow, buff, or brown colors made as an integral part of the cast concrete.

Split block is especially handsome as a low fence. By laying it up in lattice-like fashion, it makes a handsome carport, keeping out the weather but allowing air and light to pass through (Fig. 5–4).

Fig. 5–4. Split block laid up in a latticelike pattern adds a nice touch.

Slump Block

When the processor uses a mix that slumps slightly when the block is removed from the mold, it takes on an irregular appearance like old-fashioned hand molding. Slump block strongly resembles adobe or weathered stone, and it too is available with integral colors and is excellent for ranch-style homes, fireplaces, and garden walls (Fig. 5–5).

Grille Blocks

Some of the most attractive of the new concrete blocks are grille blocks, which come in a wide variety of patterns, a few of which are

Fig. 5–5.   A wall of slump block adds rustic charm and is especially suitable for ranch-type homes.

shown in Fig. 5–6. In addition to its beauty, grille blocks provide the practical protection of a concrete wall, yet allow some sun and light to enter (Fig. 5–7). They are especially useful in cutting the effects of heavy winds without blocking all circulation of air for ventilation. They are usually 4 to 6 inches thick and have faces that are 12 to 16 inches square.

Screen Block

Similar to grille block but lighter in weight and more open, screen block is being used more and more as a facing for large window areas. In this way, beauty and some temperature control is added. They protect the large window panes from icy winter blasts and strong summer

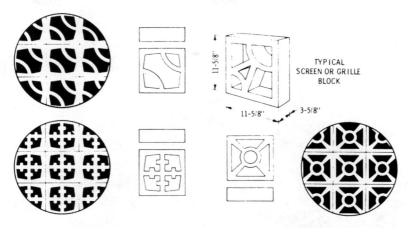

Fig. 5–6. Examples of grille blocks. The actual patterns available vary with processors in different areas.

sun, yet provide privacy and beauty to the home. Special designs are available, or they may be made up by laying single-core standard block on its side as shown in Fig. 5–8.

Patterned Block

Solid block may also be obtained with artistic patterns molded in for unusual effects both indoors and out. Some carry the trade names of Shadowal and Hi-Lite. The first has depressed diagonal recessed sections, and the second has raised half-pyramids. Either can be placed to form patterns of outstanding beauty.

Special Finishes

Concrete block is produced by some manufacturers with a special bonded-on facing to give it special finishes. Some are made with a thermosetting resinous binder and glass silica sand, which give a smooth-faced block. A marbelized finish is produced by another manufacturer, a vitreous glaze by still another. Some blocks may be obtained with a striated bark-like texture. Blocks with special aggregate can be found—some ground down smooth for a terrazo effect.

Fig. 5–7. Outside and inside views of a grille block wall.

Fig. 5–8. A large expanse of glass in a home can be given some privacy without cutting off all light.

Standard Concrete Block

Standard concrete blocks with hollow cores can make handsome walls, depending on how they are laid and on the sizes chosen. Fig. 5–9 shows a solid high wall of standard block which provides maximum privacy. A few are laid hollow core out for air circulation. Fig. 5–10 is standard block with most of the blocks laid sideways to expose the cores. Fig. 5–11 shows single-cored, thin-edged concrete blocks used to support a sloping ground level. It prevents earth runoff and adds an unusual touch of beauty.

In Fig. 5–12, precast concrete block of standard size is laid like brickwork for use as a patio floor. Level the earth and provide a gentle slope to allow for rain runoff. Put in about a 2-inch layer of sand, and

Fig. 5–9. An attractive garden wall with a symmetrical pattern.

lay the blocks in a two-block crisscross pattern. Leave a thin space between blocks and sweep sand into the spaces after the blocks are down. Fig. 5–13 shows a walk made of precast concrete blocks of extra large size. These are not standard, but they are available.

Wall Thickness

Garden walls under 4 feet high can be as thin as 4 inches, but it is best to make them 8 inches thick. Walls over 4 feet high must be at least 8 inches thick to provide sufficient strength.

A wall up to 4 feet high needs no reinforcement. Merely build up the fence from the foundation with block or brick, and mortar. Over 4 feet, however, reinforcement will be required, and the fence should

Fig. 5–10. Standard double-core block laid with cores exposed for ventilation.

be of block, not brick. As shown in the sketch of Fig. 5–14, set ½-inch-diameter steel rods in the poured concrete foundation at 4-foot centers. When you have laid the blocks (with mortar) up to the level of the top of the rods, pour concrete into the hollow cores around the rods. Then continue on up with the rest of the layers of blocks.

In areas subject to possible earthquake shocks or extra high winds, horizontal reinforcement bars should also be used in high walls. Use No. 2 (¼-inch) bars or special straps made for the purpose. Fig. 5–15 is a photograph of a block wall based on the sketch of Fig. 5–14. The foundation is concrete poured in a trench dug out of the ground. Horizontal reinforcement is in the concrete, with vertical members bent up at intervals. High column blocks are laid (16″ × 16″ × 8″) at the vertical rods. The columns are evident in the finished wall of Fig. 5–16.

Fig. 5–11.   Single-core corner blocks used in a garden slope.

Fig. 5–12.   Solid concrete blocks used as a patio floor.

Fig. 5–13. Extra-large precast concrete slabs used to make an attractive walk.

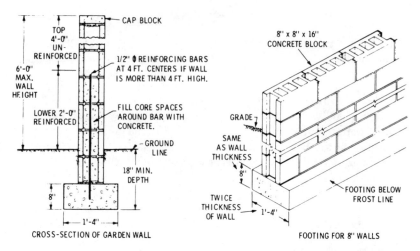

CROSS-SECTION OF GARDEN WALL

FOOTING FOR 8" WALLS

Fig. 5–14. Cross-section and view of a simple block wall. Vertical reinforcement rods are placed in the hollow cores are various intervals.

Fig. 5–15. Vertical reinforcement rods through double-thick column blocks.

Load-bearing walls are those used as exterior and interior walls in residential and industrial buildings. Not only must the wall support the roof structure, but it must bear its own weight. The greater the number of stories in the building, the greater the thickness the lower stories must be to support the weight of the concrete blocks above it, as well as roof structure.

Foundations

Any concrete-block or brick wall requires a good foundation to support its weight and prevent any position shift that may produce

Fig. 5–16. A newly finished concrete-block wall. Note reinforcement columns at various intervals.

cracks. Foundations or footings are concrete poured into forms or trenches in the earth. Chapter 4 describes forms and their construction for footings. For non-load-bearing walls, such as yard fences for example, an open trench with smooth sides is often satisfactory. Recommended footing depth is 18 inches below the grade level. In areas of hard freezes, the footing should start below the frost line.

Footings or foundations must be steel reinforced, with reinforcing rods just above the bed level. Reinforcing rods should be bent to come up vertically at regular intervals into the open cores of column sections of walls, where used, or regular sections of the wall if columns are not used. Columns of double thickness are recommended where the wall height exceeds 6 feet. Even lower height walls are better if double-thick columns are included (Fig. 5–17). The earth bed below the foundation must be well-tamped and include a layer of sand for drainage.

Mortar

Although concrete and mortar contain the same principal ingredients, the purposes they serve, and their physical requirements, are vastly different. Many architects, contractors, and engineers mistakenly

Fig. 5–17. A garden wall of concrete blocks. The columns are spaced 15 feet apart.

believe that the greater the mortar's compressive strength, the better it will perform.

Concrete is a structural element requiring great compressive strength. Mortar serves to bond two masonry units, so its tensile (stretching) bond and flexural bond strength is more important than compressive strength.

The strength of concrete is largely determined by the water/cement ratio: the less water the stronger the concrete. Mortar, however, requires a maximum amount of water consistent with workability, to provide the maximum tensile bond strength. Mortar will stiffen because of evaporation. It can be retempered, to make it more workable, by adding more water.

There are 8 types of portland cement, three of which, IA, IIA, and IIIA, are air-entrained. Because the bonding requirements of mortar differ from the requirements of concrete, not all of these cements are suitable for masonry construction.

ASTM C270–89 is the standard for masonry mortar. The 1990 BOCA *National Building Code* and the 1988 UBC reference ASTM C270. There are two types of mortar cements: portland cement/lime mix, and a masonry cement blend. Portland cement/lime mix is a mixture of portland cement, hydrated lime, sand aggregate. Masonry cement is a preblended, bagged compound, with the type of mortar marked on the bag.

There are five types of cement:

Type M is a high-strength, durable mix recommended for masonry subjected to high compressive loads, severe frost, earth pressure, hurricanes and earthquakes. Its long lasting qualities make it an excellent choice for below-grade foundations, retaining walls, manholes, sewers.

Type S is a medium-high strength mortar, used where high flexural bond strength is required, in structures subjected to normal compressive loads.

Type N is a general purpose mortar for use above in above-grade masonry. This medium-strength mortar is well suited for masonry veneers, and for interior walls and partitions.

Type O is a high-lime, low-strength mortar for use in non-load-bearing walls and partitions. It can be used in exterior veneer that will not be subjected to freezing when wet, and in load-bearing walls subjected to compressive loads of less than 100 psi. It is a very workable mortar, often used in one- and two-story residential structures.

Type K is a very low compressive, low tensile-bond mortar. Its use is limited to non-load-bearing partitions carrying only their own dead weight.

In addition to the five types, there are refractory mortars used in fireplaces and high-heat industrial boilers, chemical-resistant mortars, extra-high-strength mortars, and grouts.

Building with Concrete Blocks

Proper construction of concrete-block walls, whether for yard fencing or building structures, requires proper planning. Standard concrete blocks are made in 4-inch modular sizes. Their size allows for a ⅜-inch-thick mortar joint. By keeping this in mind, the width of a wall and openings for windows and doors may be planned without the need for cutting any of the blocks to fit.

Fig. 5–18 shows the right and wrong way to plan for openings. The illustration on the top did not take into account the 4-inch modular concept, and a number of blocks must be cut to fit the window and door opening. The illustration on the bottom shows correct planning, and no blocks need to be cut for the openings.

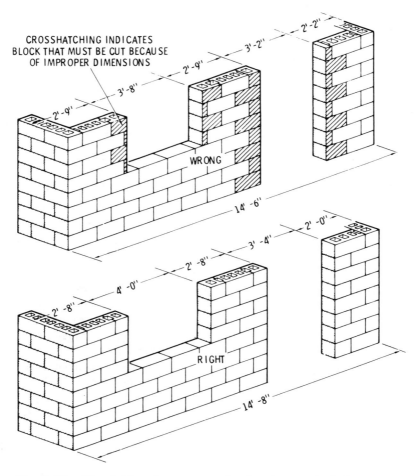

Fig. 5–18. The right and wrong way to plan door and window openings in block walls. *(Courtesy Portland Cement Association)*

Having established the length of a wall on the basis of the 4-inch modular concept (total length should be some multiple of 4 inches), actual construction begins with the corners. Stretcher blocks are then laid between the corners.

Laying Block at Corners

In laying up corners with concrete masonry blocks, place a taut line all the way around the foundation with the ends of the string tied together. It is customary to lay up the corner blocks, three or four courses high, and use them as guides in laying the walls.

A full width of mortar is placed on the footing, as shown in Fig. 5–19, and the first course is built two or three blocks long each way from the corner. The second course is half a block shorter each way than the first course; the third, half a block shorter than the second, etc. Thus, the corners are stepped off until only the corner block is laid. Use a line, and level frequently to see that the blocks are laid straight and that the corners are plumb. It is customary that such special units as corner blocks, door and window jamb blocks, fillers, and veneer blocks be provided prior to commencing the laying of the blocks.

Building the Wall Between Corners

In laying walls between corners, a line is stretched tightly from corner to corner to serve as a guide (Fig. 5–20). The line is fastened to nails or wedges driven into the mortar joints so that, when stretched, it just touches the upper outer edges of the block laid in the corners. The blocks in the wall between corners are laid so that they will just touch the cord in the same manner. In this way, straight horizontal joints are secured. Prior to laying up the outside wall, the door and window

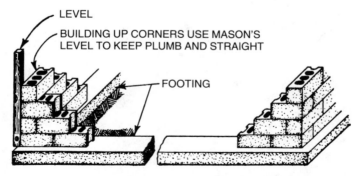

Fig. 5–19. Laying up corners when building with concrete masonry block units.

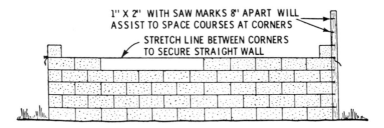

Fig. 5–20. Procedure in laying concrete block walls.

frames should be on hand to set in place as guides for obtaining the correct opening.

Applying Mortar to Blocks

The usual practice is to place the mortar in two separate strips, both for the horizontal or bed joints and for the vertical or end joints, as shown in Fig. 5–21. The mortar is applied only on the face shells of the

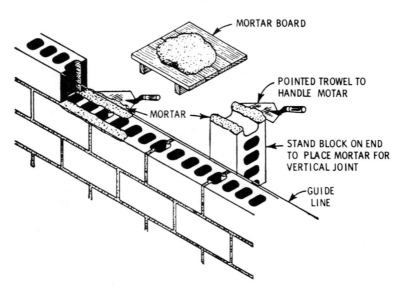

Fig. 5–21. The usual practice in applying mortar to concrete blocks.

block. This is known as *face-shell bedding*. The air spaces thus formed between the inner and outer strips of mortar help produce a dry wall.

Masons often stand the block on end and apply mortar for the end joint as shown in Fig. 5–21. Sufficient mortar is put on to make sure that all joints will be well filled. Some masons apply mortar on the end of the block previously laid as well as on the end of the block to be laid next to it to make certain that the vertical joint will be completely filled.

Placing and Setting Blocks

In placing, the block that has mortar applied to one end is picked up, as shown in Fig. 5–22, and placed firmly against the block previously placed. Note that mortar is already in place in the bed or horizontal joints.

Mortar squeezed out of the joints is carefully scraped off with the

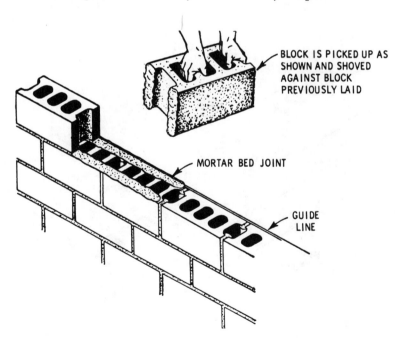

BLOCK IS PICKED UP AS SHOWN AND SHOVED AGAINST BLOCK PREVIOUSLY LAID

MORTAR BED JOINT

GUIDE LINE

Fig. 5–22. **Common method used in picking up and setting concrete blocks.**

trowel and applied on the other end of the block or thrown back onto the mortar board for later use. The blocks are laid to touch the line and are tapped with the trowel to get them straight and level as shown in Fig. 5–23. In a well-constructed wall, mortar joints will average ⅜ inch thick. Fig. 5–24 shows a mason building up a concrete-block wall.

Building Around Door and Window Frames

There are several acceptable methods of building door and window frames in concrete masonry walls. One method used is to set the frames in the proper position in the wall. The frames are then plumbed and carefully braced, after which the walls are built up against them on both sides. Concrete sills may be poured later.

The frames are often fastened to the walls with anchor bolts passing through the frames and embedded in the mortar joints. Another method of building frames in concrete masonry walls is to build open-

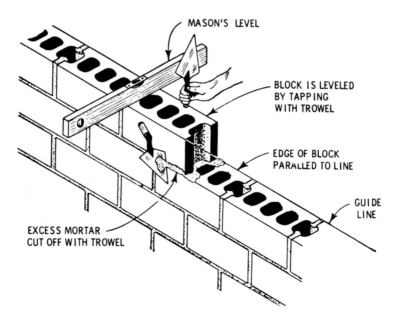

Fig. 5–23. A method of laying concrete blocks. Good workmanship requires straight courses with the face of the wall plumb and true.

(A) Several blocks are receiving mortar on the end.

(B) Blocks are tapped into position.

Fig. 5–24.   Construct of a concrete-block wall.

(C) Excess mortar is removed and alignment checked.

Fig. 5–24.   Construction of a concrete block wall (cont.).

ings for them, using special jamb blocks as shown in Fig. 5–25. The frames are inserted after the wall is built. The only advantage of this method is that the frames can be taken out without damaging the wall, should it ever become necessary.

Placing Sills and Lintels

Building codes require that concrete-block walls above openings shall be supported by arches or lintels of metal or masonry (plain or reinforced). Arches and lintels must extend into the walls not less than 4 inches on each side. Stone or other nonreinforced masonry lintels should not be used unless supplemented on the inside of the wall with iron or steel lintels. Fig. 5–26 illustrates typical methods of inserting concrete reinforced lintels to provide for door and window openings. These are usually prefabricated, but may be made up on the job if de-

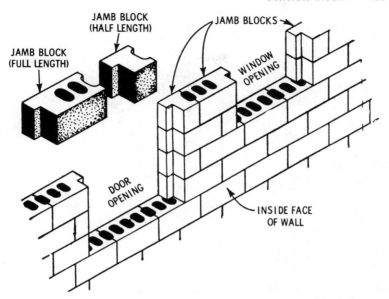

Fig. 5–25. A method of laying openings for doors and windows.

sired. Lintels are reinforced with steel bars placed 1½ inches from the lower side. The number and size of reinforcing rods depend upon the width of the opening and the weight of the load to be carried.

Sills serve the purpose of providing watertight bases at the bottom of wall openings. Since they are made in one piece, there are no joints for possible leakage of water into walls below. They are sloped on the top face to drain water away quickly. They are usually made to project 1½ to 2 inches beyond the wall face, and are made with a groove along the lower outer edge to provide a drain so that water dripping off the sill will fall free and not flow over the face of the wall causing possible staining.

Slip sills are popular because they can be inserted after the wall proper has been built, and therefore require no protection during construction. Since there is an exposed joint at each end of the sill, special care should be taken to see that it is completely filled with mortar and the joints packed tight.

Lug sills project into the masonry wall (usually 4 inches at each

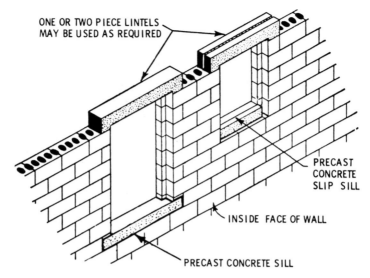

Fig. 5–26. A method of inserting precast concrete lintels and sills in concrete-block wall construction.

end.) The projecting parts are called *lugs*. There are no vertical mortar joints at the juncture of the sills and the jambs. Like the slip sill, lug sills are usually made to project from 1½ to 2 inches over the face of the wall. The sill is provided with a groove under the lower outer edge to form a drain. Frequently, they are made with washes at either end to divert water away from the juncture of the sills and the jambs. This is in addition to the outward slope on the sills.

At the time lug sills are set, only the portion projecting into the wall is bedded in mortar. The portion immediately below the wall opening is left free of contact with the wall below. This is done in case there is minor settlement or adjustments in the masonry work during construction, thus avoiding possible damage to the sill during the construction period.

Basement Walls

Basement walls shall not be less in thickness than the walls immediately above them, and not less than 12 inches for unit masonry walls.

Solid cast-in-place concrete walls are sometimes reinforced with at least one ⅜-inch deformed bar (spaced every 2 feet) continuous from the footing to the top of the foundation wall. Basement walls with 8-inch hollow concrete blocks frequently prove very troublesome. All hollow block foundation walls should be capped with a 4-inch solid concrete block, or else the core should be filled with concrete.

Building Interior Walls

Interior walls are built in the same manner as exterior walls. Load-bearing interior walls are usually made 8 inches thick; partition walls that are not load bearing are usually 4 inches thick. The recommended method of joining interior load-bearing walls to exterior walls is illustrated in Fig. 5–27.

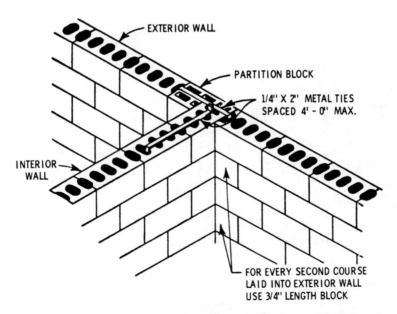

Fig. 5–27.   Detail of joining an interior and exterior wall in concrete-block construction.

Building Techniques

Sills and plates are usually attached to concrete block walls by means of anchor bolts, as shown in Fig. 5–28. These bolts are placed in the cores of the blocks, and the cores filled with concrete. The bolts are spaced about 4 inches apart under average conditions. Usually ½-inch bolts are used and should be long enough to go through two courses of blocks and project through the plate about 1 inch to permit use of a large washer and anchor bolt nut.

Installation of Heating and Ventilating Ducts

These are provided for as shown on the architect's plans. The placement of the heating ducts depends on the type of wall—whether it is load-bearing or not. A typical example of placing the heating or ventilating ducts in an interior concrete masonry wall is shown in Fig. 5–29.

Interior concrete-block walls which are not load-bearing, and which are to be plastered on both sides, are frequently cut through to provide for the heating duct, the wall being flush with the ducts on either side. Metal lath is used over the ducts.

Electrical Outlets

These are provided for by inserting outlet boxes in the walls, as shown in Fig. 5–30. All wiring should be installed to conform with the requirements of the *National Electrical Code (NEC)* and local codes in the area.

Fill Insulation

Concrete masonry is a good conductor of heat, and a poor insulator. In hot arid regions, uninsulated concrete masonry is one means of keeping a house cool during the hot day, and warm during the cold nights. But in northern climates the R value of such a wall is too low, and must be increased.

A concrete masonry block wall may be insulated on the exterior, on the interior, or the block cavities may be filled with insulation. Special foam plastic inserts are made for this purpose, but they do not com-

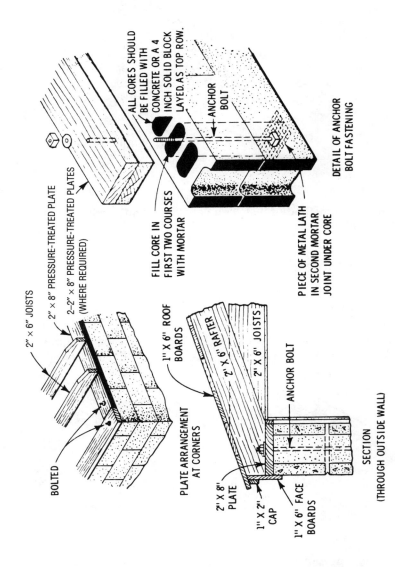

ALL CORES SHOULD BE FILLED WITH CONCRETE OR A 4 INCH SOLID BLOCK LAYED AS TOP ROW.

ANCHOR BOLT

DETAIL OF ANCHOR BOLT FASTENING

FILL CORE IN FIRST TWO COURSES WITH MORTAR

PIECE OF METAL LATH IN SECOND MORTAR JOINT UNDER CORE

2" × 6" JOISTS

2" × 8" PRESSURE-TREATED PLATE

2–2" × 8" PRESSURE-TREATED PLATES (WHERE REQUIRED)

BOLTED

PLATE ARRANGEMENT AT CORNERS

1" X 6" ROOF BOARDS

2" X 6" RAFTER

2" X 6" JOISTS

ANCHOR BOLT

2" X 8" PLATE

1" X 2" CAP

1" X 6" FACE BOARDS

SECTION (THROUGH OUTSIDE WALL)

Fig. 5–28. Details of methods used to anchor sills and plates to concrete-block walls.

127

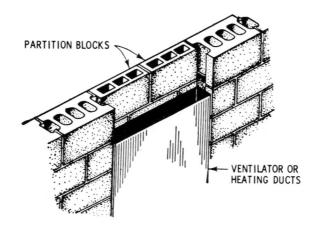

Fig. 5–29. A method of installing ventilating and heating ducts in concrete-block walls.

pletely fill the cavity. A better method is to fill the cavities with a loose fill insulation such as ZONOLITE™ (Fig. 5–31) manufactured by W. R. Grace. Zonolite is actually vermiculite, which is made from expanded mica. It does not burn, rot, settle, and is treated to repel water. Unfortunately, neither the plastic inserts nor Zonolite will stop the heat loss through the webs of the blocks. Another insulation that can be used is a cement-based foam called AIR-KRETE™. It is nontoxic, does

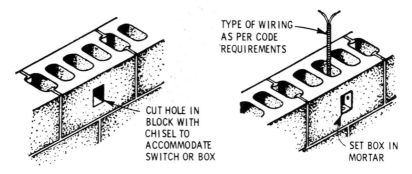

Fig. 5–30. A method of installing electrical switches and outlet boxes in concrete-block walls.

Fig. 5–31. Zonolite™ insulation used to fill cavity in concrete block.
(Courtesy W. R. Grace)

not burn or rot, and is insectproof. At its rated density, it has an R-value of 3.9 per inch. Tripolymer foam, with an R-4.8 per inch, can also be used. However, check with your local building official as to whether or not it is permitted in your state. Do not use urea formaldehyde foam insulation (UFFI).

Flashing

Adequate flashing with rust- and corrosion-resisting material is of the utmost importance in masonry construction, since it prevents water from getting through the wall at vulnerable points. Points requiring protection by flashing are:

1. Tops and sides of projecting trim under coping and sills,
2. At the intersection of a wall and the roof,
3. Under built-in gutters,
4. At the intersection of a chimney and the roof,
5. At all other points where moisture is likely to gain entrance.

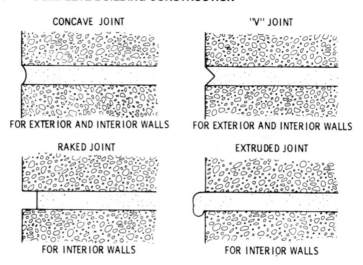

CONCAVE JOINT

"V" JOINT

FOR EXTERIOR AND INTERIOR WALLS

FOR EXTERIOR AND INTERIOR WALLS

RAKED JOINT

EXTRUDED JOINT

FOR INTERIOR WALLS

FOR INTERIOR WALLS

Fig. 5–32.   Four joint styles popular in block wall construction.

Fig. 5–33.   Brick wall with extruding joint construction.

Fig. 5–34. Block with V-joints. Block can be painted—pad painter works well on it.

Flashing material usually consists of No. 26 gauge (14-oz.) copper sheets or other approved noncorrodible material.

Types of Joints

The concave and V-joints are best for most areas. Fig. 5–32 shows four popular joints. While the raked and the extruded styles are recommended for interior walls only, they may be used outdoors in warm climates where rains and freezing weather are at a minimum. In climates where freezes can take place, it is important that no joint permits water to collect.

In areas where the raked joint can be used, you may find it looks handsome with slump block. The sun casts dramatic shadows on this type of construction. Standard blocks with extruded joints have a rustic look, and make a good background for ivy and other climbing plants (Fig. 5–33).

Tooling the Joints

Tooling consists of compressing the squeezed-out mortar of the joints back tight into the joints and taking off the excess mortar. The tool should be wider than the joint itself (wider than ½ inch). You can make an excellent tooling device from ¾-inch copper tubing bent into an S-shape. By pressing the tool against the mortar, you will make a concave joint—a common joint but one of the best. Tooling not only affects appearance but it makes the joint watertight, which is the most important function. It helps to compact and fill voids in the mortar. Fig. 5–34 shows V-joint tooling of block.

CHAPTER 6

Chimneys and Fireplaces

The term *chimney* generally includes both the chimney proper and (in house construction) the fireplace. There is no part of a house that is more likely to be a source of trouble than a chimney that is improperly constructed. Accordingly, it should be built so that it will be strong and designed and proportioned so that it gives adequate draft.

For strength, chimneys should be built of solid brick work and should have no openings except those required for the heating apparatus. If a chimney fire occurs, considerable heat may be engendered in the chimney, and the safety of the house will then depend on the integrity of the flue wall. A little intelligent care in the construction of fireplaces and chimneys will prove to be the best insurance. As a first precaution, all wood framing of floor and roof must be kept at least 2 inches away from the chimney and no woodwork of any kind should be projected into the brickwork surrounding the flues (Fig. 6–1).

When it is understood that the only power available to produce a natural draft in a chimney is that due to the small difference in weight of the column of hot gases in the chimney and of a similar column of cold air outside, the necessity of properly constructing the chimney so that the flow of gases will encounter the least resistance should be clear.

The intensity of chimney draft is measured in inches of a water column sustained by the pressure produced and depends on:

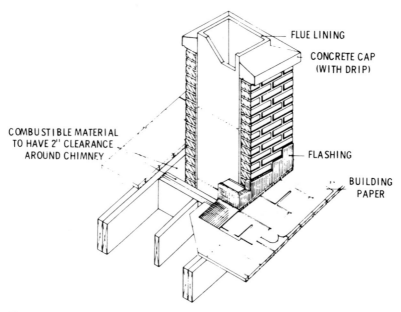

Fig. 6–1. Chimney construction above the roof.

1. The difference in temperature inside and outside the chimney,
2. The height of the chimney.

Theoretical draft in inches of water at sea level is as follows: Let,

D = Theoretical draft,
H = Distance from top of chimney to grates,
T = Temperature of air outside the chimney,
T_1 = Temperature of gases in the chimney.

Then,

$$D = 7.00 \, H \, \frac{1}{461 + T} - \frac{1}{461 + T_1}$$

The results obtained represent the theoretical draft at sea level.

For higher altitudes the calculations are subject to correction as follows:

For altitudes (in feet) of	Multiply by
1,000	0.966
2,000	0.932
3,000	0.900
5,000	0.840
10,000	0.694

A frequent cause of poor draft in house chimneys is that the peak of the roof extends higher than the chimney. In such case the wind sweeping across or against the roof will form eddy currents that drive down the chimney or check the natural rise of the gases as shown in Fig. 6–2. To avoid this, the chimney should be extended at least 2 feet higher than the roof, as shown in Fig. 6–3.

In order to reduce to a minimum the resistance or friction due to the chimney walls, the chimney should run as near straight as possible from bottom to top. This not only gives better draft but facilitates cleaning. If, however, offsets are necessary from one story to another, they should be very gradual. The offset should never be displaced so much that the center of gravity of the upper portion falls outside the area of the lower portion. In other words, the center of gravity must fall within the width and thickness of the chimney below the offset.

Flues

A chimney serving two or more floors should have a separate flue for every fireplace. The flues should always be lined with some fire-proof material. In fact, the building laws of large cities provide for this. The least expensive way to build these is to make the walls 4 inches thick, lined with burned clay flue lining. With walls of this thickness, never omit the lining and never replace the lining with plaster. The expansion and contraction of the chimney would cause the plaster to crack and an opening from the interior of the flue would be formed. See that all joints are completely filled with fire clay or refractory mortar.

Flue Lining

When coal burning furnaces were common, they were vented into unlined chimneys. The soot lined and protected the chimney walls. Coal-burning furnaces have been replaced with oil- and gas-fired

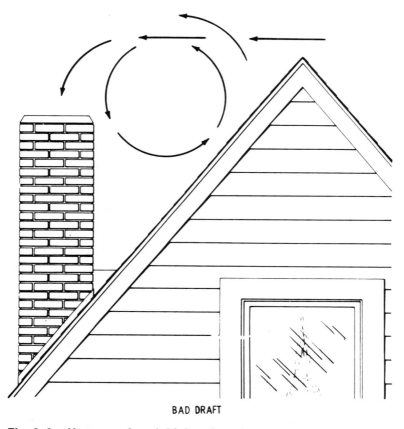

BAD DRAFT

Fig. 6–2. How a roof peak higher than the top of the chimney can cause downdrafts.

furnaces. The impurities in the gas turn into acid which attacks regular mortar. Many manufacturers of gas appliances warn against venting them into masonry chimneys. The *1990 BOCA National Mechanical Code* requires that the ". . . fire clay liner . . . be . . . carefully bedded one on the other in medium-duty refractory mortar. . . ." Although the 1988 *UBC CODE* allows unlined chimneys if the walls are 8 inches thick, they have no protection against acid attack. Do not depend on

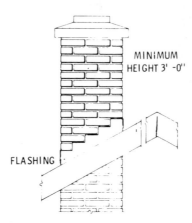

IF 10' OR LESS, CHIMNEY
MUST BE 2'-0'' HIGHER
THAN PEAK OF GABLE

MINIMUM
HEIGHT 3'-0''

FLASHING

Fig. 6–3. Ample clearance is needed between peak of roof and top of chimney. *(Courtesy Structural Clay Products Inst.)*

the building inspector enforcing the use of refractory mortar to bond flues. Verify that fireclay or refractory mortar will be used and is being used.

Clay lining for flues also follows the modular system of sizes. Table 6–1 lists currently available common sizes. The flue lining should extend the entire height of the chimney, projecting about 4 inches above the cap and a slope formed of cement to within 2 inches of the top of the lining, as shown in Fig. 6–3. This helps to give an upward direction to the wind currents at the top of the flue and tends to prevent rain and snow from being blown down inside the chimney.

The information given here is intended primarily for chimneys on residential homes. They will usually carry temperatures under 600°F. Larger chimneys, used for schools and other large buildings, have a temperature range between 600°F and 800°F. Industrial chimneys with temperatures above 800°F often are very high and require special engineering for their planning and execution. High-temperature brick chimneys must include steel reinforcing rods to prevent cracking due to expansion and contraction from the changes in temperature.

Table 6–1.   Standard Sizes of Modular Clay Flue Linings

Minimum Net inside Area (sq. in.)	Nominal[1] Dimensions (in.)	Outside[2] Dimensions (in.)	Minimum Wall Thickness (in.)	Approximate Maximum Outside Corner Radius (in.)
15	4 × 8	3.5 × 7.5	0.5	1
20	4 × 12	3.5 × 11.5	0.625	1
27	4 × 16	3.5 × 15.5	0.75	1
35	8 × 8	7.5 × 7.5	0.625	2
57	8 × 12	7.5 × 11.5	0.75	2
74	8 × 16	7.5 × 15.5	0.875	2
87	12 × 12	11.5 × 11.5	0.875	3
120	12 × 16	11.5 × 15.5	1.0	3
162	16 × 16	15.5 × 15.5	1.125	4
208	16 × 20	15.5 × 19.5	1.25	4
262	20 × 20	19.5 × 19.5	1.375	5
320	20 × 24	19.5 × 23.5	1.5	5
385	24 × 24	23.5 × 23.5	1.625	6

(1) Cross section of flue lining shall fit within rectangle of dimension corresponding to nominal size.
(2) Length in each case shall be 24 ± 0.5 inch.

Chimney Construction

Every possible thought must be given to providing good draft, leakproof mortaring, and protection from heat transfer to combustible material. Good draft means a chimney flue without obstructions. The flue must be straight from the source to the outlet. Metal pipes from the furnace into the flue must end flush with the inside of the chimney and not protrude into the flue, as shown in Fig. 6–4. The flue must be straight from the source to the outlet, without any bends, if at all possible. When two sources, such as a furnace and a fireplace, feed the one chimney, they each must have separate flues.

To prevent leakage of smoke and gas fumes from the chimney into the house, and to improve the draft, a special job of careful mortaring must be observed. The layer of mortar on each course of brick must be even and completely cover the bricks. End buttering must be complete. However, it is best to mortar the flue lining lightly between the lining and the brick. Use just enough to hold the lining securely. The air space that is left acts as additional air insulation between the lining and the brick and reduces the transfer of heat.

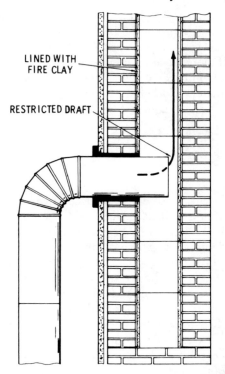

LINED WITH FIRE CLAY

RESTRICTED DRAFT

Fig. 6–4. Furnace pipes must not project into the flue of a chimney.

No combustible material, such as the wood of roof rafters or floor joists, must abut the chimney itself. There should be at least a 2-inch space between the wood and the brick of the chimney, as shown in Fig. 6–1. Brick and flue lining are built up together. The lining clay is placed first and the bricks built up around it. Another section of lining is placed and brick built up, etc.

Chimneys carrying away the exhaust of oil- and coal-burning furnaces, where still used, need a cleanout trap. An airtight cast-iron door is installed at a point below the entrance of the furnace smoke pipe.

Because of the heavier weight of the brick in a chimney, the base must be built to carry the load. A foundation for a residential chimney should be about 4 inches thick. If a fireplace is included, the foundation thickness should be increased to about 8 inches.

After the chimney has been completed, it should be tested for leaks. Build a smudge fire in the bottom and wait for smoke to come out of the top. Cover the top and carefully inspect the rest of the chimney for leaks. If there are any, add mortar at the points of leakage.

Builders should become acquainted with local codes for the construction of chimneys. Consult the applicable building and mechanical codes as well as the National Fire Protection Association (NFPA) codes.

Fireplaces

For many years fireplaces were decorative conversation pieces. With the advent of the OPEC crises, they began to be used as supplemental heat sources. Because of the increased use for longer periods of time and at higher operating temperatures, more care in their design is necessary. Higher temperatures could lead to trouble.

Centuries of trial and error has led to the standardizing of the damper and flue sizes, size of the firebox, the relation of the flue area to the area of the fireplace opening, and so on. Although there are hundreds of individual designs, there are actually three basic types of residential fireplaces: single-face, Rumford, and multi-face.

Single-face fireplaces have existed for centuries, and are the most common type of fireplace in use.

Rumford fireplaces are single-face fireplaces that differ from the deeper, almost straight-sided conventional fireplaces. In conventional fireplaces, roughly 75 percent of the heat is used to heat up exhaust gases, smoke, and the walls of the fireplace. Although this results in a hotter fire, it does not allow much of the radiant heat to enter the room and warm it. The Rumford has widely splayed or flared sides, a shallow back, and a high opening. By making a shallower firebox and flaring the sides, the Rumford fireplace allows more of the radiant heat to enter the room.

Multi-faced fireplaces (Fig. 6–5) are sometimes considered to be contemporary. But they are ancient: for example, corner fireplaces have been in use in Scandinavia for centuries. All the faces, or the opposite or adjacent faces may be opened. Because of the openings, multi-faced fireplaces are not as energy efficient as single-face fireplaces because there is less mass—brick—around the fire to hold and

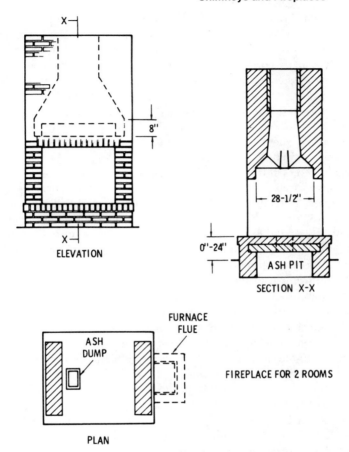

X—|

8"

X—|
ELEVATION

28-1/2"

0'-24"

ASH PIT

SECTION X-X

FURNACE
FLUE

ASH
DUMP

FIREPLACE FOR 2 ROOMS

PLAN

Fig. 6–5. A two-sided fireplace is ideal as a room divider.

radiate heat. They have the advantge of being located within a structure, so less heat is lost to the outside. Their performance may be improved by adding an outside air intake, and glass fire screens, which
should be kept closed when the fireplace is not used.

The efficiency of the common fireplace is somewhere between 0
and 15 percent. And there is little that can be done to increase that
efficiency. Adding glass fire screens and providing an outside air intake
do little to improve the performance. Glass fire screens (doors) do stop

some of the flow of indoor air up the chimney. Keeping the doors closed at night or whenever the fire is dying is important. Unfortunately, while stopping air flow up the chimney, they also stop heat flow into the room. According to researchers at Lawrence Berkeley Laboratory who measured secondary losses from fireplaces, the real efficiency of a fireplace is about 5 percent of the heating value of the wood. It is the most wasteful heating device sold in America. Catherine Beecher, in her 1869 *American Woman's Home*, eliminated all fireplaces because she saw them as dirty and inefficient.

Russian fireplaces or brick masonry heaters, have been in use in northern and eastern Europe for centuries. Figs. 6–6 and 6–7 show two examples of brick masonry fireplaces. The design of a Russian or Finnish fireplace is much too involved to deal with here. Woodburning efficiency of 80 to 90 percent is achieved by controlling the air intake into the firebox, and circulating the hot gases through a system of baffles. The hot combustion gases heat the walls of the heater, which in turn heats the room. For more information see Brick Institute of America, *Technical Notes* 19D and 19E, 1988.

Probably in no other country have so many types and styles of fireplaces been constructed as in the United States. Although the ornamental mantel facings of fireplaces may be of other materials than brick, the chimney and its foundation are invariably of masonry construction. Fig. 6–8 shows a cross section of a fireplace and chimney stack suitable for the average home.

Fireplaces are generally built in the living room or family room. Whichever you choose, the location of the fireplace should allow the maximum of heat to be radiated into the room, with consideration given to making it the center of a conversation area. At one time its location was dictated by the location of the furnace, to make use of a common chimney. Today, with compact heating systems, the chimney is often a metal pipe from the furnace flue, straight through to the roof, not of the brick construction type. The brick fireplace chimney can then be placed to suit the best fireplace location in the home and room.

Fireplace styles vary considerably from a rather large one with a wide opening (Fig. 6–9) to a smaller corner fireplace. You can also get a variety of prefabricated units that the average do-it-yourselfer can install (Fig. 6–10).

Although large pieces of wood can be burned in the larger fireplaces, regardless of size, experience has indicated certain ratios of

Fig. 6–6. Brick masonry heater. *(Courtesy Masonry Heater Association of North America)*

Fig. 6–7. Brick masonry heater. *(Courtesy Masonry Heater Association of North America)*

height, width, depth, etc., should be maintained for best flow of air under and around the burning wood. Recommended dimensions are shown in Table 6–2, which are related to the sketches in Fig. 6–11.

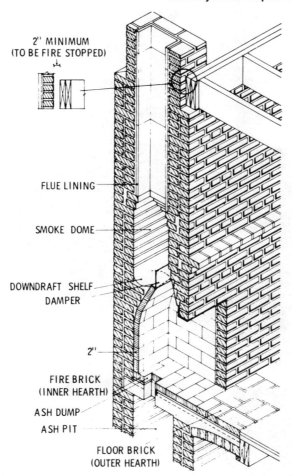

2" MINIMUM
(TO BE FIRE STOPPED)

FLUE LINING

SMOKE DOME

DOWNDRAFT SHELF
DAMPER

2"

FIRE BRICK
(INNER HEARTH)

ASH DUMP

ASH PIT

FLOOR BRICK
(OUTER HEARTH)

Fig. 6–8. Cross section of a typical fireplace and chimney of modern design.

Fireplace Construction

Brick masonry is nearly always used for fireplace construction. Sometimes brick masonry is used around a metal fireplace form trademarked HEATILATOR. While any type of brick may be used for the

Fig. 6–9. **A large fireplace gives off heat and adds a homeyness to a room that nothing else can equal.** *(Courtesy Armstrong)*

Fig. 6–10. Today there are many types of prefab fireplaces.

outside of the fireplace, the fire pit must be lined with a high-tempera-ture fire clay or fire brick.

The pit is nearly always sloped on the back and the sides. This is to reflect forward as much of the heat as possible. The more surface expo-sure that is given to the hot gases given off by the fire, the more heat will be radiated into the room. Fig. 6–8 is a cutaway view of an all-brick fireplace for a home with a basement. The only nonbrick item is the adjustable damper. A basement makes possible a very large ash storage before cleanout is necessary. The ash dump opens into the basement cavity. A cleanout door at the bottom opens inward into the basement.

Fig. 6–12 is a side cutaway view of a typical fireplace for a home built on a concrete slab. It uses the metal form mentioned. The ash pit is a small metal box which can be lifted out, as shown in Fig. 6–13. In some slab home construction, the ash pit is a cavity formed in the con-crete foundation with an opening for cleanout at the rear of the house. A metal grate over the opening prevents large pieces of wood from dropping into the ash pit as shown in Fig. 6–14.

Table 6–2. Recommended Sizes of Fireplace Openings

Opening			Minimum back (hori- zontal) c	Vertical back wall, a	Inclined back wall, b	Outside dimen- sions of standard rectan- gular flue lining	Inside diameter of stan- dard round flue lining
Width, w	Height, h	Depth, d					
Inches	Inches	Inches	Inches	Inches	Inches	Inches	Inches
24	24	16–18	14	14	16	8½ by 8½	10
28	24	16–18	14	14	16	8½ by 8½	10
24	28	16–18	14	14	20	8½ by 8½	10
30	28	16–18	16	14	20	8½ by 13	10
36	28	16–18	22	14	20	8½ by 13	12
42	28	16–18	28	14	20	8½ by 18	12
36	32	18–20	20	14	24	8½ by 18	12
42	32	18–20	26	14	24	13 by 13	12
48	32	18–20	32	14	24	13 by 13	15
42	36	18–20	26	14	28	13 by 13	15
48	36	18–20	32	14	28	13 by 18	15
54	36	18–20	38	14	28	13 by 18	15
60	36	18–20	44	14	28	13 by 18	15
42	40	20–22	24	17	29	13 by 13	15
48	40	20–22	30	17	29	13 by 18	15
54	40	20–22	36	17	29	13 by 18	15
60	40	20–22	42	17	29	18 by 18	18
66	40	20–22	48	17	29	18 by 18	18
72	40	22–28	51	17	29	18 by 18	18

Importance of a Hearth

Every fireplace should include a brick area in front of it where hot wood embers may fall with safety. The plan view of Fig. 6–15 shows a brick hearth built 16 inches out from the fireplace itself as required by the 1990 *BOCA Mechanical Code*. The side of the hearth must extend a minimum of 8 inches on each side of the hearth. These extensions of the hearth are for a fireplace opening of less than 6 square feet. The hearths of larger fireplaces must extend out 20 inches and a minimum of 12 inches on each side of the opening. The hearth and the hearth extension must be constructed of no less than 4 inches of solid masonry. This should be about the minimum distance. Most often the hearth is

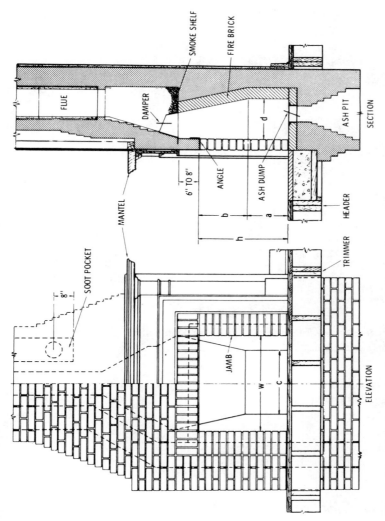

Fig. 6–11. Sketch of a basic fireplace. Letters refer to sizes recommended in Table 6–2.

149

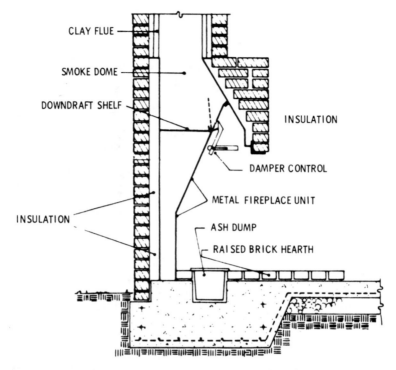

Fig. 6–12. Fireplace built on a concrete slab. *(Courtesy Structural Clay Products Inst.)*

raised several inches above the floor level. This raises the fireplace itself, all of which makes for easier tending of the fire.

In addition to the protection of the floor by means of a hearth, every wood-burning fireplace should have a screen to prevent flying sparks from being thrown beyond the hearth distance and onto a carpeted or plastic tile floor.

Ready-Built Fireplace Forms

There are a number of metal forms available, which make fireplace construction easier. Like the prefab units, they make a good starting point for the handy homeowner who can build a fireplace addition in

Fig. 6–13. Metal lift-out ash box used in many fireplaces built on a concrete slab.

Fig. 6–14. A cast-iron grate over the ash box to keep large pieces of burning wood from falling into the ash box.

ASH DUMP   TERRACOTA FLUE LINING

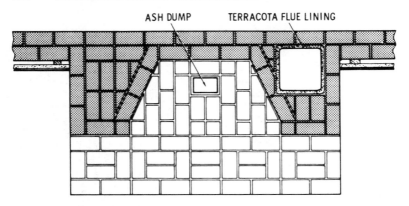

Fig. 6–15. A brick hearth in front of the fireplace catches hot embers that may fall out of the fire.

the home (Fig. 6–16). Many brands are available from fireplace shops and building supply dealers.

These units are built of heavy metal or boiler-plate steel and designed to be set into place and concealed by the usual brickwork, or other construction, so that no practical change in the fireplace mantel design is required by their use. One claimed advantage for modified fireplace units is that the correctly designed and proportioned fire box manufactured with throat, damper, smoke shelf, and chamber provides a form for the masonry, thus reducing the risk of failure and assuring a smokeless fireplace.

There is, however, no excuse for using incorrect proportions; the desirability of using a form, as provided by the modified unit, is not necessary merely to obtain good proportions. Each fireplace should be designed to suit individual requirements and if correct dimensions are adhered to, a satisfactory fireplace will be obtained.

Prior to selecting and erecting a fireplace, several suitable designs should be considered and a careful estimate of the cost of each one should be made. Remember that even though the unit of a modified fireplace is well designed, it will not operate properly if the chimney is inadequate. Therefore, for satisfactory operation, the chimney must be made in accordance with the rules for correct construction to give satisfactory operation with the modified unit as well as with the ordinary fireplace.

Manufacturers of modified units also claim that labor and materi-

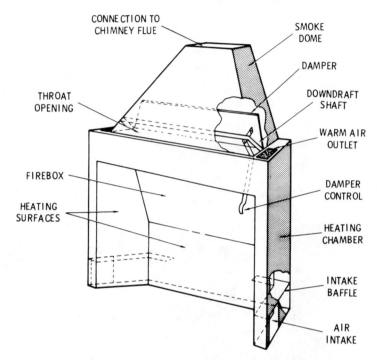

Fig. 6–16. A prefabricated metal form that makes fireplace construction easier.

als saved tend to offset the purchase price of the unit and that the saving in fuel tends to offset the increase in first cost. A minimum life of 20 years is usually claimed for the type and thickness of metal commonly used in these units.

As illustrated in Figs. 6–17 and 6–18, and the sketches of Figs. 6–19 and 6–20, the brick work is built up around a metal fireplace form. The back view of Fig. 6–17 shows a layer of fireproof insulation between the metal form and brick. The layer of fireproof wool batting should be about 1 inch thick. Note the ash door, which gives access to the ash pit for the removal of ashes from the outside of the house.

Fig. 6–18 shows a partially-built front view. By leaving a large air cavity on each side of the metal form and constructing the brick work with vents, some of the heat passing through the metal sides will be

Fig. 6–17. Brick work around the back of a fireplace form.

Fig. 6–18. Front view of the brick work around a metal fireplace form.

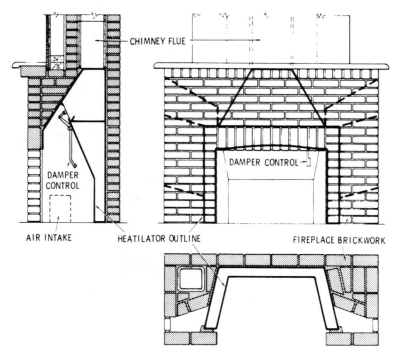

CHIMNEY FLUE

DAMPER CONTROL

DAMPER CONTROL

AIR INTAKE

HEATILATOR OUTLINE

FIREPLACE BRICKWORK

Fig. 6–19.   Sketch of a typical fireplace built around a metal form.

returned to the room. The rowlock-stacked brick with no mortar, but an air space, permits cool air to enter below and warmed air to come out into the room from the upper outlet.

The front of the form includes a linetel for holding the course of brick just over the opening. A built-in damper is part of the form. Even with the use of a form, a good foundation is necessary for proper support as there is still quite a bit of brick weight. Chimney construction following the illustrations and descriptions previously given is still necessary.

Other Fireplace Styles

There are a number of other fireplace styles available. One such is the hooded type which permits the construction of the fireplace out into the room, rather than into the wall (Fig. 6–21).

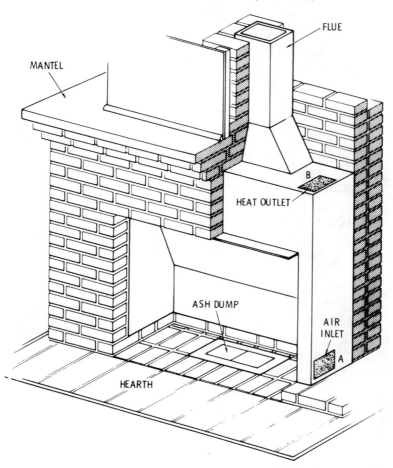

Fig. 6–20. Cutaway sketch of fireplace using a metal form.

Another style is two-sided, similar to that shown in the sketch of Fig. 6–5. It is used for building into a semi-divider-type wall, such as between a living area and a dining area. Thus, the fire can be enjoyed from either room, or both at once, and what heat is given off is divided between the two areas.

Important to successful wood burning is good circulation of air

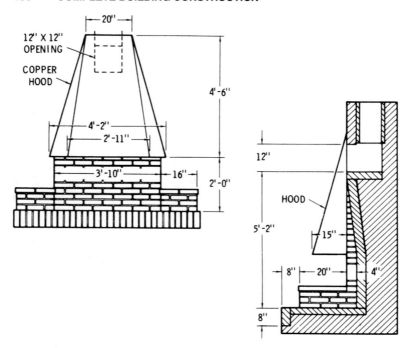

Fig. 6–21. A hood projecting out from the wall carries flue gases up through chimney.

under and around the sides. A heavy metal grate which lifts the burning logs above the floor of the fireplace is essential (Fig. 6–22).

Smoky Fireplaces

When a fireplace smokes, it should be examined to make certain that the essential requirements of construction as previously outlined have been fulfilled. If the chimney has not been stopped up with fallen brick and the mortar joints are in good condition, a survey should be made to ascertain that nearby trees or tall buildings do not cause eddy currents down the flue.

To determine whether the fireplace opening is in incorrect proportion to the flue area, hold a piece of sheet metal across the top of the

Fig. 6–22.   A grate is used to hold logs above the base, allowing air to move under and through the burning wood.

fireplace opening and then gradually lower it, making the opening smaller until smoke does not come into the room. Mark the lower edge of the metal on the sides of the fireplace.

The opening may then be reduced by building in a metal shield or hood across the top of the fireplace so that its lower edge is at the marks made during the test. Trouble with smoky fireplaces can also usually be remedied by increasing the height of the flue.

Uncemented flue-lining joints cause smoke to penetrate the flue joints and descend out of the fireplace. The best remedy is to tear out the chimney and join linings properly.

Where flue joints are uncemented and mortar in surrounding brick work disintegrated, there is often a leakage of air in the chimney causing poor draft. This prevents the stack from exerting the draft possibilities which its height would normally ensure.

Another cause of poor draft is wind being deflected down the chimney. The surroundings of a home may have a marked bearing on fireplace performance. Thus, for example, if the home is located at the foot of a bluff or hill or if there are high trees close at hand the result may be to deflect wind down the chimney in heavy gusts. A most common and efficient method of dealing with this type of difficulty is to provide a hood on the chimney top.

Carrying the flue lining a few inches above the brick work with a bevel of cement around it can also be used as a means of promoting a clean exit of smoke from the chimney flue. This will effectively prevent wind eddies. The cement bevel also causes moisture to drain from the top and prevents frost troubles between lining and masonry.

CHAPTER 7

Woods

Wood is our most versatile, most useful building material, and a general knowledge of the physical characteristics of various woods used in building operations is important for carpenter and handyman alike.

Wood may be classified:

1. Botanically. All trees which can be sawed into lumber or timbers belong to the division called Spermatophytes. This includes softwoods as well as hardwoods.
2. With respect to its density, as
   a. Soft,
   b. Hard.
3. With respect to its leaves, as
   a. Needle- or scale-leaved, botanically Gymnosperms, or conifers, commonly called softwoods. Most of them, but not all, are evergreens;
   b. Broad-leaved, botanically Angiosperms, commonly called hardwoods. Most are deciduous, shedding their leaves in the fall. Only one broad-leaved hardwood, the Chinese Ginkgo, belongs to the subdivision Gymnosperms.
4. With respect to its shade or color, as
   a. White or very light,
   b. Yellow or yellowish,

    c. Red,

    d. Brown,

    e. Black, or nearly black.

5. In terms of grain, as

    a. Straight,

    b. Cross,

    c. Fine,

    d. Coarse,

    e. Interlocking.

6. With respect to the nature of the surface when dressed, as

    a. Plain—example, white pine;

    b. Grained—example, oak;

    c. Figure or marked—example, bird's-eye maple.

A section of a timber tree, as shown in Figs. 7–1 and 7–2, consists of:

1. Outer bark—living and growing only at the cambium layer. In most trees, the outside continually sloughs away.
2. In some trees, notably hickories and basswood, there are long tough fibers, called bast fibers, in the inner bark. In some trees such as the beech, they are notably absent.
3. Cambium layer—Sometimes this is only one cell thick. Only these cells are living and growing.
4. Medullary rays or wood fibers, which run radially.
5. Annual rings, or layers of wood.
6. Pith.

Around the pith, the wood substance is arranged in approximately concentric rings. The part nearest the pith is usually darker than the parts nearest the bark and is called the heartwood. The cells in the heartwood are dead. Nearer the bark is the sapwood, where the cells are living but not growing.

As winter approaches, all growth ceases, and thus each annual ring is separate and in most cases distinct. The leaves of the deciduous trees, or those which shed their leaves, and the leaves of some of the conifers, such as cypress and larch, fall, and the sap in the tree may freeze hard. The tree is dormant but not dead. With the warm days of the next spring, growth starts again strongly, and the cycle is repeated. The width of the annual rings varies greatly, from 30 to 40 or more per inch

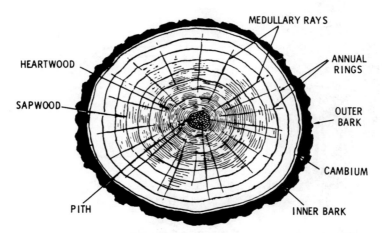

MEDULLARY RAYS

HEARTWOOD

ANNUAL RINGS

SAPWOOD

OUTER BARK

CAMBIUM

PITH

INNER BARK

Fig. 7–1. Cross section of a 9-year-old oak showing pith, concentric rings comprising the woody part, cambium layer, and bark. The arrangement of the wood in concentric rings is due to the fact that one layer was formed each year. These rings, or layers, are called annual rings. That portion of each ring formed in spring and early summer is softer, lighter colored, and weaker than that formed during the summer and is called spring wood. The denser, stronger wood formed later is called summer wood. The cells in the heartwood of some species are filled with various oils, tannins, and other substances, which make these timbers rot-resistant. There is practically no difference in the strength of heartwood and sapwood if they weigh the same. In most species, only the sapwood can be readily impregnated with preservatives.

in some slow-growing species, to as few as 3 or 4 per inch in some of the quick-growing softwoods. The woods with the narrowest rings, because of the large percentage of summer wood, are generally strongest, although this is not always the case.

Cutting at the Mill

When logs are taken to the mill, they may be cut in a variety of ways. One way of cutting is quartersawing. Fig. 7–3 shows four varia-

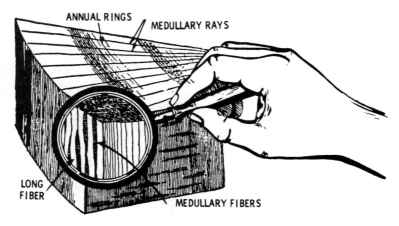

ANNUAL RINGS

MEDULLARY RAYS

LONG
FIBER

MEDULLARY FIBERS

Fig. 7–2. A piece of wood magnified slightly to show its structure. The wood is made up of long, slender cells called fibers, which usually lie parallel to the pith. The length of these cells is often 100 times their diameter. Transversely, bands of other cells, elongated but much shorter, serve to carry sap and nutrients across the trunk radially. Also, in the hardwoods, long vessels or tubes, often several feet long, carry liquids up to the tree. There are no sap-carrying vessels in the softwoods, but spaces between the cells may be filled with resins.

tions of this method. Each quarter is laid on the bark and ripped into quarters, as shown in the figure. Quartersawing is rarely done this way, though, because only a few wide boards are yielded; there is too much waste. More often, when quartersawed stock is required, the log is started as shown in Figs. 7–4 and 7–5, sawed until a good figure (pattern) shows, then turned over and sawed. This way, there is little waste, and the boards are wide. In other words, most quartersawed lumber is resawed out of plain sawed stock.

The plain sawed stock, as shown, is simply flat sawn out of quartersawed. Quartersawed stock has its uses. Boards shrink most in a direction parallel with the annual rings, and door stiles and rails are often made of quartersawed material.

Lumber is sold by the board foot, meaning one 12-inch 1-inch-thick square of wood. Any stock under 2 inches thick is known as lum-

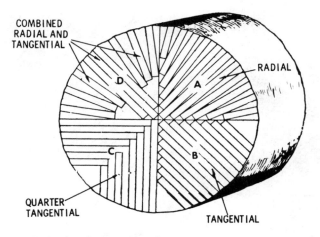

Fig. 7–3. Methods of quartersawing.

ber; over 4 inches, it is timber. The terms have become interchangeable, however, and are used interchangeably in this book.

Lumber, of course, is sold in nominal and actual size, the actual size being what the lumber is after being milled. As the years have gone

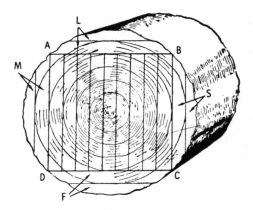

Fig. 7–4. Plain or bastard sawing, called flat or slash sawing. The log is first squared by removing boards *M, S, L,* and *F,* giving the rectangular section *ABCD.* This is necessary to obtain a flat surface on the log.

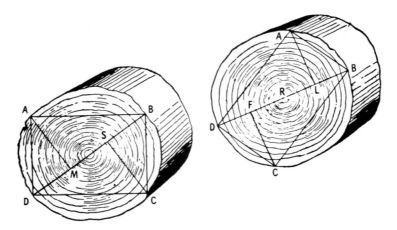

Fig. 7–5. Obtaining beams from a log.

by, the actual size has gotten smaller. A 2″ × 4″, for example, used to be an actual size of 1⁹⁄₁₆″ × 3⁹⁄₁₆″. It is now 1½″ × 3½″, and other boards go up or down in size in half-inch increments.

Defects

The defects found in manufactured lumber, as shown in Figs. 7–6 and 7–7, are grouped in several classes:

1. Those found in the natural log, as
   a. Snakes,
   b. Knots,
   c. Pitchpockets.
2. Those due to deterioration, as
   a. Rot,
   b. Dote.
3. Those due to imperfect manufacture, as
   a. Imperfect machining,
   b. Wane,
   c. Machine burn,
   d. Checks and splits from imperfect drying.

Heart shakes, as shown in Fig. 7–6, are radial cracks that are wider at the pith of the tree than at the outer end. This defect is most commonly found in those trees which are old, rather than in young vigorous saplings; it occurs frequently in hemlock.

A wind or cup shake is a crack following the line of the porous part of the annual rings and is curved by a separation of the annual rings. A wind shake may extend for a considerable distance up the trunk. Other explanations for wind shakes are expansion of the sapwood and wrenchings received due to high winds (hence the name). Brown ash is especially susceptible to wind shakes.

A star shake resembles the wind shake but differs in that the crack extends across the center of the trunk without any appearance of decay at that point; it is larger at the outside of the tree.

Dry rot, to which timber is so subject, is due to fungi; the name is misleading as it only occurs in the presence of moisture and the absence of free air circulation.

Selection of Lumber

A variety of factors must be considered when picking lumber for a particular project. For example, is it seasoned or not, that is, has it been dried naturally—the lumber is stacked up with air spaces between, as shown in Fig. 7–8—or artificially, as it is when dried in kilns? The idea

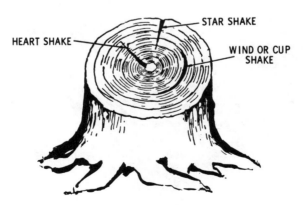

STAR SHAKE

HEART SHAKE

WIND OR CUP SHAKE

Fig. 7–6. Various defects that can be found in lumber.

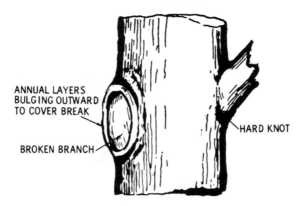

ANNUAL LAYERS
BULGING OUTWARD
TO COVER BREAK

HARD KNOT

BROKEN BRANCH

Fig. 7–7. Hard knot and broken branch show nature's way of covering the break.

is to produce lumber with a minimum amount of moisture that will warp least on the job. If your project requires nonwarping material, ask for kiln-dried lumber. If not, so-called green lumber will suffice. Green lumber is often used for framing (outside) where the slight warpage

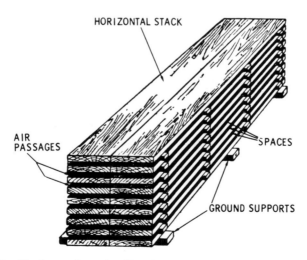

HORIZONTAL STACK

AIR
PASSAGES

SPACES

GROUND SUPPORTS

Fig. 7–8. Horizontal stack of lumber for air drying.

that occurs after it is nailed in place is not a problem. Green lumber is less costly than kiln-dried, of course.

Another factor to consider is the grade of the lumber. The best lumber you can buy is Clear, which means the material is free of defects. Following this is Select, which has three subdivisions—Nos. 1, 2, and 3—with No. 1 the best of the Select with only a few blemishes on one side of the board and few, if any, on the other. Last is Common. This is good wood, but it will have blemishes and knots that can interfere with a project if you want to finish it with a clear material.

Rough lumber also has a grading system, and it reflects the material, which comes green or kiln-dried.

Lumber has two grading systems, numerical and verbal. Numerically there are Nos. 1, 2, and 3 with No. 1 the best and No. 3 the least desirable. Roughly corresponding to these numbers are Construction, Standard, and Utility.

Wood may also be characterized as hardwood or softwood. These designations do not refer to the physical hardness of the wood, although hardwoods are normally harder than softwoods. The designation refers to the kinds of trees the wood comes from: cone-bearing trees are softwoods, leaf-bearing trees are hardwoods. By far the most valuable softwood is pine, a readily available material in all sections of the country. Of course the type of pine will depend on the particular area. Hardwoods, which come in random lengths from 8 to 16 feet long and 4 to 16 inches wide are all Clear—these would be mahogany, birch, oak, or maple. Hardwoods are usually much more expensive than softwoods.

In addition to the above, there are overall characteristics of the particular wood to consider. What follows is a round-up of individual characteristics that will make the material more or less suitable for your project.

Brown Ash—Not a framing timber, but an attractive trim wood. Brown heart, lighter sapwood. The trees often wind shake so badly that the heart is entirely loose. Attractive veneers are sliced from stumps and forks.

Northern White Cedar—Light-brown heart, sapwood thin and nearly white. Light, weak, soft, decay resistant, holds paint well.

Western Red Cedar—Also called canoe cedar or shinglewood. Light, soft, straight-grained, small shrinkage, holds paint well. Heart is light brown, extremely rot resistant. Sap quite narrow, nearly white. Used for shingles, siding, boat building.

Eastern Red Cedar, or Juniper—Pungent aromatic odor said to repel moths. Red or brown heartwood, extremely rot-resistant white sapwood. Used for lining clothes closets and chests and for fence posts.

Cypress—Probably our most durable wood for contact with the soil. Wood moderately light, close-grained, heartwood red to nearly yellow, sapwood nearly white. Does not hold paint well, but otherwise desirable for siding and outside trim. Attractive for inside trim.

Red Gum—Moderately heavy, interlocking grain; warps badly in seasoning; heart is reddish brown, sapwood nearly white. The sapwood may be graded out and sold as white gum, the heartwood as red gum, or together as unselected gum. Cuts into attractive veneers.

Hickory—A combination of hardness, weight, toughness, and strength found in no other native wood. A specialty wood, almost impossible to nail when dry. Not rot-resistant.

Eastern Hemlock—Heartwood is pale brown to reddish, sapwood not distinguishable from heart. May be badly wind shaken. Brittle, moderately weak, not at all durable. Used for cheap, rough framing veneers.

Western Hemlock—Heartwood and sapwood almost white with purplish tinge. Moderately strong, not durable, mostly used for pulpwood.

Black Locust—Heavy, hard, strong; the heartwood is exceptionally durable. Not a framing timber. Used mostly for posts and poles.

Hard Maple—Heavy, strong, hard, and close grained; color light brown to yellowish. Used mostly for wear-resistant floors, and furniture. Circularly growing fibers cause the attractive "birds-eye" grain in some trees. One species, the Oregon Maple, occasionally contains the attractive "quilted" grain.

Soft Maple—Softer and lighter than hard maple; lighter colored. Box elder is sometimes marketed with soft maple. Used for much the same purposes as hard maple, but not nearly so desirable.

White Oak—Several species are marketed together, but the woods are practically identical. Hard, heavy, tough, strong, and somewhat rot-resistant. Brownish heart, lighter sapwood. Desirable for trim and flooring, and one of our best hardwood framing timbers.

Red Oak—Several species are marketed together. They cannot be distinguished one from the other, but can be distinguished from the white oaks. Good framing timber, but not rot-resistant.

Western White Pine—also called Idaho White Pine. Creamy or light-brown heartwood, sapwood thick and white. Used mostly for mill-work and siding. Moderately light, moderately strong, easy to work, holds paint well.

Red or Norway Pine—Resembles the lighter weight specimens of southern yellow pine. Moderately strong and stiff, moderately soft, heartwood pale red to reddish brown. Used for millwork, siding, fram-ing, and ladder rails.

Long-Leaf Southern Yellow Pine—Not a species but a grade. All southern yellow pine that has six or more annual rings per inch is marketed as long-leaf, and it may contain lumber from any of the sev-eral species of southern pine. Heavy, hard, and strong, but not espe-cially durable in contact with the soil. The sapwood takes creosote well. One of our most useful timbers for light framing.

Short-Leaf Southern Yellow Pine—Contains timber from any of several related species of southern pine having less than six annual rings per inch. Quite satisfactory for light framing, and the sapwood is attractive as an interior finish.

Douglas Fir—Our most plentiful commercial timber. Varies greatly in weight, color, and strength. Strong, moderately heavy, splintery, splits easily. Used in all kinds of construction; much is rotary cut for plywood.

Yellow or Tulip Poplar—Our easiest-working native wood. Old growth has a yellow to brown heart. Sapwood and young trees are tough and white. Not a framing lumber, but used for siding.

Redwood—One of our most durable and rot-resistant timbers. Light, soft, moderately high strength, heartwood reddish brown, sap-wood white. Does not paint exceptionally well, as it oftentimes "bleeds" through. Used mostly for siding and outside trim, decks, furniture (Figs. 7–9 and 7–10).

Sitka Spruce—Light, soft, medium strong, heart is light reddish brown, sapwood is nearly white, shading into the heartwood. Usually cut into boards, planing-mill stock, and boat lumber.

Eastern Spruce—Stiff, strong, hard, and tough. Moderately lightweight, light color, little difference between heart and sapwood. Commercial eastern spruce includes wood from three related species. Used for pulpwood, framing lumber, millwork, etc.

Engelmann Spruce—Color, white with red tint. Straight grained, light weight, low strength. Used for dimension lumber and boards, and for pulpwood. Extremely low rot-resistance.

Fig. 7–9. For building outdoor furniture and items such as planters, redwood is great.

Tamarack, or Larch—Small to medium sized trees; not much is sawed into framing lumber, but much is cut into boards. Yellowish brown heart, sapwood white. Much is cut into posts and poles.

Black Walnut—Our most attractive cabinet wood. Heavy, hard, and strong, heartwood is a beautiful brown, sap nearly white. Mostly used for fine furniture, but some is used for fine interior trim. Somewhat rot-resistant. Used also for gunstocks.

White Walnut or Butternut—Sapwood light to brown, heart light chestnut brown with an attractive sheen. The cut is small, mostly going into cabinet work and interior trim. Moderately light, rather weak, not rot-resistant.

Fig. 7–10. Redwood is a favorite, but expensive, material for deck building.

Decay of Lumber

Five conditions are necessary for the decay—rotting—of wood: (1) fungi; (2) moisture; (3) air; (4) favorable temperatures; (5) food. Fungi are plants that are unable to make their own food, and must depend on a host plant or plant products such as wood. The fungi get their food from the cell wall in wood.

They can live and grow on wood *only* if there is moisture, air, and the right temperature. At a moisture content below 20 percent, fungi either die or become dormant. Properly-dried wood will not decay unless it gets wet. Dry-rot fungi can carry water from a distance to wood, but the moisture conditions must be correct for the rot to develop. Too low a temperature stops the growth of the fungi. Kiln-drying temperatures will kill the fungi, if the heat can penetrate to where the fungi are located. Fungi do best at temperatures between 68 and 96.8 degrees Fahrenheit ($20°C$ to $36°C$). However, the temperature range varies for different fungi species.

If any one of the five conditions is missing, the wood will not decay. Wood completely submerged in water is deprived of air, and will not rot. When fungi, moisture, air, and temperatures cannot be controlled, the food source can be shut off. Pressure-treating wood with a preservative such as Chromated Copper Arsenate (CCA) poisons the wood and deprives the fungi of food.

There is no such thing as "dry rot"; however, the term is rather loosely used sometimes when speaking about any rot or any dry and decayed wood. Although rotten wood may be dry when observed, it was wet while decay was progressing. This kind of decay is often found inside living, growing trees, but it occurs only in the presence of water.

Other Materials

Plywood

In addition to boards and lumber, carpenters and handymen have come to rely on other materials. Among the most important is plywood (Fig. 7–11).

The most familiar plywood used in the United States is made from douglas fir. Short logs are chucked into a lathe and a thin, continuous layer of wood is peeled off. This thin layer is straightened, cut to convenient sizes, covered with glue, laid up with the grain in successive plies crisscrossing, and subjected to heat and pressure. This is the plywood of commerce, one of our most useful building materials.

All plywood has an odd number of plies, allowing the face plies to

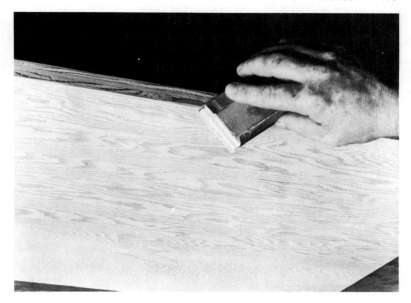

Fig. 7–11. Plywood is indispensable to the builder. It comes in forms from utilitarian to elegant.

have parallel grain while the lay-up is "balanced" on each side of a center ply. This process equalizes stresses set up when the board dries or when it is subsequently wetted and dried.

Grade Designations—Structural panel grades are identified according to the veneer grade used on the face and back of the panel, or by a name such as APA RATED SHEATHING, APA RATED STURD-I-FLOOR. See Fig. 7–12 for veneer grades. The highest quality grades are N and A. The minimum grade of veneer permitted in Exterior plywood is C-grade. D-grade veneer is used in panels to be used indoors or in areas protected from permanent exposure to weather.

Sanded, Unsanded, and Touch-Sanded Panels—Grade B, or better, panels are always sanded smooth because they are intended for use in cabinets, furniture, shelving. APA RATED SHEATHING is unsanded. APA UNDERLAYMENT, STURD-I-FLOOR, C-D PLUGGED, and C-C PLUGGED are touch-sanded to make the panel thickness more uniform.

N	Smooth surface "natural finish" veneer. Select, all heartwood or all sapwood. Free of open defects. Allows not more than 6 repairs, wood only, per 4 x 8 panel, made parallel to grain and well matched for grain and color.
A	Smooth, paintable. Not more than 18 neatly made repairs, boat, sled, or router type, and parallel to grain, permitted. May be used for natural finish in less demanding applications. Synthetic repairs permitted.
B	Solid surface. Shims, circular repair plugs and tight knots to 1 inch across grain permitted. Some minor splits permitted. Synthetic repairs permitted.
C Plugged	Improved C veneer with splits limited to 1/8-inch width and knotholes and borer holes limited to 1/4 x 1/2 inch. Admits some broken grain. Synthetic repairs permitted.
C	Tight knots to 1-1/2 inch. Knotholes to 1 inch across grain and some to 1-1/2 inch if total width of knots and knotholes is within specified limits. Synthetic or wood repairs. Discoloration and sanding defects that do not impair strength permitted. Limited splits allowed. Stitching permitted.
D	Knots and knotholes to 2-1/2-inch width across grain and 1/2 inch larger within specified limits. Limited splits are permitted. Stitching permitted. Limited to Exposure 1 or Interior panels.

Fig. 7–12. Veneer grades. *(Courtesy American Plywood Association)*

Unsanded and touch-sanded panels, and those of Grade B or better veneer on one side only, usually have the APA trademark on the panel back. Panels with both sides of B-grade or better veneer have the APA trademark on the panel edge.

Exposure Durability—Plywood panels are produced in 4 exposure durabilities: Exterior, Exposure 1, Exposure 2, and Interior.

Exterior panels have a fully waterproof bond and may be used where continual exposure to both weather and moisture is possible.

Exposure 1 panels are made with the same glues used in Exterior panels. These panels may be used on roofs, for example, where some time may pass before they are protected from weathering or moisture. They also may be used outdoors where only one side is exposed to the weather, such as on open soffits.

Caution: Exposure 1 panels are commonly called CDX. Too many builders assume that because the glues are X—exterior— these panels can be left continually exposed to weather and moisture. CDX panels are not designed to withstand prolonged exposure to weather and elements because they lack the C-grade veneers. They should be covered as soon as possible to prevent excessive moisture absorption. Although direct rain on the surface of the panel will not weaken the panel, the moisture raises the grain and causes checking and other minor surface problems. There have been reports of these moisture-caused defects telegraphing through thin fiberglass shingles.

Exposure 2 panels are manufactured with intermediate glue. They are designed for use with only moderate exposure to weather or moisture. Exposure 2 is not a common lumberyard stock item.

Interior panels are manufactured with interior glues and are to be used indoors *only*.

Figs. 7–13(A), (B), and (C) are guides to APA-related panels.

In selecting plywood, the rule is simple. Just pick what is right for the job at hand. If one side is going to be hidden, for example, you do not need a high grade.

APA RATED SHEATHING Typical Trademark	**APA** RATED SHEATHING 24/16 7/16 INCH SIZED FOR SPACING EXPOSURE 1 000 NER-QA397 PRP-108		Specially designed for subflooring and wall and roof sheathing. Also good for a broad range of other construction and industrial applications. Can be manufactured as conventional veneered plywood, as a composite, or as a nonveneer panel. EXPOSURE DURABILITY CLASSIFICATIONS: Exterior, Exposure 1, Exposure 2. COMMON THICKNESSES: 5/16, 3/8, 7/16, 15/32, 1/2, 19/32, 5/8, 23/32, 3/4.
APA STRUCTURAL I RATED SHEATHING(3) Typical Trademark	**APA** RATED SHEATHING STRUCTURAL I 32/16 15/32 INCH SIZED FOR SPACING EXPOSURE 1 000 PS 1-83 C-D NER-QA397 PRP-108	**APA** RATED SHEATHING 32/16 15/32 INCH SIZED FOR SPACING EXPOSURE 1 000 STRUCTURAL I RATED DIAPHRAGMS · SHEAR WALLS PANELIZED ROOFS NER-QA397 PRP-108	Unsanded grade for use where shear and cross-panel strength properties are of maximum importance, such as panelized roofs and diaphragms. Can be manufactured as conventional veneered plywood, as a composite, or as a nonveneer panel. EXPOSURE DURABILITY CLASSIFICATIONS: Exterior, Exposure 1. COMMON THICKNESSES: 5/16, 3/8, 7/16, 15/32, 1/2, 19/32, 5/8, 23/32, 3/4.
APA RATED STURD-I-FLOOR Typical Trademark	**APA** RATED STURD-I-FLOOR 24 OC 23/32 INCH SIZED FOR SPACING T&G NET WIDTH 47-1/2 EXPOSURE 1 000 NER-QA397 PRP-108		Specially designed as combination subfloor-underlayment. Provides smooth surface for application of carpet and pad and possesses high concentrated and impact load resistance. Can be manufactured as conventional veneered plywood, as a composite, or as a nonveneer panel. Available square edge or tongue-and-groove. EXPOSURE DURABILITY CLASSIFICATIONS: Exterior, Exposure 1, Exposure 2. COMMON THICKNESSES: 19/32, 5/8, 23/32, 3/4, 1-1/8.
APA RATED SIDING Typical Trademark	**APA** RATED SIDING 24 OC 15/32 INCH SIZED FOR SPACING EXTERIOR 000 NER-QA397 PRP-108	**APA** RATED SIDING 303-18-S/W 16 OC 11/32 INCH SIZED FOR SPACING EXTERIOR 000 PS 1-83 FHA-UM-64 NER-QA397 PRP-108	For exterior siding, fencing, etc. Can be manufactured as conventional veneered plywood, as a composite or as a nonveneer siding. Both panel and lap siding available. Special surface treatment such as V-groove, channel groove, deep groove (such as APA Texture 1-11), brushed, rough sawn and texture-embossed (MDO). Span Rating (stud spacing for siding qualified for APA Sturd-I-Wall applications) and face grade classification (for veneer-faced siding) indicated in trademark. EXPOSURE DURABILITY CLASSIFICATION: Exterior. COMMON THICKNESSES: 11/32, 3/8, 7/16, 15/32, 1/2, 19/32, 5/8.

(1) Specific grades, thicknesses and exposure durability classifications may be in limited supply in some areas. Check with your supplier before specifying.
(2) Specify Performance Rated Panels by thickness and Span Rating. Span Ratings are based on panel strength and stiffness. Since these properties are a function of panel composition and configuration as well as thickness, the same Span Rating may appear on panels of different thickness. Conversely, panels of the same thickness may be marked with different Span Ratings.
(3) All plies in Structural I plywood panels are special improved grades and panels marked PS 1 are limited to Group 1 species. Other panels marked Structural I Rated qualify through special performance testing.

Structural II plywood panels are also provided for, but rarely manufactured. Application recommendations for Structural II plywood are identical to those for RATED SHEATHING plywood.

Fig. 7–13. (A) APA performance-rated panels. *(Courtesy American Plywood Association)*

APA A-A Typical Trademark **A-A · G-1 · EXPOSURE1-APA · 000 · PS1-83**	Use where appearance of both sides is important for interior applications such as built-ins, cabinets, furniture, partitions; and exterior applications such as fences, signs, boats, shipping containers, tanks, ducts, etc. Smooth surfaces suitable for painting. EXPOSURE DURABILITY CLASSIFICATIONS: Interior, Exposure 1, Exterior. COMMON THICKNESSES: 1/4, 11/32, 3/8, 15/32, 1/2, 19/32, 5/8, 23/32, 3/4.
APA A-B Typical Trademark **A-B · G-1 · EXPOSURE1-APA · 000 · PS1-83**	For use where appearance of one side is less important but where two solid surfaces are necessary. EXPOSURE DURABILITY CLASSIFICATIONS: Interior, Exposure 1, Exterior. COMMON THICKNESSES: 1/4, 11/32, 3/8, 15/32, 1/2, 19/32, 5/8, 23/32, 3/4.
APA A-C Typical Trademark APA A-C GROUP 1 EXTERIOR 000 PS 1-83	For use where appearance of only one side is important in interior applications, such as paneling, built-ins, shelving, partitions, flow racks, etc. EXPOSURE DURABILITY CLASSIFICATION: Exterior. COMMON THICKNESSES: 1/4, 11/32, 3/8, 15/32, 1/2, 19/32, 5/8, 23/32, 3/4.
APA A-D Typical Trademark APA A-D GROUP 1 EXPOSURE 1 000 PS 1-83	For use where appearance of only one side is important in interior applications, such as paneling, built-ins, shelving, partitions, flow racks, etc. EXPOSURE DURABILITY CLASSIFICATIONS: Interior, Exposure 1. COMMON THICKNESSES: 1/4, 11/32, 3/8, 15/32, 1/2, 19/32, 5/8, 23/32, 3/4.
APA B-B Typical Trademark **B-B · G-2 · EXPOSURE1-APA · 000 · PS1-83**	Utility panels with two solid sides. EXPOSURE DURABILITY CLASSIFICATIONS: Interior, Exposure 1, Exterior. COMMON THICKNESSES: 1/4, 11/32, 3/8, 15/32, 1/2, 19/32, 5/8, 23/32, 3/4.
APA B-C Typical Trademark APA B-C GROUP 1 EXTERIOR 000 PS 1-83	Utility panel for farm service and work buildings, boxcar and truck linings, containers, tanks, agricultural equipment, as a base for exterior coatings and other exterior uses or applications subject to high or continuous moisture. EXPOSURE DURABILITY CLASSIFICATION: Exterior. COMMON THICKNESSES: 1/4, 11/32, 3/8, 15/32, 1/2, 19/32, 5/8, 23/32, 3/4.
APA B-D Typical Trademark APA B-D GROUP 2 EXPOSURE 1 000 PS 1-83	Utility panel for backing, sides of built-ins, industry shelving, slip sheets, separator boards, bins and other interior or protected applications. EXPOSURE DURABILITY CLASSIFICATIONS: Interior, Exposure 1. COMMON THICKNESSES: 1/4, 11/32, 3/8, 15/32, 1/2, 19/32, 5/8, 23/32, 3/4.
APA UNDERLAYMENT Typical Trademark APA UNDERLAYMENT GROUP 1 EXPOSURE 1 000 PS 1-83	For application over structural subfloor. Provides smooth surface for application of carpet and pad and possesses high concentrated and impact load resistance. EXPOSURE DURABILITY CLASSIFICATIONS: Interior, Exposure 1. COMMON THICKNESSES[4]: 11/32, 3/8, 1/2, 19/32, 5/8, 23/32, 3/4.
APA C-C PLUGGED Typical Trademark APA C-C PLUGGED GROUP 2 EXTERIOR 000 PS 1-83	For use as an underlayment over structural subfloor, refrigerated or controlled atmosphere storage rooms, pallet fruit bins, tanks, boxcar and truck floors and linings, open soffits, and other similar applications where continuous or severe moisture may be present. Provides smooth surface for application of carpet and pad and possesses high concentrated and impact load resistance. EXPOSURE DURABILITY CLASSIFICATION: Exterior. COMMON THICKNESSES[4]: 11/32, 3/8, 1/2, 19/32, 5/8, 23/32, 3/4.
APA C-D PLUGGED Typical Trademark APA C-D PLUGGED GROUP 2 EXPOSURE 1 000 PS 1-83	For built-ins, cable reels, separator boards and other interior or protected applications. Not a substitute for Underlayment or APA Rated Sturd-I-Floor as it lacks their puncture resistance. EXPOSURE DURABILITY CLASSIFICATIONS: Interior, Exposure 1. COMMON THICKNESSES: 3/8, 1/2, 19/32, 5/8, 23/32, 3/4.

(1) Specific plywood grades, thicknesses and exposure durability classifications may be in limited supply in some areas. Check with your supplier before specifying.

(2) Sanded exterior plywood panels, C-C Plugged and Underlayment grades can also be manufactured in Structural I (all plies limited to Group 1 species).

(3) Some manufacturers also produce plywood panels with premium N-grade veneer on one or both faces. Available

only by special order. Check with the manufacturer.

(4) Panels 1/2 inch and thicker are Span Rated and do not contain species group number in trademark.

Fig. 7–13. (B) APS sanded and touch-sanded plywood panels.
(Courtesy American Plywood Association)

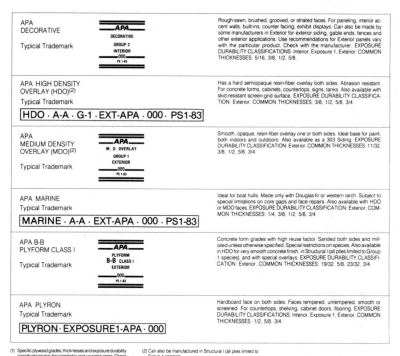

APA DECORATIVE Typical Trademark	APA DECORATIVE GROUP 2 INTERIOR 000 PS 1-83	Rough-sawn, brushed, grooved, or striated faces. For paneling, interior accent walls, built-ins, counter facing, exhibit displays. Can also be made by some manufacturers in Exterior for exterior siding, gable ends, fences and other exterior applications. Use recommendations for Exterior panels vary with the particular product. Check with the manufacturer. EXPOSURE DURABILITY CLASSIFICATIONS: Interior, Exposure 1, Exterior. COMMON THICKNESSES: 5/16, 3/8, 1/2, 5/8.
APA HIGH DENSITY OVERLAY (HDO)[2] Typical Trademark HDO · A-A · G-1 · EXT-APA · 000 · PS1-83		Has a hard semiopaque resin-fiber overlay both sides. Abrasion resistant. For concrete forms, cabinets, countertops, signs, tanks. Also available with skid-resistant screen-grid surface. EXPOSURE DURABILITY CLASSIFICATION: Exterior. COMMON THICKNESSES: 3/8, 1/2, 5/8, 3/4.
APA MEDIUM DENSITY OVERLAY (MDO)[2] Typical Trademark	APA M. D. OVERLAY GROUP 1 EXTERIOR 000 PS 1-83	Smooth, opaque, resin-fiber overlay one or both sides. Ideal base for paint, both indoors and outdoors. Also available as a 303 Siding. EXPOSURE DURABILITY CLASSIFICATION: Exterior. COMMON THICKNESSES: 11/32, 3/8, 1/2, 5/8, 3/4.
APA MARINE Typical Trademark MARINE · A-A · EXT-APA · 000 · PS1-83		Ideal for boat hulls. Made only with Douglas-fir or western larch. Subject to special limitations on core gaps and face repairs. Also available with HDO or MDO faces. EXPOSURE DURABILITY CLASSIFICATION: Exterior. COMMON THICKNESSES: 1/4, 3/8, 1/2, 5/8, 3/4.
APA B-B PLYFORM CLASS I Typical Trademark	APA PLYFORM B-B CLASS I EXTERIOR 000 PS 1-83	Concrete form grades with high reuse factor. Sanded both sides and mill-oiled unless otherwise specified. Special restrictions on species. Also available in HDO for very smooth concrete finish. In Structural I (all plies limited to Group 1 species), and with special overlays. EXPOSURE DURABILITY CLASSIFICATION: Exterior. COMMON THICKNESSES: 19/32, 5/8, 23/32, 3/4.
APA PLYRON Typical Trademark PLYRON · EXPOSURE1-APA · 000		Hardboard face on both sides. Faces tempered, untempered, smooth or screened. For countertops, shelving, cabinet doors, flooring. EXPOSURE DURABILITY CLASSIFICATIONS: Interior, Exposure 1, Exterior. COMMON THICKNESSES: 1/2, 5/8, 3/4.

(1) Specific plywood grades, thicknesses and exposure durability classifications may be in limited supply in some areas. Check with your supplier before specifying.

(2) Can also be manufactured in Structural I (all plies limited to Group 1 species)

Fig. 7–13. (C) APA specialty plywood panels. *(Courtesy American Plywood Association)*

Particleboard

Particleboard is made from wood chips or particles combined with synthetic resin binders and pressed in a hot-plate press to form a flat sheet. Because the particles are so small, they are not visible: a sheet of the material looks as though it were made from compressed sawdust. Structural particleboard is made from graded small particles, arranged in layers according to particle size. It is generally used for carrying loads such as floors. The panels range in thickness from 1/4 inch to 1 1/2 inch, in widths of 3 feet to 8 feet, up to 24 feet long. Particleboard is used in kitchen countertops and cabinets, underlayment for carpets or resilient floor coverings, furniture, core material for large tabletops and many other uses. Particleboard is not waferboard.

Waferboard, or flakeboard, and particleboard are similar, and often confused. Aspenite, for example, is waferboard, not particleboard, as it is often called. Waferboard is made from large wafer-shaped pieces of wood.

Strandboard is made of long, narrow particles. The particles may be random or oriented, as in Oriented Strand Board (OSB). During manufacturing the wafers are oriented—pointing in the same direction—mechanically or electronically to align the grain in each layer (Fig. 7–14). The layers are cross-laminated, so that each layer is perpendicular—at right angles—to each other. The OSB face strands run the 8-foot length of the panel. The strands are compressed into a three-to-five layer board with phenolic resin or isocyanurate adhesives. As a result, an ¾-inch thick panel of OSB weighs about 10 percent more than the same thickness of plywood.

Builders are often confused about the difference between OSB and waferboard. Until recently, it was easy to tell the difference: OSB had long narrow strands pointing—oriented—in the same direction. Waferboard had square-shaped pieces that pointed left, right, diagonally and so on. Now most waferboard panels have the strands oriented to make the panels stronger. The shape of the strand no longer matters. Therefore, make certain the panel is APA performance-rated. (See Fig. 7–15, How to Read an APA Label.)

OSB comes in thicknesses of ⅜, ⁷⁄₁₆, ½, ¹⁹⁄₃₂, ⅝, ²³⁄₃₂, and ¾-inch. The 4′ × 8′ panel is standard, but lengths up to 16 feet are available. Most tongue-and-groove panels are made with a 47½-inch net width on the top surface. But this varies with the manufacturer. OSB is stronger than some plywood. For the same thickness, OSB is 10 percent stronger across its surface than 4-ply plywood or composite-core plywood, and 20 percent stronger than 3-ply plywood. It is not surprising, then, that more and more builders are using OSB in place of more expensive plywood. In some areas OSB is replacing plywood as the choice for roof sheathing and subflooring.

Hardboard

Hardboard is an all-wood material manufactured from exploded-wood chips, using either the wet or dry process. In the wet process, the

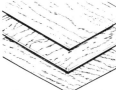

APA
PERFORMANCE RATED
PANEL COMPOSITIONS

PLYWOOD

Plywood is the original structural wood panel. It is composed of thin sheets of veneer, or plies, arranged in layers to form a panel. Plywood always has an odd number of layers, each one consisting of one or more plies, or veneers.

In plywood manufacture, a log is turned on a lathe and a long knife blade peels the veneer. The veneers are clipped to a suitable width, dried, graded, and repaired if necessary. Next the veneers are laid up in cross-laminated layers. Sometimes a layer will consist of two or more plies with the grain running in the same direction, but there will always be an odd number of layers, with the face layers typically having the grain oriented parallel to the long dimension of the panel.

Adhesive is applied to the veneers which are laid up. Laid-up veneers are then put in a hot press where they are bonded to form panels.

Wood is strongest along its grain, and shrinks and swells most across the grain. By alternating grain direction between adjacent layers, strength and stiffness in both directions are maximized, and shrinking and swelling are minimized in each direction.

ORIENTED STRAND BOARD

Nonveneer panels manufactured with various techniques have been marketed with such names as waferboard, oriented strand board, and structural particleboard. Today, most nonveneer structural wood panels are manufactured with oriented strands or wafers, and are commonly called oriented strand board (OSB).

OSB is composed of compressed strands arranged in layers (usually three to five) oriented at right angles to one another. The orientation of layers achieves the same advantages of cross-laminated veneers in plywood. Since wood is stronger along the grain, the cross-lamination distributes wood's natural strength in both directions of the panel. Whether a reconstituted panel is composed of strands or wafers, nearly all manufacturers orient the material to achieve maximum performance.

Most OSB panels are textured on one side to reduce slickness.

COM-PLY

COM-PLY is an APA product name for composite panels that are manufactured by bonding reconstituted wood cores between wood veneer. By combining reconstituted wood with conventional wood veneer, COM-PLY panels allow for more efficient resource use while retaining the wood grain appearance on the panel face and back.

COM-PLY panels are manufactured in a three- or five-layer arrangement. A three-layer panel has a reconstituted wood core and a veneer face and back. The five-layer panel has a wood veneer in the center as well as on the face and back. When manufactured in a one-step pressing operation, voids in the veneers are filled automatically by the particles as the panel is pressed in the bonding process.

Fig. 7–14. Performance-rated panel composition. *(Courtesy American Plywood Association)*

fibers are bound together by lignin, which is a natural resin found in trees and plants. In the dry process, lignin gets a boost from a phenolic resin that is added during the manufacture. Different additives are used to increase the hardboard's stability, and to reduce its rate of moisture absorption. The chips are steamed under high pressure which is suddenly released causing the fibers to explode and separate. The

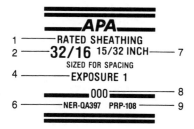

1 Panel grade
2 Span Rating
3 Tongue-and-groove
4 Exposure durability classification
5 Product Standard
6 Code recognition of APA as a quality assurance agency
7 Thickness
8 Mill number
9 APA's Performance Rated Panel Standard
10 Siding face grade
11 Species group number
12 FHA recognition

Fig. 7–15. How to read an APA label. *(Courtesy American Plywood Association)*

fibers are recombined under heat and pressure into large sheets. The weight and density of the sheets depends on the amount of the pressure. After pressing, the hardboard is kiln-dried and moisture is added to bring the moisture content to between 2 percent and 9 percent.

Hardboard is made in three basic types: *standard, tempered,* and *service. Standard* hardboard is given no further treatment after manu-

CHAPTER 8

Framing

Knowledge of Lumber

The basic construction material in carpentry is lumber. There are many kinds of lumber varying greatly in structural characteristics. Here we deal with the lumber common to construction carpentry, its application, the standard sizes in which it is available, and the methods of computing lumber quantities in terms of *board feet*.

Standard Sizes of Bulk Lumber

Lumber is usually sawed into standard lengths, widths, and thicknesses. This permits uniformity in planning structures and in ordering material. Table 8–1 lists the common widths and thicknesses of wood in rough and dressed dimensions in the United States. Standards have been established for dimension differences between nominal size and the standard size (which is actually the reduced size when dressed). It is important that these dimension differences be taken into consideration when planning a structure. A good example of the dimension difference may be illustrated by the common 2 × 4. As may be seen in the table, the familiar quoted size (2 × 4) refers to a rough or nominal dimension, but the actual standard size to which the lumber is dressed is 1½″ × 3½″.

Table 8–1. Your Guide to Sizes of Lumber

WHAT YOU ORDER	WHAT YOU GET		WHAT YOU USED TO GET
	* Dry or Seasoned	** Green or Unseasoned	Seasoned or Unseasoned
1 × 4	¾ × 3½	25⁄32 × 3⁹⁄16	25⁄32 × 3⅝
1 × 6	¾ × 5½	25⁄32 × 5⅝	25⁄32 × 5½
1 × 8	¾ × 7¼	25⁄32 × 7½	25⁄32 × 7½
1 × 10	¾ × 9¼	25⁄32 × 9½	25⁄32 × 9½
1 × 12	¾ × 11¼	25⁄32 × 11½	25⁄32 × 11½
2 × 4	1½ × 3½	1⁹⁄16 × 3⁹⁄16	1⅝ × 3⅝
2 × 6	1½ × 5½	1⁹⁄16 × 5⅝	1⅝ × 5½
2 × 8	1½ × 7¼	1⁹⁄16 × 7½	1⅝ × 7½
2 × 10	1½ × 9¼	1⁹⁄16 × 9½	1⅝ × 9½
2 × 12	1½ × 11¼	1⁹⁄16 × 11½	1⅝ × 11½
4 × 4	3½ × 3½	3⁹⁄16 × 3⁹⁄16	3⅝ × 3⅝
4 × 6	3½ × 5½	3⁹⁄16 × 5⅝	3⅝ × 5½
4 × 8	3½ × 7¼	3⁹⁄16 × 7½	3⅝ × 7½
4 × 10	3½ × 9¼	3⁹⁄16 × 9½	3⅝ × 9½
4 × 12	3½ × 11¼	3⁹⁄16 × 11½	3⅝ × 11½

*19% Moisture Content or under.
**Over 19% Moisture Content.

Grades of Lumber

Lumber as it comes from the sawmill is divided into three main classes: yard lumber, structural material, and factory or shop lumber. In keeping with the purpose of this book, only yard lumber will be considered. Yard lumber is manufactured and classified on a quality basis into sizes, shapes, and qualities required for ordinary construction and general building purposes. It is then further subdivided into classifications of select lumber and common lumber.

Select Lumber—Select lumber is of good appearance and finished or dressed. It is identified by the following grade names:

Grade A. Grade A is suitable for natural finishes, of high quality, and is practically clear.
Grade B. Grade B is suitable for natural finishes, of high quality, and is generally clear.

Grade C. Grade C is adapted to high-quality paint finish.

Grade D. Grade D is suitable for paint finishes and is between the higher finishing grades and the common grades.

Common Lumber—Common lumber is suitable for general construction and utility purposes and is identified by the following grade names:

No. 1 common. No. 1 common is suitable for use without waste. It is sound and tight-knotted, and may be considered watertight material.

No. 2 common. No. 2 common is less restricted in quality than No. 1, but of the same general quality. It is used for framing, sheathing, and other structural forms where the stress or strain is not excessive.

No. 3 common. No. 3 common permits some waste with prevailing grade characteristics larger than in No. 2. It is used for footings, guardrails, and rough subflooring.

No. 4 common. No. 4 common permits waste, and is of low quality, admitting the coarsest features such as decay and holes. It is used for sheathing, subfloors, and roof boards in the cheaper types of construction. The most important industrial outlet for this grade is for boxes and shipping crates.

Framing Lumber

The frame of a building consists of the wooden form constructed to support the finished members of the structure. It includes such items as posts, girders (beams), joists, subfloor, sole plate, studs, and rafters. Softwoods are usually used for light wood framing and all other aspects of construction carpentry considered in this book. One of the classifications of softwood lumber cut to standard sizes is called yard lumber which is manufactured for general building purposes. It is cut into the standard sizes required for light framing, including 2 × 4, 2 × 6, 2 × 8, 2 × 10, 2 × 12, and all other sizes required for framework, with the exception of those sizes classed as structural lumber.

Although No. 1 and No. 3 common are sometimes used for framing, No. 2 common is most often used, and is therefore most often stocked and available in retail lumber yards in the common sizes for various framing members. However, the size of lumber required for any specific structure will vary with the design of the building, such as light-frame or heavy-frame, and the design of the particular members, such as beams or girders.

Exterior walls traditionally consisted of three layers—sheathing, building paper, and siding. Plywood sheathing replaced board sheathing. Tyvek and other breathable air retarders replaced building paper. Rigid non-structural foam boards are also used as sheathing.

Computing Board Feet

The arithmetic method of computing the number of board feet in one or more pieces of lumber is by the use of the following formula:

$$\frac{\text{Pieces} \times \text{Thickness (inches)} \times \text{Width (inches)} \times \text{Length (feet)}}{12}$$

Example—Find the number of board feet in a piece of lumber 2 inches thick, 10 inches wide, and 6 feet long.

$$\frac{1 \times 2 \times 10 \times 6}{12} = 10 \text{ board feet.}$$

Example—Find the number of board feet in 10 pieces of lumber 2 inches thick, 10 inches wide, and 6 feet long.

$$\frac{10 \times 2 \times 10 \times 6}{12} = 100 \text{ board feet.}$$

Example—Find the number of board feet in a piece of lumber 2 inches thick, 10 inches wide, and 18 inches long.

$$\frac{2 \times 10 \times 18}{144} = 2\frac{1}{2} \text{ board feet}$$

(***Note:*** If all three dimensions are expressed in inches, the same formula applies except the divisor is changed to 144.)

Board feet can also be calculated by use of a table normally found on the back of the tongue of a steel framing square. The inch graduations on the outer edge of the square are used in combination with the values in the table to get a direct indication of the number of board feet in a particular board. Complete instructions come with the square.

Methods of Framing

Good material and workmanship will be of very little value unless the underlying framework of a building is strong and rigid. The resistance of a house to such forces as tornadoes and earthquakes, and control of cracks due to settlement depends on a good framework.

Although it is true that no two buildings are put together in exactly the same manner, disagreement exists among architects and carpenters as to which method of framing will prove most satisfactory for a given condition. *Light-framed construction* may be classified into three distinct types known as:

1. Balloon frame,
2. Plank and beam,
3. Western frame (also identified as platform frame).

Balloon-Frame Construction

The principal characteristic of *balloon framing* is the use of studs extending in one piece from the foundation to the roof, as shown in Fig. 8–1. The joists are nailed to the studs and also supported by a ledger board set into the studs. Diagonal sheathing may be used instead of wall board to eliminate corner bracing.

Plank-and-Beam Construction

The *plank-and-beam construction* is said to be the oldest method of framing in the country, having been imported from England in colonial times. Although in a somewhat modified form, it is still being used in certain states, notably in the east. Originally, this type of framing

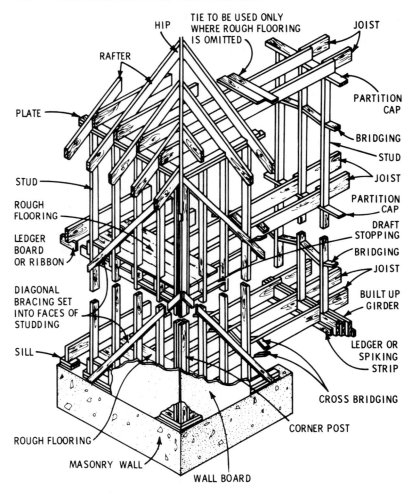

HIP

TIE TO BE USED ONLY
WHERE ROUGH FLOORING
IS OMITTED

JOIST

RAFTER

PLATE

STUD

ROUGH
FLOORING

LEDGER
BOARD
OR RIBBON

DIAGONAL
BRACING SET
INTO FACES OF
STUDDING

SILL

ROUGH FLOORING

MASONRY WALL

WALL BOARD

PARTITION
CAP

BRIDGING

STUD

JOIST

PARTITION
CAP

DRAFT
STOPPING

BRIDGING

JOIST

BUILT UP
GIRDER

LEDGER OR
SPIKING
STRIP

CROSS BRIDGING

CORNER POST

Fig. 8–1. Details of balloon-frame construction.

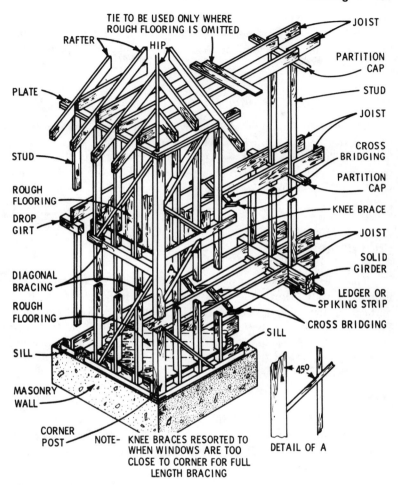

TIE TO BE USED ONLY WHERE
ROUGH FLOORING IS OMITTED

RAFTER

HIP

JOIST

PARTITION
CAP

PLATE

STUD

JOIST

STUD

CROSS
BRIDGING

ROUGH
FLOORING

PARTITION
CAP

DROP
GIRT

KNEE BRACE

JOIST

DIAGONAL
BRACING

SOLID
GIRDER

ROUGH
FLOORING

LEDGER OR
SPIKING STRIP

CROSS BRIDGING

SILL

SILL

MASONRY
WALL

45°

CORNER
POST

NOTE- KNEE BRACES RESORTED TO
WHEN WINDOWS ARE TOO
CLOSE TO CORNER FOR FULL
LENGTH BRACING

DETAIL OF A

Fig. 8–2.   Details of plank-and-beam construction.

was characterized by heavy timber posts at the corners, as shown in Fig. 8–2, and often with intermediate posts between, which extended continuously from a heavy foundation sill to an equally heavy plate at the roof line.

Western-Frame Construction

This type of framing is characterized by platforms independently framed, the second or third floor being supported by the studs from the first floor, as shown in Fig. 8–3. The chief advantage in this type of framing (in all-lumber construction) lies in the fact that if there is any settlement due to shrinkage, it will be uniform throughout and will not be noticeable.

Foundation Sills

The foundation sill consists of a plank or timber resting on the foundation wall. It forms the support or bearing surface for the outside of the building and, as a rule, the first floor joists rest upon it. Shown in Fig. 8–4 is the balloon-type construction of first and second floor joist and sills. In Fig. 8–5 is shown the joist and sills used in plank-and-beam framing, and in Fig. 8–6 is shown the western-type construction.

Size of Sills

The size of sills for small buildings of light frame construction may be as small as a 2 × 4, and as large as a 2 × 12 on a 10-inch-thick foundation wall. They may be 2 × 10s or 2 × 12s placed on edge and topped with a 2 × 4 laid flat on top. Sill plate may be flush with the outside of the foundation wall, or flush with the inside foundation edge. In general, sill plates are 2 × 6s. For two-story buildings, and especially in locations subject to earthquakes or tornadoes, a double sill is desirable, as it affords a larger nailing surface for diagonal sheathing brought down over the sill, and ties the wall framing more firmly to its sills. In cases where the building is supported by posts or piers, it is necessary to increase the sill size, since the sill supported by posts acts as a girder. In balloon framing, for example, it is customary to build up the sills with two or more planks 2 or 3 inches thick which are nailed together.

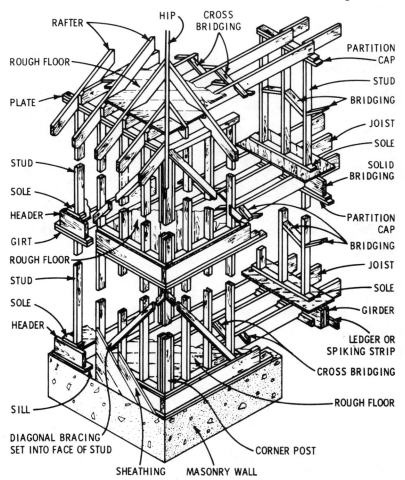

Fig. 8–3. Details of western-frame construction.

In most types of construction, since it is not necessary that the sill be of great strength, the foundation will provide uniform solid bearing throughout its entire length. The main requirements are: resistance to crushing across the grain; ability to withstand decay and attacks of in-

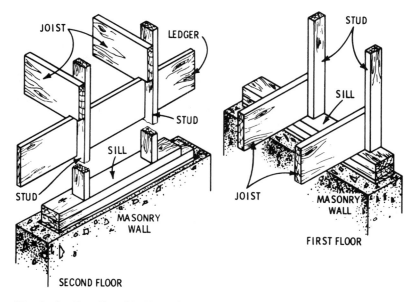

Fig. 8–4. Details of balloon framing of sill plates and joists.

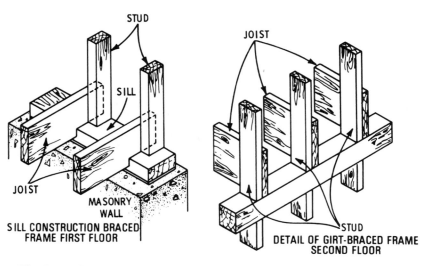

Fig. 8–5. Details of plank-and-beam framing of sill plates and joists.

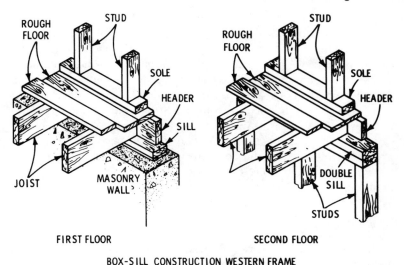

ROUGH FLOOR • STUD • SOLE • HEADER • SILL • JOIST • MASONRY WALL

FIRST FLOOR

ROUGH FLOOR • STUD • SOLE • HEADER • DOUBLE SILL • STUDS

SECOND FLOOR

BOX-SILL CONSTRUCTION WESTERN FRAME

Fig. 8–6. Details of western framing of sill plates and joists.

sects; and ability to furnish adequate nailing area for studs, joists, and sheathing.

Length of Sill

The length of the sill is determined by the size of the building, and hence the foundation should be laid out accordingly. Dimension lines for the outside of the building are generally figured from the outside face of the subsiding or sheathing, which is about the same as the outside finish of unsheathed buildings.

Anchorage of Sill

It is important, especially in locations of strong winds, that buildings be thoroughly anchored to the foundation. Sill plates may be secured to the foundation with ½-inch diameter anchor bolts, embedded no less than 8 inches into the concrete. In concrete masonry blocks they must be embedded no less than 15 inches. The bolts must be spaced 12 inches from the end of the plate, and spaced 8 feet on center maximum. There must be a minimum of 2 bolts per plate. These are the basic requirements of both the *BOCA* and the *UBC* codes.

Sometimes anchor bolts are embedded too deeply, improperly located, and the holes in the plate made too large. The washer, and sometimes the nut, are not installed. Galvanized steel anchor straps are an alternative to the bolts. They too are embedded in concrete, and their arms wrap around the sill plate. Simpson Strong-Tie® Company manufactures an MA Mudsill Anchor, Models MA4 and MA6. Panel-clip Company also manufactures a Y-shaped metal anchor, the end of which is embedded in the concrete wall. These anchors should be spaced 12 inches in from the end of the plate, and 4 feet on center maximum, for the Simpson MA4, and 4½ feet for the MA6 (Fig. 8–7).

Splicing of Sill

As previously stated, a 2 × 6 sill is large enough for small buildings under normal conditions, if properly bedded on the foundations. In order to properly accomplish the splicing of a sill, it is necessary that special precaution be taken. A poorly fitted joint weakens rather than strengthens the sill frame. Where the sill is built up of two planks, the joints in the two courses should be staggered.

Placing of Sill

The traditional method of installing sill or mud plates and sealing them to the top of the foundation was to grout them in. A bed of mortar (cement and sand) was trowelled on the wall, the sill plate placed over the bolts, and the nuts tightened. Tightening continued until a small amount of the mortar squeezed out on both sides. Using a level, the nuts were further tightened until the mud plate was level. Excess mortar was removed, and the edges of the mortar bed were angled to about 45 degrees. Thus the mortar bed could compensate for an nonlevel rough foundation. Although the mortar provided an air seal, in time it did crack and loosen. Because accepted practice is not to provide a capillary break between the top of the footing and the bottom of the concrete foundation wall, the sill plate is subjected to moisture contact. Use pressure-treated wood.

Grouting is a lost art and is rarely seen today. Grouting has been replaced with fiberglass, plastic, or cellulose wrapped in plastic, *sill*

SIMPSON
Strong-Tie® CONNECTORS

Patent No. 3,889,441

MA MUDSILL ANCHORS

- A low-labor, high-value method to secure mudsills to monolithic slabs or foundation walls
- Replaces anchor bolts and washers
- Eliminates drilling the sill
- Includes depth gauges for easy, perfect installation
- No special tools required
- Can be installed before sill placement or attached to sill (see illustration)
- Arrowhead design is ideal for inserting into screeded surface

MATERIAL: 16 gauge
FINISH: Galvanized
INSTALLATION: ▪ Use all specified fasteners. See General Notes.
- Place anchors not more than 1' from the end of each sill. Maximum anchor spacing for the MA4 and MA6 is 4' and 4½' on center, respectively.
- Not for use where (a) a horizontal cold joint exists between the slab and foundation wall or footing beneath, or (b) they are installed in slabs poured over foundation walls formed of concrete block.

CODE NUMBERS: ICBO No. 1211. Dade County, FL No. 89-0131.2.
City of L.A. No. RR 22086.

MA4 and MA6

MODEL NO.	SILL SIZE	W	FASTENERS			UPLIFT AVG ULT	ALLOWABLE LOADS[1]		
			SIDES TOTAL	TOP			UPLIFT	PARALLEL TO PLATE	PERPEN-DICULAR TO PLATE
MA4	2×4	3⅝	2-10dx1½	2-10dx1½		2655	830	550	1180
	3×4		4-10dx1½	2-10dx1½		—	1060	680	1180
MA6	2×6	5⅝	2-10dx1½	4-10dx1½		4020	1060	680	1180
	3×6		4-10dx1½	4-10dx1½		—	1290	680	1180

1. Loads may not be increased for short-term loading.

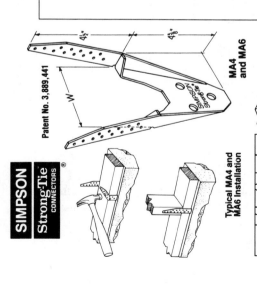

Typical MA4 and MA6 Installation

Optional method with mudsill anchors in place for positioning into screeded concrete.

Fig. 8–7. Simpson mud anchors. (Courtesy Simpson Strong-Tie® Connectors)

197

sealers. Sill sealers prevent direct contact between the foundation and the plate, and the plastic sill sealer is a capillary break. However, sill sealer will not make the sill plate tight to the foundation. Gaps between the foundation and sill plate are either left open or plugged with wood shims. This widespread common practice among builders, and permitted by too many building inspectors, should be stopped. The crushing of the shims leads to spongy, nonlevel floors that can vibrate and squeak. The gaps between the sill plate and foundation permit wind and rain to enter.

Even if the foundation is level and smooth, sill sealer should not be totally depended on to seal the sill plate. The inner and outer edges of the sill plate should be caulked around the entire perimeter of the foundation. An alternative to time-consuming caulking is to use EPDM gaskets.

Girders

A girder in small-house construction consists of a large beam at the first-story line which takes the place of an interior foundation wall and supports the inner ends of the floor joists. In a building where the space between the outside walls is more than 14 to 15 feet, it is generally necessary to provide additional support near the center to avoid the necessity of excessively heavy floor joists. When a determination is made as to the number of girders and their location, consideration should be given to the required length of the joists, to the room arrangement, as well as to the location of the bearing partitions.

Length of Joists

The length of floor joists depends on: (a) span; (b) size of floor joist; (c) live and dead loads; (d) spacing between joists; (e) the F_b and the modulus E; (f) the wood species and wood grade; (g) whether or not the subfloor is glued to the joists.

Bridging Between Joists

Floor bridging has long been the subject of controversy. More than 20 years ago, the National Association of Home Builders (NAHB) and the Forest Products Laboratory (FPL) (operated by the University of Wisconsin for the United States Department of Agriculture) had shown that bridging, as normally applied, added little to the stiffness of the floor in resisting static (non-moving) loads. The *BOCA* code does not require midspan bridging in Use Groups R-2 and R-3—multiple-family dwellings, boarding houses, and one- and two-family dwelling units—unless the live load exceeds 40 psf, or the depth of the floor joist exceeds 12 inches nominal. The Canadian Building Code permits a piece of 1×3 strapping or furring to be nailed to the bottom of the joist at midspan for the length of the building, in lieu of bridging. The *CABO One And Two Family Dwelling Code* does not require bridging when floor joists are 12 inches deep or less.

Interior Partitions

An interior partition differs from an outside partition, in that it seldom rests on a solid wall. Its supports therefore require careful consideration, making sure they are large enough to carry the required weight. The various interior partitions may be bearing or nonbearing, and may run at either right angles or parallel to the joints upon which they rest.

Partitions Parallel to Joists

Here the entire weight of the partition will be concentrated upon one or two joists, which perhaps are already carrying their full share of the floor load. In most cases, additional strength should be provided. One method is to provide double joists under such partitions—to put an extra joist beside the regular ones. Computation shows that the average partition weighs nearly three times as much as a single joist should be expected to carry. The usual (and approved method) is to double the joists under nonbearing partitions. An alternative method is to place a joist on each side of the partition.

Partitions at Right Angles to Joists

For nonbearing partitions, it is not necessary to increase the size or number of the joists. The partitions themselves may be braced, but even without bracing, they have some degree of rigidity.

Framing Around Openings

It is necessary that some parts of the studs be cut out around windows or doors in outside walls or partitions. It is *imperative* to insert some form of a header to support the lower ends of the top studs that have been cut off. There is a member that is termed a rough sill at the bottom of the window openings. This sill serves as a nailer, but does not support any weight.

Headers

Headers are of two classes, namely:

1. Nonbearing headers which occur in the walls which are parallel with the joists of the floor above, and carry only the weight of the framing immediately above;
2. Load-bearing headers in walls which carry the end of the floor joists either on plates or rib bands immediately above the openings, and must therefore support the weight of the floor or the floors above.

Size of Headers

The determining factor in header sizes is whether they are load-bearing or not. In general, it is considered good practice to use a double 2×4 header placed on edge unless the opening in a nonbearing partition is more than 3 feet wide. In cases where the trim inside and outside is too wide to prevent satisfactory nailing over the openings, it may become necessary to double the header to provide a nailing base for the trim.

Corner Studs

These are studs which occur at the intersection of two walls at right angles to each other. Fig. 8–8 shows one way to frame a 3-stud corner.

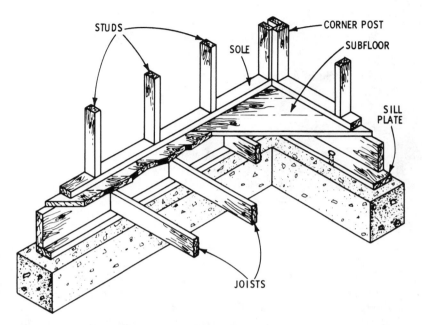

STUDS

SOLE

CORNER POST

SUBFLOOR

SILL
PLATE

JOISTS

Fig. 8–8. Detailed view of a corner stud.

Roofs

Generally, it may be said that rafters serve the same purpose for a roof as joists do for the floors. They provide a support for sheathing and roof material. Among the various kinds of rafters used, regular rafters extending without interruption from the eave to the ridge are the most common.

Spacing of Rafters

Spacing of rafters is determined by the stiffness of the sheathing between rafters, by the weight of the roof, and by the rafter span. In most cases, the rafters are spaced 16 or 24 inches on center.

Size of Rafters

The size of the rafters will depend upon the following factors:

1. The span,
2. The weight of the roof material,
3. The snow and wind loads.

Span of Rafters

In order to avoid any misunderstanding as to what is meant by the span of a rafter, the following definition is given:

The *rafter span* is the horizontal distance between the supports and not the overall length from end to end of the rafter. The span may be between the wall plate and the ridge, or from outside wall to outside wall, depending on the type of roof.

Length of Rafters

Length of rafters must be sufficient to allow for the necessary cut at the ridge and to allow for the protection of the eaves as determined by the drawings used. This length should not be confused with the span as used for determining strength.

Collar Beams

Collar beams may be defined as ties between rafters on opposite sides of a roof. If the attic is to be utilized for rooms, collar beams may be lathed as ceiling rafters, providing they are spaced properly. In general, collar beams should not be relied upon as ties. The closer the ties are to the top, the greater the leverage action. There is a tendency for the collar beam nails to pull out, and also for the rafter to bend if the collar beams are too low. The function of the collar beam is to stiffen the roof. These beams are often, although not always, placed at every rafter. Placing them at every second or third rafter is usually sufficient.

Size of Collar Beams

If the function of a collar beam were merely to resist a thrust, it would be unnecessary to use material thicker than 1-inch lumber. But,

as stiffening the roof is their real purpose, the beams must have sufficient body to resist buckling, or else must be braced to prevent bending.

Hip Rafters

The hip roof is built in such a shape that its geometrical form is that
of a pyramid, sloping down on all four sides. A hip is formed where two
adjacent sides meet. The hip rafter is the one which runs from the corner of the building upward along the same plane of the common rafter.
Where hip rafters are short and the upper ends come together at the
corner of the roof (lending each other support), they may safely be of
the same size as that of the regular rafters.

For longer spans, however, and particularly when the upper end of
the hip rafter is supported vertically from below, an increase of size is
necessary. The hip rafter will necessarily be slightly wider than the
jacks, in order to give sufficient nailing surface. A properly-sheathed
hip roof is nearly self-supporting. It is the strongest roof of any type of
framing in common use.

Dormers

The term *dormer* is given to any window protruding from a roof.
The general purpose of a dormer may be to provide light or to add to
the architectural effect.

In general construction, there are three types of dormers, as follows:

1. Dormers with flat sloping roofs, but with less slope than the roof in
   which they are located, as shown in Fig. 8–9;
2. Dormers with roofs of the gable type at right angles to the roof
   (Fig. 8–10);
3. A combination of the above types, which gives the hip-type dormer.

When framing the roof for a dormer window, an opening is provided in which the dormer is later built in. As the spans are usually
short, light material may be used.

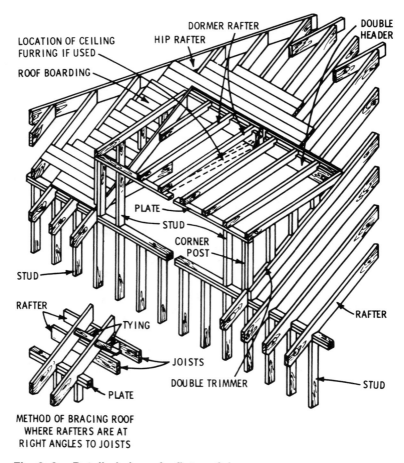

DORMER RAFTER

DOUBLE HEADER

HIP RAFTER

LOCATION OF CEILING FURRING IF USED

ROOF BOARDING

PLATE

STUD

CORNER POST

STUD

RAFTER

TYING

RAFTER

JOISTS

DOUBLE TRIMMER

STUD

PLATE

METHOD OF BRACING ROOF WHERE RAFTERS ARE AT RIGHT ANGLES TO JOISTS

Fig. 8–9. Detailed view of a flat-roof dormer.

Stairways

The well-built stairway is something more than a convenient means of getting from one floor to another. It must be placed in the right location in the house. The stairs themselves must be so designed that traveling up or down can be accomplished with the least amount of discomfort.

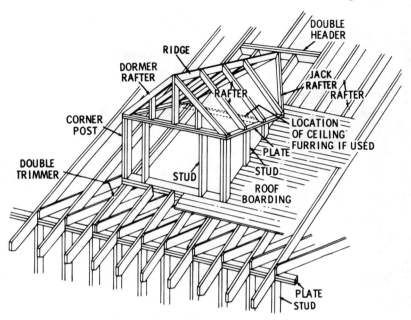

Fig. 8–10. Detailed view of gable-roof dormer.

The various terms used in stairway building (Fig. 8–11) are as follows:

1. The *rise* of a stairway is the height from the top of the lower floor to the top of the upper floor;
2. The *run* of the stairs is the length of the floor space occupied by the construction;
3. The *pitch* is the angle of inclination at which the stairs run;
4. The *tread* is that part of the horizontal surface on which the foot is placed;
5. The *riser* is the vertical board under the front edge of the tread;
6. The *stringer* is the framework on either side which is cut to support the treads and risers.

A commonly followed rule in stair construction is that the *tread* should not measure less than 9 inches deep, and the *riser* should not be more than 8 inches high. The width measurement of the tread and

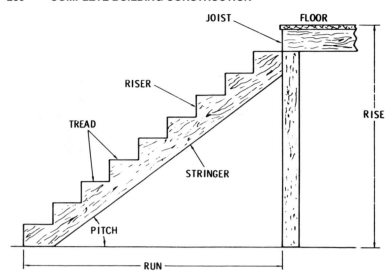

Fig. 8–11. Parts of a stairway.

height of the riser combined should not exceed 17 inches. Measurements are for the cuts of the stringers, not the actual width of the boards used for risers and treads. Treads usually have a projection, called a *nosing*, beyond the edge of the riser.

Fire and Draft Stops

It is known that many fires originate on lower floors. It is therefore important that fire stops be provided in order to prevent a fire from spreading through the building by way of air passages between the studs. Similarly, fire stops should be provided at each floor level to prevent flames from spreading through the walls and partitions from one floor to the next. Solid blocking should be provided between joists and studs to prevent fire from passing across the building.

In the platform frame and plank-and-beam framing, the construction itself provides stops at all levels. In this type of construction, therefore, fire stops are needed only in the floor space over the bearing partitions. Masonry is sometimes utilized for fire stopping, but is usually

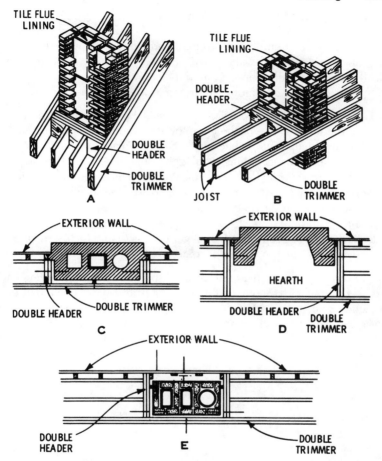

Fig. 8–12. Framing around chimneys and fireplaces: (A) roof framing around chimney; (B) floor framing around chimney; (C) framing around chimney above fireplace; (D) floor framing around fireplace; (E) framing around concealed chimney above fireplace.

adaptable in only a few places. Generally, obstructions in the air passages may be made of 2-inch lumber, which will effectively prevent the rapid spread of fire. Precautions should be made to insure the proper fitting of fire stops throughout the building.

Chimney and Fireplace Construction

Although carpenters are ordinarily not concerned with the building of the chimney, it is necessary, however, that they be acquainted with the methods of framing around the chimney.

The following minimum requirements are recommended:

1. No wooden beams, joists, or rafters shall be placed within 2 inches of the outside face of the chimney. No woodwork shall be placed within 4 inches of the back wall of any fireplace.
2. No studs, furring, lathing, or plugging should be placed against any chimney or in the joints thereof. Wooden construction shall either be set away from the chimney or the plastering shall be directly on the masonry or on metal lathing or on incombustible furring material.
3. The walls of fireplaces shall never be less than 8 inches thick if of brick or 12 inches if built of stone.

Formerly, it was advised to pack all spaces between chimneys and wood framing with incombustible insulation. It is now known that this practice is not as fire-resistant as the empty air spaces, since the air may carry away dangerous heat while the insulation may become so hot that it becomes a fire hazard itself. Fig. 8–12 shows typical framing around chimneys and fireplaces.

CHAPTER 9

Girders

Chapter 8 gave an overall idea of several kinds of framing. The details of each vary greatly. There are many ways of constructing each part, and in this connection, buyers or those contemplating having a house built should be acquainted not only with the right construction methods but also with the methods that are not good.

It is poor economy to specify inexpensive and inferior construction as houses so built are often not satisfactory. In this and following chapters the various parts of the frame, such as *girders, sills, corner posts,* and *studding* are considered in detail, showing the numerous ways in which each part is treated.

Girders

By definition, a *girder* is a principal beam extended from wall to wall of a building affording support for the joists or floor beams where the distance is too great for a single span. Girders may be either solid or built up.

Construction of Girders

Girders may be of steel, solid wood, or built up from 2x lumber. A center-bearing wall may be substituted for a girder. Commercially manufactured glue-laminated (glulams) beams may also be used, especially if left exposed in finished basements.

The joints on the outside of the girder should fall directly over the post or lally column. However, when the girder is continuous over three or more supports, the joints may be located between ⅙ and ¼ the span length from the intermediate lally column (Fig. 9–1). The nails should be long enough to penetrate all layers and be clinched. Use 20d nails at the ends, driven at an angle, and 20d nails at the top and bottom of the girder, spaced 32 inches on center staggered. Place them so that they are not opposite the nails on the other side. The beam/girder should have 4 inches minimum bearing in the beam pocket.

Caution—Do not use wood shims under the girder. Because the wood shim is in compression, it will begin to creep with time. As it creeps—sags—the girder begins to sag and eventually the floor sags. Install a ⅜- to 1-inch-thick iron-bearing plate, 2 inches wider than the beam, in the beam pocket before the girder or solid beam is set in place. The plate serves two purposes: it provides a capillary break to prevent what many incorrectly call *dry rot;* and it helps distribute the load of the girder over a larger wall area.

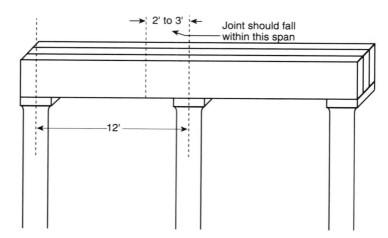

Fig. 9–1. Location of joint when girder is continous over three or more supports.

Why Beam Ends Rot

Fungi live on wood or other plants because they cannot make their own food. Unless the moisture content of the wood is above 20 percent, fungi cannot live on the wood. Therefore, kiln-dried lumber is used in construction. Yet the dry lumber of a beam/built-up girder gets wet at the ends, leaving it wide open for rotting. There are two reasons for this.

Although the outside of the foundation wall is dampproofed to provide a capillary break, the top of the footing is rarely ever dampproofed. The footing, sitting on damp ground, sucks up moisture which works its way up to the top of the foundation, even if this foundation were 6 miles high. Because the dry wood of the girder rests in an untreated beam pocket, the dry wood is exposed to the moisture in the concrete.

Water in the form of vapor is present in the basement. The amount of moisture vapor in moist air depends on the temperature of the air. The warmer the air, the more moisture vapor it can hold. The cooler the air, the less moisture it can hold. When at a given temperature the air holds all the moisture it can, it is said to be *saturated*. Its relative humidity (RH) is said to be 100 percent. Perfectly dry air has an RH of 0 percent. For example, a cubic yard of air at 68 degrees Fahrenheit can hold $\frac{1}{2}$ ounce of water vapor. If the air temperature cools down to 40 degrees Fahrenheit, the air can hold only $\frac{1}{10}$ ounce of water. The other $\frac{4}{10}$ condenses out as liquid water. Moisture condensation happens when the air is cooled down below a certain critical *temperature* called the *dew point temperature*. This is a common occurrence on hot humid summer days: a cold bottle taken out of the refrigerator starts *sweating* as the air in contact with it reaches the dew point temperature. This is also seen in winter when moist air condenses on single-pane glass, as the outside temperature begins to drop. These cool surfaces are called the *first condensing surfaces*.

The mass of the concrete foundation becomes cool as it sits on cool damp earth. Warm basement air touching the foundation—the first condensing surface—cools down, condenses as liquid water, and deposits itself in the beam pocket. The ends of the dry wood beam or built-up girder soak up the water and increase the moisture content of the wood. The five necessary conditions are present, and the wood eventually rots. Dampproof the beam pocket while coating the exterior of the foundation. Or line the beam pocket with bituthene or EPDM. The ends of the girder can be wrapped with TU-TUF, which serves two functions: it provides a capillary break and protects the end of the girder. It also prevents air, infiltrating into the beam pocket, from moving horizontally between the wood layers and into the basement. Without this protection the girder acts like duct work to bring cold air into the basement. The *BOCA* code requires a ½-inch air space on the top, sides, and ends of a girder sitting in a concrete beam pocket.

As a rule, the size of a solid beam, or the size and number of layers of 2x lumber for a built-up girder, will be listed on the drawings by the architect. Tables such as Tables 9–1 and 9–2 can be used to find the size of the girder for one- and two-story structures, or the girder size may be calculated.

Calculating the Size of A Girder

To calculate girder size the following facts are needed:

1. Length and width of house,
2. The total psf load of joists and bearing partitions,
3. Girder tributary area,
4. The load per lineal foot on the girder,
5. The spacing between the lally columns,
6. The total load on the girder.

Assume a single story house 24 feet wide and 40 feet long, with a total dead load (DL) and live load (LL) of 50 psf. Fig. 9–2 is a foundation plan showing the 40 foot girder and the location and spacing of the lally columns. In order to calculate the load per lineal foot on the girder, the half-width or tributary area of the joist span must be found.

Tributary Area

The total loading area of a building is carried by both the foundation walls and the girder. But the girder carries more weight than the foundation walls. A simple example will show why. A person carrying a 60-pound 12-foot ladder supports all the weight of the ladder. However, two persons, one at each end of the ladder, will carry only 30 pounds each. Suppose we have two 12-foot ladders on the ground in line with each other, and three persons to carry them. One individual

Table 9–1. Girder Size and Allowable Spans—One-Story Floor Loads

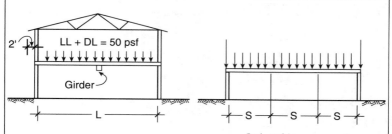

End splits may not
exceed one girder
depth.

In order that stresses not be exceeded,
in the following table the girder may
not be offset from centerline of house
by more than one foot. However, girders
may be located to suit design conditions
provided unit stresses conform to industry
standards for grade and species.

Nominal Lumber Sizes	Girder Spans = S in Feet					
	House Widths = L in Feet					
	22	24	26	28	30	32
Lumber having an allowable bending stress not less than 1000 psi						
2 - 2 × 6	4′ - 0″	—	—	—	—	—
3 - 2 × 6	5′ - 3″	5′ - 0″	4′ - 10″	4′ - 8″	4′ - 5″	4′ - 2″
2 - 2 × 8	5′ - 3″	4′ - 10″	4′ - 5″	4′ - 2″	—	—
3 - 2 × 8	6′ - 11″	6′ - 7″	6′ - 4″	6′ - 2″	5′ - 10″	5′ - 5″
2 - 2 × 10	6′ - 9″	6′ - 2″	5′ - 8″	5′ - 3″	4′ - 11″	4′ - 8″
3 - 2 × 10	8′ - 10″	8′ - 5″	8′ - 1″	7′ - 10″	7′ - 5″	6′ - 11″
2 - 2 × 12	8′ - 2″	7′ - 6″	6′ - 11″	6′ - 5″	6′ - 0″	5′ - 8″
3 - 2 × 12	10′ - 9″	10′ - 3″	9′ - 10″	9′ - 6″	9′ - 0″	8′ - 5″
Lumber having an allowable bending stress not less than 1500 psi						
2 - 2 × 6	4′ - 10″	4′ - 5″	4′ - 1″	—	—	—
3 - 2 × 6	6′ - 5″	6′ - 2″	5′ - 11″	5′ - 8″	5′ - 3″	4′ - 11″
2 - 2 × 8	6′ - 4″	5′ - 10″	5′ - 4″	5′ - 0″	4′ - 8″	4′ - 4″
3 - 2 × 8	8′ - 6″	8′ - 1″	7′ - 9″	7′ - 5″	7′ - 0″	6′ - 6″
2 - 2 × 10	8′ - 1″	7′ - 5″	6′ - 10″	6′ - 4″	5′ - 11″	5′ - 7″
3 - 2 × 10	10′ - 10″	10′ - 4″	9′ - 11″	9′ - 6″	8′ - 11″	8′ - 4″
2 - 2 × 12	9′ - 10″	9′ - 0″	8′ - 4″	7′ - 9″	7′ - 2″	6′ - 9″
3 - 2 × 12	13′ - 2″	12′ - 7″	12′ - 1″	11′ - 7″	10′ - 10″	10′ - 2″

Table 9–2. Girder Size and Allowable Spans—Two-Story Floor Loads

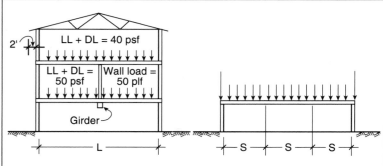

End splits may not exceed the girder depth.

In order that stresses not be exceeded, in the following table the girder may not be offset from centerline of house by more than one foot. However, girders may be located to suit design conditions provided unit stresses conform to industry standards for grade and species.

Nominal Lumber Sizes	Girder Spans = S in Feet					
	House Widths = L in Feet					
	22	24	26	28	30	32
Lumber having an allowable bending stress not less than 1000 psi						
3 - 2 × 8	4′ - 2″	—	—	—	—	—
2 - 2 × 12	4′ - 4″	4′ - 0″	—	—	—	—
3 - 2 × 10	5′ - 4″	4é - 11″	4é - 7″	4é - 3″	4é - 0″	—
3 - 2 × 12	6′ - 6″	6′ - 0″	5′ - 6″	5′ - 2″	4′ - 10″	4′ - 6″
Lumber having an allowable bending stress not less than 1500 psi						
2 - 2 × 10	4′ - 3″	—	—	—	—	—
3 - 2 × 8	5′ - 0″	4′ - 7″	4′ - 3″	—	—	—
2 - 2 × 12	5′ - 2″	4′ - 9″	4′ - 5″	4′ - 1″	—	—
2 - 2 × 10	6′ - 5″	7′ - 6″	5′ - 6″	5′ - 1″	4′ - 9″	4′ - 6″
3 - 2 × 12	7′ - 9″	7′ - 2″	6′ - 8″	6′ - 2″	5′ - 9″	5′ - 5″

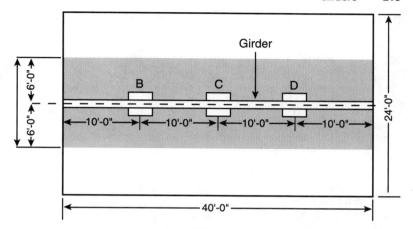

Fig. 9–2. Foundation plan showing location of girder and lally columns.

each is at the outer ends of the two ladders. The third person is in the middle between the two inner ends of the ladders. Lifting them off the ground, the three persons support a total weight of 120 pounds. The two persons on the outer ends still carry only 30 pounds each. But the middle individual supports 60 pounds, one-half of the weight of each of the two ladders. The two outer persons represent the two outside foundation walls. The middle person represents the beam or built-up girder. The girder thus supports one-half of the weight, while the other half is divided equally between the outside walls.

The design of the structure might require that the beam be offset from the center of the foundation. If the girder in the one-story house (Fig. 9–2) was located at 14 feet from the end wall, neither the total length of the floor joists nor their weight changes. Using the ladder example, assume one ladder is 14 feet long and the other 10 feet long, each weighing 5 pounds per foot. The person on the outside end of the 10-foot ladder supports one-half of the 10-foot ladder, or 25 pounds, while the person on the outside of the 14-foot ladder supports one-half, or 35 pounds. The middle person supports one-half of each ladder or a total of 60 pounds—the same weight as with the 12-foot ladders.

The beam/girder is not supporting all of the floor area, but only a part of it. The floor area the girder supports is called the *tributary area,* or the *contributing area.* Tributary refers to the weight contributed to each load-carrying member. The tributary area of the girder is the area from which it receives all of its load. It is also called the *half-width,* or *girder loading area.* In order to calculate the load per lineal foot on the girder, we must find the tributary area.

A general rule to find the girder loading area: the girder will carry the weight of the floor on each side to the middle of the joists—span—that rest on it. Fig. 9–2 shows the 40-foot long girder located exactly in the middle of the 24-foot-wide end wall. One-half the length of the floor joists—they are 12 feet long—on each side of the girder is 6 feet. Six feet plus six feet is 12 feet. This is the tributary width, the half-width, the midpoint of the floor joists, and is the shaded area shown in Fig. 9–2. When the girder is centered, the girder width, or the half-width, is exactly half of the width of the building. In this example, 24⁄2 or 12 feet. But what if the girder is offset 14 feet from one wall and 10 feet from the other? Take one-half of each distance—14 feet/2 + 10 feet/2—and add them together—7 feet + 5 feet = 12 feet.

Load Distribution

In the example we have been using, the girder carries one-half the total floor weight. This is because the girder is supporting the inner ends of the floor joists, and one-half the weight of every joist resting on it. The other half is divided equally between the two foundation walls. We have assumed that the floor joists are butted or overlapped over the girder. Lapped or butted joists have little resistance to bending when they are loaded, and tend to sag between the lally columns.

However, when the floor joists are continuous—one piece—they are better able to resist the bending over the girder. As a result, the girder now has to support more weight than when the joists are in two pieces. A girder under continuous joists now carries 5⁄8 instead of one-half the load.

Now that we have found the tributary area, we must calculate the total live and dead loads (LL+DD) on the girder. However, we have assumed a LL+DD of (10 psf + 40 psf) 50 psf floor load for the 24′ × 40′ single-story ranch. The roof is trussed; therefore, there are no load-bearing partitions.

Example—Using a tributary width of 12 feet, girder length of 40 feet, and LL+DD of 50 psf, find (a) the tributary area; (b) the girder load per lineal foot; (c) the girder load between the lally columns in Fig. 9–2; (d) the total load on the girder.

Tributary area = $12' \times 40' = 480$ ft^2.
Girder load per lineal foot = 50 psf $\times 12' = 600$ pounds.
Girder load between lally columns = 600 pounds $\times 10' = 6000$ pounds.
Total load on girder = 600 pounds $\times 40' = 24,000$ pounds.

Of course, multiplying the tributary area by the floor load gives us the total load on the girder: 480 ft$^2 \times 50$ psf = 24,000 pounds. Dividing 24,000 pounds by 40 feet = 600 pounds, the load per lineal foot on the girder.

Columns and Column Footings

Girders are supported by wooden posts, concrete or brick piers, hollow pipe columns, or lally columns. A *lally column* is a circular steel shell filled with a special concrete that is carefully vibrated to eliminate all voids. Although hollow cylindrical columns and adjustable columns are used throughout the country, the lally column is the most commonly used girder support in the Northeast. They do not burn, rot, or shrink. Whatever column is used, it must rest a concrete footing that is large enough and deep enough to support the imposed loads. As a general rule, the size of the column footing, the column size and spacing, are specified on the foundation plan. Occasionally, a builder receives an incomplete set of plans, and must either hire an architect to complete them, or attempt it himself. Here we will discuss the basics of column support design, lally column selection, and spacing.

In Chapter 3, **Foundations,** footing design based on footing load and soil bearing capacity was explained. The total load on the footing had to be found. Again the average load per square foot acting on the column footing must be calculated. But the tributary area of the column is needed in order to find the total load carried by the column footing.

Fig. 9–2 shows the lally columns spaced at 10 feet oc. The beam is located in the center of the 24-foot gable end wall. We know the beam tributary width is 12 feet, but what is the half-width of the column? The column supports half the weight of the girder to the midpoint of the span on both sides of the girder. In others words, the midpoint is 5 feet on each side of the column or a total of 10 feet (Fig. 9–3). Therefore, the tributary area of the column is $12' \times 10' = 120$ ft^2.

To calculate the load on the lally column footing we need to know:

1. Ceiling load,
2. Floor load,
3. Partition load.

The roof is trussed. There is no attic storage space, and no load-bearing partitions. Therefore, the floor load is 50 psf. The total footing load is equal to the floor load per square foot multiplied by the column tributary area:

50 psf $\times$ 120 ft^2 = 6000 pounds.

Next, the size of the column footing must be calculated. Our soil bearing capacity is 2000 psf.

$$\text{Footing area (ft}^2) = \frac{\text{footing load (lb/ft)}}{q_a \text{ (lb/ft}^2)}$$

Area = 6000 pounds/2000 psf = 3 ft^2

The standard $2' \times 2' \times 1'$ (4 ft^2) concrete footing is more than adequate. An 8-foot, 3½-inch lightweight lally column will support 21,000 pounds (kips), and is more than adequate to support 6000 pounds.

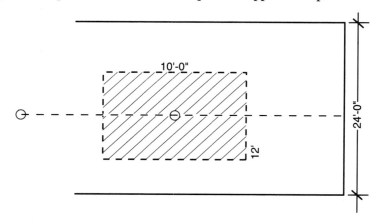

Fig. 9–3. Tributary area of column.

Selecting the Girder

There are a number of ways to find the size of the girder necessary to support the floor loads: consult tables such as Tables 9–1 and 9–2, use the data in *Wood Structural Design Data,* or calculate it using structural engineering formulas. These calculations are beyond the scope of this book. However, before we can use these or other tables, we must understand what is meant by F_b and how to use it.

F_b (pronounced "eff sub-b") is the allowable bending stress, or extreme fiber stress in bending. It tells how strong the wood fibers are. When a beam is loaded it starts bending. As the beam bends, the upper wood fibers try to shorten, causing them to be in compression. At the same time, the lower wood fibers try to get longer, putting them in tension. These two opposing forces meet head on like a pair of scissors, creating a shear force in the beam. Horizontal shear in a beam would be the upper fibers sliding over the lower fibers (Fig. 9–4).

Beams do not break easily because wood is very strong in tension parallel to the grain. But a beam can fail when the lower fibers tear apart from tension. In selecting a beam or laminating a girder, choose the wood whose fibers are strong enough to resist this kind of failure. Tables 9–1 and 9–2 call for a wood with an F_b of 1000 psi or 1500 psi. For example, Southern yellow pine (SYP) No. 2, Hem-fir No. 2, or Douglas fir south No. 2, will meet the F_b requirements. Because Tables 9–1 and 9–2 do not list which woods have which F_b, a table of Working Stresses for Joists and Rafters, Visual Graded Lumber, must be used. Such tables can be found in the *CABO One and Two Family Dwelling Code,* in the *Uniform Building Code,* and in many books on carpentry or framing. Span tables are not published in the *BOCA* code.

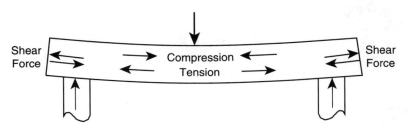

Fig. 9–4. Horizontal shear.

Selecting Girder Lumber and F_b

Table 9–3 lists different species and grades of lumber, and some of their various properties. Because we want the F_b, the other properties will be ignored.

The girder in the single-story ranch spans 10 feet between columns, and supports a load of 6000 pounds, at this spacing. According to Table 9–1, a structure 24 feet wide whose beam spans 10'–3", can use a built-up girder made from three 2 × 12s, with an F_b of 1000 psi minimum.

In the south, Southern pine is the common building material; F_b's of 1200 to 1400 and higher are common. In the west, Douglas fir is the common building lumber, and F_b's from 1250 to 1700 and higher are common. In the northeast, Spruce-pine-fir and Hem-fir No. 2 and better are the common framing lumber. Southern yellow pine and Douglas fir are not stock items. Spruce-pine-fir has an F_b of 875 psi, and Hem-fir, 1000 psi. Select Structural and No. 1 with F_b's of 1050 to 1400 psi can be ordered. A higher rating for the F_b is allowed when the lumber is used repetitively as floor joists, studs, or rafters. Some specialty yards stock Douglas-fir beams. Occasionally, steel beams or engineered wood beams are used. In general, however, built-up girders using Spruce-pine-fir are common in the northeast. After having consulted the table to find which woods have which F_b, the girder material can be selected.

Solid Beams versus Built-up Beams

The tables in *Wood Structural Design Data* and Table 9–4 are for solid wood beams of rectangular cross section, surfaced 4 sides to standard dress dimensions. A 6 × 18 beam would actually be 5½" × 17½". Built-up girders cannot support as much weight as solid girders because they are smaller. A dressed 6 × 10 built-up girder is only 4½ inches × 9¼ inches. A dressed solid 6 × 10 beam is 5½ inches × 9½ inches. Therefore, to find the actual carrying capacity of a built-up girder based on a solid beam's dimensions, a correction factor must be applied:

Multiply by 0.897 when 4" girder is made of (2) 2" pieces.
Multiply by 0.887 when 6" girder is made of (3) 2" pieces.
Multiply by 0.867 when 8" girder is made of (4) 2" pieces.
Multiply by 0.856 when 10" girder is made of (5) 2" pieces.

Table 9-3. Strength Properties of Common Species and Grades of Structural Lumber

Species and commercial grade	Size classification	Design values in pounds per square inch						
		Extreme fiber in bending "f"		Tension parallel to grain	Horizontal shear "H"	Compression perpendicular to grain	Compression parallel to grain	Modulus of elasticity "E"
		Single-member uses	Repetitive-member uses*					
BALSAM FIR (Surfaced dry or surfaced green. Used at 19% max. m.c.)								
Select Structural	2" to 4" thick 2" to 4" wide	1350	1550	800	60	170	1050	1,200,000
No. 1		1150	1300	675	60	170	825	1,200,000
No. 2		950	1100	550	60	170	650	1,100,000
No. 3		525	600	300	60	170	400	900,000
Appearance		1000	1150	650	60	170	1000	1,200,000
Stud	2" to 4" thick 2" to 4" wide	525	600	300	60	170	400	900,000
Construction	2" to 4" thick 4" wide	675	800	400	60	170	750	900,000
Standard		375	450	225	60	170	625	900,000
Utility		175	200	100	60	170	400	900,000
Select Structural	2" to 4" thick 5" and wider	1150	1350	775	60	170	925	1,200,000
No. 1		1000	1150	650	60	170	825	1,200,000
No. 2		825	950	425	60	170	700	1,100,000
No. 3		475	550	250	60	170	450	900,000
Appearance		1000	1150	650	60	170	1000	1,200,000
Stud		475	550	250	60	170	450	900,000

*Repetitive members include joists, studs, rafters, trusses and built-up beams with at least three members spaced no more than 24" apart.

Table 9–3. (cont.)

Species and commercial grade	Size classification		Design values in pounds per square inch						
			Extreme fiber in bending "f"		Tension parallel to grain	Horizontal shear "H"	Compression perpendicular to grain	Compression parallel to grain	Modulus of elasticity "E"
			Single-member uses	Repetitive-member uses*					
DOUGLAS FIR-LARCH (Surfaced dry or surfaced green. Used at 19% max. m.c.)									
Dense Select Structural			2450	2800	1400	95	455	1850	1,900,000
Select Structural			2100	2400	1200	95	385	1600	1,800,000
Dense No. 1			2050	2400	1200	95	455	1450	1,900,000
No. 1	2" to 4"		1750	2050	1050	95	385	1250	1,800,000
Dense No. 2	thick		1700	1950	1000	95	455	1150	1,700,000
No. 2	2" to 4"		1450	1650	850	95	385	1000	1,700,000
No. 3	wide		800	925	475	95	385	600	1,500,000
Appearance			1750	2050	1050	95	385	1500	1,800,000
Stud			800	925	475	95	385	600	1,500,000
Construction	2" to 4"		1050	1200	625	95	385	1150	1,500,000
Standard	thick		600	675	350	95	385	925	1,500,000
Utility	4" wide		275	325	175	95	385	600	1,500,000
Dense Select Structural			2100	2400	1400	95	455	1650	1,900,000
Select Structural			1800	2050	1200	95	385	1400	1,800,000
Dense No. 1	2" to 4"		1800	2050	1200	95	455	1450	1,900,000
No. 1	thick		1500	1750	1000	95	385	1250	1,800,000
Dense No. 2	5" and		1450	1700	775	95	455	1250	1,700,000

Grade	Size							
No. 2	wider	1250	1450	650	95	385	1050	1,700,000
No. 3		725	850	375	95	385	675	1,500,000
Appearance		1500	1750	1000	95	385	1500	1,800,000
Stud		725	850	375	95	385	675	1,500,000

DOUGLAS FIR SOUTH (Surfaced dry or surfaced green. Used at 19% max. m.c.)

Grade	Size							
Select Structural	2" to 4"	2000	2300	1150	90	335	1400	1,400,000
No. 1	thick	1700	1950	975	90	335	1150	1,400,000
No. 2	2" to 4"	1400	1600	825	90	335	900	1,300,000
No. 3	wide	775	875	450	90	335	550	1,100,000
Appearance		1700	1950	975	90	335	1350	1,400,000
Stud		775	875	450	90	335	550	1,100,000
Construction	2" to 4"	1000	1150	600	90	335	1000	1,100,000
Standard	thick	550	650	325	90	335	850	1,100,000
Utility	4" wide	275	300	150	90	335	550	1,100,000
Select Structural	2" to 4"	1700	1950	1150	90	335	1250	1,400,000
No. 1	thick	1450	1650	975	90	335	1150	1,400,000
No. 2	5" and	1200	1350	625	90	335	950	1,300,000
No. 3	wider	700	800	350	90	335	600	1,100,000
Appearance		1450	1650	975	90	335	1350	1,400,000
Stud		700	800	350	90	335	600	1,100,000

Table 9–3. (cont.)

Species and commercial grade	Size classification	Extreme fiber in bending "f"		Tension parallel to grain	Horizontal shear "H"	Compression perpendicular to grain	Compression parallel to grain	Modulus of elasticity "E"
		Single-member uses	Repetitive-member uses*					
EASTERN HEMLOCK—TAMARACK (Surfaced dry or surfaced green. Used at 19% max. m.c.)								
Select Structural	2″ to 4″ thick	1800	2050	1050	85	365	1350	1,300,000
No. 1		1500	1750	900	85	365	1050	1,300,000
No. 2	2″ to 4″ wide	1250	1450	725	85	365	850	1,100,000
No. 3		700	800	400	85	365	525	1,000,000
Appearance		1300	1500	900	85	365	1300	1,300,000
Stud		700	800	400	85	365	525	1,000,000
Construction	2″ to 4″ thick	900	1050	525	85	365	975	1,000,000
Standard		500	575	300	85	365	800	1,000,000
Utility	4″ wide	250	275	150	85	365	525	1,000,000
Select Structural	2″ to 4″ thick	1550	1750	1050	85	365	1200	1,300,000
No. 1		1300	1500	875	85	365	1050	1,300,000
No. 2	5″ and wider	1050	1200	575	85	365	900	1,100,000
No. 3		625	725	325	85	365	575	1,000,000
Appearance		1300	1500	875	85	365	1300	1,300,000
Stud		625	725	325	85	365	575	1,000,000

Design values in pounds per square inch

EASTERN SPRUCE (Surfaced dry or surfaced green. Used at 19% max. m.c.)

	Size							
Select Structural	2" to 4" thick	1500	1750	875	65	255	1150	1,400,000
No. 1		1300	1500	750	65	255	900	1,400,000
No. 2		1050	1200	625	65	255	700	1,200,000
No. 3	2" to 4" wide	575	675	325	65	255	425	1,100,000
Appearance		1100	1250	750	65	255	1050	1,400,000
Stud		575	675	325	65	255	425	1,100,000
Construction	2" to 4" thick	775	875	450	65	255	800	1,100,000
Standard		425	500	250	65	255	675	1,100,000
Utility	4" wide	200	225	100	65	255	425	1,100,000
Select Structural	2" to 4" thick	1300	1500	875	65	255	1000	1,400,000
No. 1		1100	1250	750	65	255	900	1,400,000
No. 2	5" and	900	1000	475	65	255	750	1,200,000
No. 3	wider	525	600	275	65	255	475	1,100,000
Appearance		1100	1250	750	65	255	1050	1,400,000
Stud		525	600	275	65	255	475	1,100,000

ENGELMANN SPRUCE—ALPINE FIR (ENGELMANN SPRUCE—LODGEPOLE PINE) (Surfaced dry or surfaced green.) Used at 19% max. m.c.

	Size							
Select Structural	2" to 4" thick	1350	1550	800	70	195	950	1,300,000
No. 1		1150	1350	675	70	195	750	1,300,000
No. 2		950	1100	550	70	195	600	1,100,000
No. 3	2" to 4" wide	525	600	300	70	195	375	1,000,000
Appearance		1150	1350	675	70	195	900	1,300,000
Stud		525	600	300	70	195	375	1,000,000

Table 9–3. (cont.)

Species and commercial grade	Size classification	Design values in pounds per square inch						
		Extreme fiber in bending "f"		Tension parallel to grain	Horizontal shear "H"	Compression perpendicular to grain	Compression parallel to grain	Modulus of elasticity "E"
		Single-member uses	Repetitive-member uses*					
Construction	2" to 4" thick 4" wide	700	800	400	70	195	675	1,000,000
Standard		375	450	225	70	195	550	1,000,000
Utility		175	200	100	70	195	375	1,000,000
Select Structural	2" to 4" thick 5" and wider	1200	1350	775	70	195	850	1,300,000
No. 1		1000	1150	675	70	195	750	1,300,000
No. 2		825	950	425	70	195	625	1,100,000
No. 3		475	550	250	70	195	400	1,000,000
Appearance		1000	1150	675	70	195	900	1,300,000
Stud		475	550	250	70	195	400	1,000,000
HEM-FIR (Surfaced dry or surfaced green. Used at 19% max. m.c.)								
Select Structural	2" to 4" thick	1650	1900	975	75	245	1300	1,500,000
No. 1		1400	1600	825	75	245	1050	1,500,000
No. 2		1150	1350	675	75	245	825	1,400,000
No. 3	2" to 4" wide	650	725	375	75	245	500	1,200,000
Appearance		1400	1600	825	75	245	1250	1,500,000
Stud		650	725	375	75	245	500	1,200,000

Construction	2″ to 4″	825	975	500	75	245	925	1,200,000
Standard	thick	475	550	275	75	245	775	1,200,000
Utility	4″ wide	225	250	125	75	245	500	1,200,000
Select Structural		1400	1650	950	75	245	1150	1,500,000
No. 1	2″ to 4″	1200	1400	800	75	245	1050	1,500,000
No. 2	thick	1000	1150	525	75	245	875	1,400,000
No. 3	5″ and	575	675	300	75	245	550	1,200,000
Appearance	wider	1200	1400	800	75	245	1250	1,500,000
Stud		575	675	300	75	245	550	1,200,000

LODGEPOLE PINE (Surfaced dry or surfaced green. Used at 19% max. m.c.)

Select Structure		1500	1750	875	70	250	1150	1,300,000
No. 1	2″ to 4″	1300	1500	750	70	250	900	1,300,000
No. 2	thick	1050	1200	625	70	250	700	1,200,000
No. 3	2″ to 4″	600	675	350	70	250	425	1,000,000
Appearance	wide	1300	1500	750	70	250	1050	1,300,000
Stud		600	675	350	70	250	425	1,000,000
Construction	2″ to 4″	775	875	450	70	250	800	1,000,000
Standard	thick	425	500	250	70	250	675	1,000,000
Utility	4″ wide	200	225	125	70	250	425	1,000,000
Select Structural		1300	1500	875	70	250	1000	1,300,000
No. 1	2″ to 4″	1100	1300	750	70	250	900	1,300,000
No. 2	thick	925	1050	475	70	250	750	1,200,000
No. 3	5″ and	525	625	275	70	250	475	1,000,000
Appearance	wider	1100	1300	750	70	250	1050	1,300,000
Stud		525	625	275	70	250	475	1,000,000

Table 9–3. (cont.)

Species and commercial grade	Size classification	Extreme fiber in bending "f"		Tension parallel to grain	Horizontal shear "H"	Compression perpendicular to grain	Compression parallel to grain	Modulus of elasticity "E"
		Single-member uses	Repetitive-member uses*					
MOUNTAIN HEMLOCK (Surfaced dry or surfaced green. Used at 19% max. m.c.)								
Select Structural	2" to 4" thick	1750	2000	1000	95	370	1250	1,300,000
No. 1		1450	1700	850	95	370	1000	1,300,000
No. 2	2" to 4" wide	1200	1400	700	95	370	775	1,100,000
No. 3		675	775	400	95	370	475	1,000,000
Appearance		1450	1700	850	95	370	1200	1,300,000
Stud		675	775	400	95	370	475	1,000,000
Construction	2" to 4" thick	875	1000	525	95	370	900	1,000,000
Standard	4" wide	500	575	275	95	370	725	1,000,000
Utility		225	275	125	95	370	475	1,000,000
Select Structural	2" to 4" thick	1500	1700	1000	95	370	1100	1,300,000
No. 1		1250	1450	850	95	370	1000	1,300,000
No. 2	5" and wider	1050	1200	550	95	370	825	1,100,000
No. 3		625	700	325	95	370	525	1,000,000
Appearance		1250	1450	850	95	370	1200	1,300,000
Stud		625	700	325	95	370	525	1,000,000

Design values in pounds per square inch

RED PINE (Surfaced dry or surfaced green. Used at 19% max. m.c.)								
Select Structural		1400	1600	800	70	280	1050	1,300,000
No. 1	2" to 4"	1200	1350	700	70	280	825	1,300,000
No. 2	thick	975	1100	575	70	280	650	1,200,000
No. 3	2" to 4"	525	625	325	70	280	400	1,000,000
Appearance	wide	1200	1350	675	70	280	925	1,300,000
Stud		525	625	325	70	280	400	1,000,000
Construction	2" to 4"	700	800	400	70	280	750	1,000,000
Standard	thick	400	450	225	70	280	600	1,000,000
Utility	4" wide	175	225	100	70	280	400	1,000,000
Select Structural	2" to 4"	1200	1350	800	70	280	900	1,300,000
No. 1	thick	1000	1150	675	70	280	825	1,300,000
No. 2	5" and	825	950	425	70	280	675	1,200,000
No. 3	wider	500	550	250	70	280	425	1,000,000
Appearance		1000	1150	675	70	280	925	1,300,000
Stud		500	550	250	70	280	425	1,000,000

SOUTHERN PINE (Surfaced dry or surfaced green. Used at 19% max. m.c.)								
Select Structural		2000	2300	1150	100	405	1550	1,700,000
Dense Select Structural		2350	2700	1350	100	475	1800	1,800,000
No. 1		1700	1950	1000	100	405	1250	1,700,000
No. 1 Dense	2" to 4" thick	2000	2300	1150	100	475	1450	1,800,000
No. 2	2" to 4"	1400	1650	825	90	405	975	1,600,000
No. 2 Dense	wide	1650	1900	975	90	475	1150	1,600,000
No. 3		775	900	450	90	405	575	1,400,000
No. 3 Dense		925	1050	525	90	475	675	1,500,000
Stud		775	900	450	90	405	575	1,400,000

Table 9–3. (cont.)

Species and commercial grade	Size classification	Design values in pounds per square inch						
		Extreme fiber in bending "f"		Tension parallel to grain	Horizontal shear "H"	Compression perpendicular to grain	Compression parallel to grain	Modulus of elasticity "E"
		Single-member uses	Repetitive-member uses*					
Construction	2" to 4" thick 4" wide	1000	1150	600	100	405	1100	1,400,000
Standard		575	675	350	90	405	900	1,400,000
Utility		275	300	150	90	405	575	1,400,000
Select Structural		1750	2000	1150	90	405	1350	1,700,000
Dense Select Structural		2050	2350	1300	90	475	1600	1,800,000
No. 1	2" to 4" thick 5" and wider	1450	1700	975	90	405	1250	1,700,000
No. 1 Dense		1700	2000	1150	90	475	1450	1,800,000
No. 2		1200	1400	625	90	405	1000	1,600,000
No. 2 Dense		1400	1650	725	90	475	1200	1,600,000
No. 3		700	800	350	90	405	625	1,400,000
No. 3 Dense		825	925	425	90	475	725	1,500,000
Stud		725	850	350	90	405	625	1,400,000
SPRUCE—PINE—FIR (Surfaced dry or surfaced green. Used at 19% max. m.c.)								
Select Structural	2" to 4" thick 2" to 4" wide	1450	1650	850	70	265	1100	1,500,000
No. 1		1200	1400	725	70	265	875	1,500,000
No. 2		1000	1150	600	70	265	675	1,300,000
No. 3		550	650	325	70	265	425	1,200,000
Appearance		1200	1400	700	70	265	1050	1,500,000
Stud		550	650	325	70	265	425	1,200,000

Grade							
Construction	725	850	425	70	265	775	1,200,000
Standard	400	475	225	70	265	650	1,200,000
Utility	175	225	100	70	265	425	1,200,000
Select Structural	1250	1450	825	70	265	975	1,500,000
No. 1	1050	1200	700	70	265	875	1,500,000
No. 2	875	1000	450	70	265	725	1,300,000
No. 3	500	575	275	70	265	450	1,200,000
Appearance	1050	1200	700	70	265	1050	1,500,000
Stud	500	575	275	70	265	450	1,200,000

WESTERN HEMLOCK (Surfaced dry or surfaced green. Used at 19% max. m.c.)

Grade							
Select Structural	1800	2100	1050	90	280	1450	1,600,000
No. 1	1550	1800	900	90	280	1150	1,600,000
No. 2	1300	1450	750	90	280	900	1,400,000
No. 3	700	800	425	90	280	550	1,300,000
Appearance	1550	1800	900	90	280	1350	1,600,000
Stud	700	800	425	90	280	550	1,300,000
Construction	925	1050	550	90	280	1050	1,300,000
Standard	525	600	300	90	280	850	1,300,000
Utility	250	275	150	90	280	550	1,300,000
Select Structural	1550	1800	1050	90	280	1300	1,600,000
No. 1	1350	1550	900	90	280	1150	1,600,000
No. 2	1100	1250	575	90	280	975	1,400,000
No. 3	650	750	325	90	280	625	1,300,000
Appearance	1350	1550	900	90	280	1350	1,600,000
Stud	650	750	325	90	280	625	1,300,000

Table 9–4. Girder Size and Allowable Spans

Allowable Uniformly Distributed Loads for Solid Wood Girders and Beams
Computed for Actual Dressed Sizes of Douglas Fir, Southern Yellow Pine
(Allowable Fiber Stress 1400 lbs. Per Square Inch)

Solid Girder Size	Span in Feet						
	6	7	8	10	12	14	16
2 × 6	1318	1124	979	774	636	536	459
3 × 6	2127	1816	1581	1249	1025	863	740
4 × 6	2938	2507	2184	1726	1418	1194	1023
6 × 6	4263	3638	3168	2504	2055	1731	1483
2 × 8	1865	1865	1760	1395	1150	973	839
3 × 8	3020	3020	2824	2238	1845	1560	1343
4 × 8	4165	4165	3904	3906	2552	2160	1802
6 × 8	6330	6330	5924	4698	3873	3277	2825
8 × 8	8630	8630	8078	6406	5281	4469	3851
2 × 10	2360	2360	2360	2237	1848	1569	1356
3 × 10	3810	3810	3810	3612	2984	2531	2267
4 × 10	5265	5265	5265	4992	4125	3500	3026
6 × 10	7990	7990	7990	6860	6261	5312	4593
8 × 10	10920	10920	10920	9351	8537	7244	6264
2 × 12	2845	2845	2845	2845	2724	2315	2006
3 × 12	4590	4590	4590	4590	4394	3734	3234
4 × 12	6350	6350	6350	6350	6075	5165	4474
6 × 12	9640	9640	9640	9640	9220	7837	6791
8 × 12	13160	13160	13160	13160	12570	10685	9260
2 × 14	3595	3595	3595	3595	3595	3199	2776

Example—Given a 10-foot span and a girder load of 6000 pounds, what size beam is required?

According to Table 9–4, at a 10-foot span, a solid 8 × 8 beam with an F_b of 1400 will support 6406 pounds. If a solid beam with an F_b of 1400 is available, it can be used. It allows for a little more head room. However, if the 8 × 8 is made from 2 × 8s, its bearing capacity must be multiplied by 0.867. This reduces its load-carrying capacity to: 6406 × 0.867 = 5554 pounds. The 6 × 10 beam with a capacity of 6860 pounds reduces to 6860 × 0.867 = 5977 pounds. The 8 × 10 beam with a capacity of 9351 × 0.867 = 8107 pounds, is more than adequate. The choice: a larger beam, or a reduced span.

Reducing the span to 8 feet would require another lally column (4 total), and another column footing. The additional labor and material costs, while not large, must be weighed against a lower cost girder. With an 8-foot span, carrying 6000 pounds, an 8 × 8 built-up girder could support 8079 pounds × 0.867 = 7004 pounds. Table 9–1 shows that for a single-story house, 24 feet wide, with a girder span of 10′–3″, carrying 6000 pounds, three 2 × 12s will easily carry this load. The choices are: keep the span at 10 feet and use 2 × 12s, or use 2 × 8s and add an extra column and column footing.

Column Spacing

In the one-story ranch the 40-foot-long girder is supported by lally columns spaced 10 feet on center. As the spacing between columns increases, the girder must be made larger. The girder can be made larger by increasing its width, depth, or both. Doubling the width of a girder doubles its strength. Doubling the depth increases its strength four times. A beam 3 inches wide by 6 inches deep will carry 4 times as much weight as one 3 inches wide by 3 inches deep. However, as the girder gets deeper, more headroom in the basement is lost. Keep this in mind when a buyer, wanting a more open basement, requests only one lally column.

Our calculations show that the 40-foot beam is carrying 24,000 pounds or 24 kips. *Kip* is derived from Ki(lo) and P(ound) and means 1000 pounds. Therefore, 24 kips equals 24,000 pounds. Assume a lally column at the center of the girder, and the span now extending 20 feet on each side.

Example—Given a 40-foot beam carrying 24 kips, a tributary width of 12 feet, and a lally column 20 feet on center, find the required beam dimensions and the size of the column. Assume an 8-foot-long lally column.

The beam carries 24,000 pounds/2 = 12,000 pounds for a 20-foot span.

Using an F_b of 1400, *Wood Structural Design Data* lists a 6 × 18 as able to support 13 kips. However if the height from the bottom of the floor joists to the slab floor is 7′–10″, the headroom under an 18-inch deep beam is only 6′–5″. Using a Douglas fir-larch No. 1 with an F_b of 1800 would bring the beam size down to a 10 × 12. This would provide 7 feet of headroom under the beam.

Flitch Beams

Before glue-laminated wood beams and rolled steel beams were readily available, a common method of making wood beams stronger was to make a sandwich of two wood planks separated by a steel plate. This composite material acted as a unit, and allowed the wood to carry considerably more weight without increasing the depth of the wood planks. Known as a *flitch beam* (Fig. 9–5), it is rarely seen today, because it is labor and material intensive, particularly in the use of steel plate, which has to be drilled and then bolted to the wood planks. Calculating the size of a flitch beam using structural engineering formulas, to replace the 6 × 18 beam, is too complex for this book. Two 2 × 12s, with a ½-inch steel plate in the middle, all bolted together, would carry for a 20-foot span, 10971 pounds; over 1000 pounds less than required. The two 2 × 12s at 20 feet with an F_b of 1400 psi can support only 3090 pounds. With the ½-inch steel plate, it will now support 10971 pounds.

Steel Beams

Another alternative to deep beams and low headroom is the steel beam. Wide-flange shapes have replaced the older American Standard I-beam shape, because they are more structurally efficient. Wide-flanges come in a variety of sizes and weights, ranging in size from 4 inches to

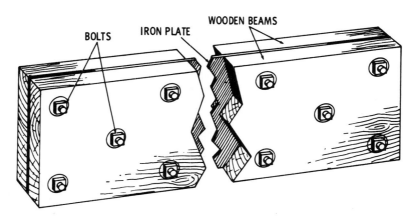

Fig. 9–5. A flitch plate girder, or flitch beam.

36 inches, and in weights of 12 pounds per foot to 730 pounds per foot. Mild structural steel, ASTM A36 Carbon Steel, is the most commonly used steel in steel building frames. It has an F_b of 22,000 psi.

The same kind of information used for calculating the size of wood beams is used with steel beams. In both cases, the engineering calculations are too complex to discuss here. As with the 6 × 18 beam, published tables will be used to select the shape and size of the steel beam.

The American Institute of Steel Construction (AISC) publishes design data for all common steel shapes, special designs, and tables giving allowable uniform loads in kips for beams laterally supported. This information is contained in the AISC *Manual of Steel Construction.* Fig. 9–6 is a sample of one of these tables.

Steel Beams and Fire

Alec Nash in his book, *Structural Design for Architects,* notes,

> *... timber is ... therefore combustible. . . . The fact that it will burn when exposed to fire, however, does not mean that it is devoid of fire resistance. The burned fibres of the timber on the outside . . . subjected to fire provide a certain degree of thermal insulation between the fire source and the timber. . . .* (reprinted by permission)

Although a building fire may not be hot enough to melt steel, the heat is able to weaken it enough to cause structural damage.

Many builders claim that wood beams are the best construction materials to structurally withstand fire. Because 2x lumber is only 1½ inches wide, it burns completely. When flames touch one surface of a heavy beam, the wood starts burning and forms a char layer on the surface (Fig. 9–7). The rate of charring depends on many things, but is slower for high-density woods and woods with a high moisture content. The temperature at the base of the char is about 550 degrees Fahrenheit. Because wood is a poor conductor of heat, the temperature drops rapidly inside the char zone. The char zone actually insulates and protects the strength of the inner wood.

Nash reminds us that,

> *There is another strange irony in that timber, an organic material and therefore combustible, does possess a measure of fire resistance by virtue of the charring process whilst steel, a metal and therefore*

F_y = 36 ksi	BEAMS W Shapes Allowable uniform loads in kips for beams laterally supported For beams laterally unsupported, see page 2-146										W 10

Designation		W 10			W 10			W 10				
Wt./ft		45	39	33	30	26	22	19	17	15	12	Deflection In.
Flange Width		8	8	8	5¾	5¾	5¾	4	4	4	4	
L_c		8.50	8.40	8.40	6.10	6.10	6.10	4.20	4.20	4.20	3.90	
L_u		22.8	19.8	16.5	13.1	11.4	9.40	7.20	6.10	5.00	4.30	

F_y = 36 ksi — Span in Feet		45	39	33	30	26	22	19	17	15	12	Deflection
	3								70	66	54	.02
	4							74	64	55	43	.04
	5				90	77	70	60	51	44	35	.06
	6			81	86	74	61	50	43	36	29	.09
	7	102	90	79	73	63	52	43	37	31	25	.12
	8	97	83	69	64	55	46	37	32	27	22	.16
	9	86	74	62	57	49	41	33	29	24	19	.20
	10	78	67	55	51	44	37	30	26	22	17	.25
	11	71	61	50	47	40	33	27	23	20	16	.30
	12	65	56	46	43	37	31	25	21	18	14	.35
	13	60	51	43	39	34	28	23	20	17	13	.42
	14	56	48	40	37	32	26	21	18	16	12	.48
	16	49	42	35	32	28	23	19	16	14	11	.63
	18	43	37	31	29	25	20	17	14	12	10	.80
	20	39	33	28	26	22	18	15	13	11	8.6	.98
	22	35	30	25	23	20	17	14	12	10	7.8	1.19
	24	32	28	23	21	18	15	12	11	9.1	7.2	1.42

Properties and Reaction Values											
S_x in.³	49.1	42.1	35.0	32.4	27.9	23.2	18.8	16.2	13.8	10.9	
V kips	51	45	41	45	39	35	37	35	33	27	For explanation of deflection, see page 2-32
R_1 kips	26.0	21.0	18.3	16.7	13.5	10.7	12.1	10.7	9.39	7.05	
R_2 kips/in.	8.32	7.48	6.89	7.13	6.18	5.70	5.94	5.70	5.46	4.51	
R_3 kips	33.3	26.3	21.0	23.9	17.9	14.4	16.0	13.8	11.7	7.74	
R_4 kips/in.	4.19	3.64	3.53	3.09	2.37	2.31	2.36	2.54	2.76	2.03	
R kips	48	39	33	35	26	22	24	23	21	15	

Load above heavy line is limited by maximum allowable web shear.

Fig. 9–6. From manual of steel construction. (*Courtesy of American Institute of Steel Construction*)

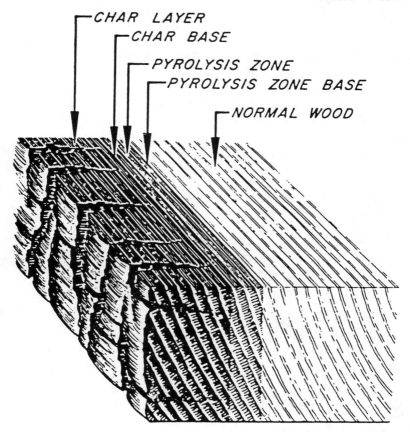

CHAR LAYER
CHAR BASE
PYROLYSIS ZONE
PYROLYSIS ZONE BASE
NORMAL WOOD

Fig. 9–7. Layers of char help to protect heavy timbers. 2x lumber is too thin and is structurally inadequate once it burns.

incombustible, has virtually no fire resistance. (reprinted by permission)

Steel certainly has a place in residential construction. But as with any material, its cost must be considered. Delivery charges, hole drilling, and the cost of hiring a crane can add substantially to the price of a steel beam. The cost of enclosing it in fire-resistant material—to prevent major structural damage during a fire—is another added cost.

Engineered Lumber

Anyone seeing glulam beams, paralam, wood I-beams with plywood webs, and OSB might think these materials are modern. However, the Egyptians in 300 BC developed engineered lumber known as veneer. The Romans developed it to a high degree, but with the advent of the Dark Ages, AD 476–1453, it became a lost art. Not until the mid-1500s did it reappear. Plywood came into existence in 1927.

Laminated Layered Products

MICRO-LAM® was developed by Trus Joist Corporation in the early 1970s. It is made up of a series of $\frac{1}{10}$- and $\frac{1}{8}$-inch thick veneers, laid up with all grain parallel. The veneer is coated with waterproof adhesives and seal-cured by pressure and heat. The resulting product is a dense board up to 2½ inches thick, 24 inches wide, up to 80 feet long.

Georgia-Pacific and Louisiana-Pacific also manufacture LVL products. All of these products are stronger than dimension lumber, dimensionally stable, resistant to splitting, crooking, shrinking, warping, and twisting. They are lighter and have more load-bearing capacity per pound than solid sawn lumber. See Figs. 10–10 and 10–11 in Chapter 10, *Floor Framing.*

Caution—Each of these manufacturers publishes Application Booklets on spans, loads, applications, and how to use their product. DO NOT USE one manufacturer's specifications with another manufacturer's product.

Glulam

Glulam (structural glue-laminated timber) was first used in Europe. An auditorium under construction in Switzerland in 1893 employed laminated arches glued with a casein glue. Today, about 30 members of the American Institute of Timber Construction produce glulams (Fig. 9–8).

Standard nominal 2-inch lumber is used as the laminations, or lams. The lams are dressed to 1½ inches and bonded together with adhesives. Glulams may be straight or curved, laminated horizontally or vertically, and available in standard or custom sizes. Glulam beams

Fig. 9–8. Glulam beam. *(Courtesy Weyerhaeuser)*

are made with the strongest lams on the top and the bottom of the beam. The beams may be balanced or unbalanced: the unbalanced beam has TOP stamped in the top. Either face may be used as the Top in a balanced beam.

Glulam beams are made in three appearance grades: Premium, Architectural, and Industrial. Like laminated veneer lumber, glulam is strong, durable, and dimensionally stable. Glulam offers better fire safety than steel. Wood ignites at about 480 degrees Fahrenheit, but charring can begin as low as 300 degrees Fahrenheit. Wood chars at a rate of about 1/40 inch (0.025″) per minute. After 30 minutes exposure to fire, only 3/4 inch of the glulam will be damaged (Fig. 9–7). As we have seen, the char insulates the wood and allows it to withstand higher fire temperatures. For more information see American Wood Systems (APA) publication, *GLULAMS*.

Choosing an LVL or Glulam Beam

Load and span tables show that a 10×12 solid girder with an F_b of 1800 psi will span 20 feet and support 12.5 kips. Consulting LVL beam applications data reveals that at a 20-foot span supporting 600 pounds/LF, at least three $1\frac{3}{4}'' \times 11\frac{7}{8}''$ beams, if not more, are necessary. Consult the manufacturer's load and span tables, or your architect. At between $4 and $5 per foot these beams are more expensive than more conventional sawn lumber. When appearance and openness are major issues, some brands of LVL beams have a more suitable *appearance*; however, all brands will meet structural requirements.

When fire resistance, appearance, great spans and loads are important, glulam beams are the clear choice. Stock glulam beams are about 3 to 4 times more expensive than LVL beams. First cost is always a consideration, but it should never be the *only* consideration.

CHAPTER 10

Floor Framing

A 2 × 10, or any piece of dimensional lumber, can vary in depth from 9¼ inches to 9⅝ inches. If the tops of the floor joists and the floor are to be level over the entire surface of the floor, the joists must be selected from those that are closest to, in this case, 9¼ inches. Check, say, every fourth joist while they are still bundled. Mark the actual dimension on the ends of the joist. Or make a GO-NO GO gage that allows quick selection of joists nearest to the required depth. Selecting the joists can eliminate having to cut end notches. *Do not use wood shims* to level floor joists. Shims, under compression, creep as the weight of the structure eventually crushes them paper thin.

Once the crowns have been identified and marked, set the joists in place, crown up, ready for nailing to the joist header, also called a rim joist, band or box sill. By butting the ends of the joists over the girder, rather than overlapping, the joists are in-line and the 1½-inch offset is eliminated. The ends of each joist are now located exactly on the 16-inch line or the 2-foot module. Although the code official may ask for a 1x lumber tie (Fig. 10–1), it is necessary on only one side of the joists. A plywood floor which is continuous over the two joist ends will tie the butted ends together (Fig. 10–2), making the metal or lumber tie unnecessary.

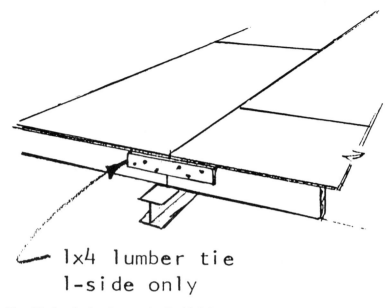

1x4 lumber tie
1-side only

Fig. 10–1.   1x lumber on butted joists.

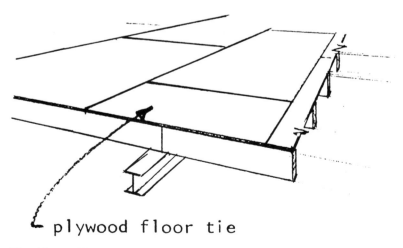

plywood floor tie

Fig. 10–2.   Plywood over butted floor joists.

Alternative Methods and Materials

Many builders believe that the code forces them to frame only one way. Unfortunately, too often the code official's ideas and the code requirements are at odds. There are two types of codes: performance and specification. No code is 100 percent one or the other. There are elements of both in each code. A performance code states what is to be accomplished: frame that wall, pour the foundation, shingle the roof, but not *how* to do it, or at what spacing or span. In other words, whatever you do, base it on accepted *standards:* ASTM, ACI, the standards listed in Appendix A of the *BOCA* code, Chapter 60 of the *Uniform Building Code,* or on good engineering practice. Do not confuse *accepted practice,* which is often wrong, with ASTM or other published standards.

Specification codes, on the other hand, lay out the methods to be followed and the specific requirements. Even a specification code, however, does not demand that a structure be designed or built in a specific way or in one particular configuration. Nor is it the intent of the code to do so. No matter how much we may dislike codes—it is often the code official who is the problem rather than the code—they do not stop or prevent new material and methods. The point of all this is that every code contains a statement, such as, it is not the intent of this code to prevent the use or alternative methods and materials provided they are equal or superior to that prescribed by the code. (See 1990 *BOCA* Section 107.4, 1988 *UBC* code Section 105, and Section R-108 of the 1989 *CABO One And Two Family Dwelling Code.*)

There is *no* 11th Commandment that says, "Thou shalt not frame any way other than 16 inches on center." The building official has the right to make rules and regulations regarding the methods and materials of construction, the right to interpret the code, and the right to adopt rules and regulations. You have the right to challenge and appeal his decisions. You have the right to use alternative methods of framing and alternative materials. He has the right to ask you to document your claims, or have your plans stamped by a registered professional licensed engineer or architect.

Avoid misunderstandings, challenges, and STOP WORK ORDERS by having a pre-framing conference with the building official. Even if you frame conventionally, and especially if you have never built in a given town before, it is a good idea to find out what you and the code official expect from each other. It is better to solve problems in the building department office than to start construction and have the inspector stop you half way through.

Perhaps it is not possible to know the codes as well as the code official. But you should be familiar enough with the codes that apply to residential construction to challenge the official when he makes a mistake in interpretation. For example, floor joists do not have to be overlapped. They can be butted, even if this is not the *usual* way of doing it. That they are secured to each other is what matters, not whether they are overlapped or butted.

Cantilevered In-Line Joist System

When two joist ends are secured together with a metal or plywood gusset, the spliced joint does not have to be located directly over the girder. They can extend beyond the girder, cantilevered, as in Fig. 10–3.

The cantilevered system is more structurally efficient than the simple 2-joist span. There is less stress on the suspended joists because the span is shorter than the joist that runs from foundation wall to the girder. Off-center spliced joists act as though they were a single unit. A reduction in joist size may be possible: a 2 × 8 rather than a 2 × 10. However, even if no reduction is allowed, the cantilevered joist combined with a plywood subfloor is stiffer than the simple 2-joist span. The offset joist can be pre-assembled and each joist simply dropped in place.

Table 10–1 lists the types of lumber, joist size for different house widths, and for 16- and 24-inch oc framing. Table 10–1 is based on the following assumptions:

1. LL + DD = 50 psf over both clear spans;
2. Joist lumber kiln dried, and according to Table 10–1;
3. Plywood splice made from 1/2-inch APA RATED SHEATHING marked PS1;
4. Splice fasteners are 10d common nails—DO NOT USE GLUE;
5. Subfloor per APA recommendations;
6. Joists must be installed so that the splices occur alternately on one side, then the other, of the girder (Fig. 10–3);

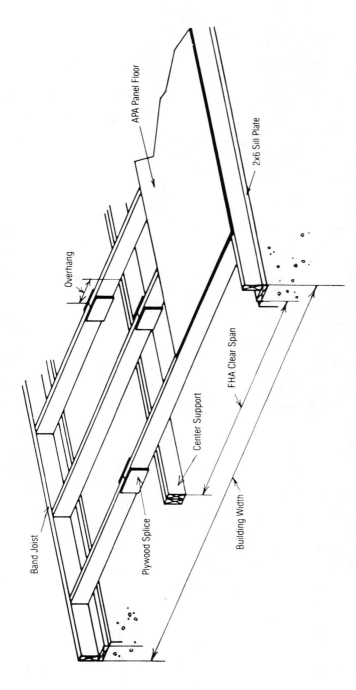

Fig. 10–3. Cantilevered joist diagram from APA. *(Courtesy American Plywood Association)*

APA Panel Floor

2x6 Sill Plate

Overhang

FHA Clear Span

Center Support

Building Width

Band Joist

Plywood Splice

Table 10–1. Joist Spans

16″ Joist Spacing

Joist Lumber	House Width						
	22′	24′	26′	28′	30′	32′	34′
	Joist Size						
Douglas fir-larch							
No. 1	2x6*	2x6*	2x8	2x8*	2x8*	2x10	2x10
No. 2	2x6*	2x8	2x8	2x8*	2x8*	2x10	2x10
No. 3	2x8	2x8*	2x10	2x10	2x10*	2x12	2x12
Southern pine							
No. 1 KD	2x6*	2x6*	2x8	2x8*	2x8*	2x10	2x10
No. 2 KD	2x6*	2x8	2x8	2x8*	2x8*	2x10	2x10
No. 3 KD	2x8	2x8*	2x10	2x10	2x10*	2x12	2x12
Hem-fir							
No. 1	2x8	2x8	2x8	2x8*	2x8*	2x10	2x10*
No. 2	2x8	2x8	2x8*	2x8*	2x10	2x10	2x10*
No. 3	2x10	2x10	2x10*	2x12	2x12	2x12*	—

24″ Joist Spacing

Joist Lumber	House Width						
	22′	24′	26′	28′	30′	32′	34′
	Joist Size						
Douglas fir-larch							
No. 1	2x8	2x8	2x10	2x10	2x10	2x10*	2x12
No. 2	2x8	2x8*	2x10	2x10	2x10	2x10*	2x12
No. 3	2x10	2x10*	2x12	2x12	2x12*	—	—
Southern pine							
No. 1 KD	2x8	2x8	2x10	2x10	2x10	2x10*	2x12
No. 2 KD	2x8	2x8*	2x10	2x10	2x10*	2x10*	2x12
No. 3 KD	2x10	2x10*	2x12	2x12	2x12*	—	—
Hem-fir							
No. 1	2x8	2x10	2x10	2x10	2x12	2x12	—
No. 2	2x8*	2x10	2x10	2x10*	2x12	2x12	—
No. 3	2x12	2x12	2x12*	—	—	—	—

*Joist size is a reduction from HUD-MPS requirement for simple spans, as shown in the NFPA Span Tables for Joists and Rafters, adopted as the standard for conventional floor construction. (See MPS 4900.1. Appendix E.)

(Courtesy American Plywood Association)

7. All interior partitions must be non-load-bearing, except over girder;
8. The girder may be in the middle of the building;
9. Subfloor must be installed per APA recommendations.

Fig. 10–4 is the joist-cutting schedule and the table of lengths for the floor joist system. Fig. 10–5 lists the splice size and nailing schedule.

Glued Floors

About 30 years ago, the APA and an adhesive manufacturer developed the glued/nailed subfloor concept. Although gluing the subfloor is a common practice, many builders omit this step. Why use glue when nails will do? With nails only, the stresses that develop in a structure concentrate around nails and screws. When glue is added, the stresses spread out through the framing. As a result of full-scale testing, the APA has shown that the stiffness of a glued and nailed floor is at least one-third greater than that of the same floor nailed-only. In Chapter 2, **Concrete,** we noted that a stronger floor could be obtained by increasing the depth of the slab. The resistance of a floor joist to bending is a function of its depth. The deeper a floor joist, the stiffer it is. Gluing the subfloor to the joist increases its depth by the thickness of the subfloor. A 9¼-inch 2 × 10 becomes a full 10 inches deep when ¾-inch plywood is glued to it.

Caution—Gluing the subfloor to the joists produces a *stiffer* floor, but not one that is much stronger than a nailed-only floor. *Strength* is not the same as *stiffness*. The span of floor joists framed with weaker wood such as Spruce-pine-fir (SPF) is controlled by the *strength* of the wood. Plywood glued to SPF joists will stiffen them, but the small increase in strength will not allow an increase in span length.

J. E. Gordon in his book, *Structures: Or Why Things Don't Fall Down,* quotes a passage from *The New Science of Materials:*

> *A biscuit is stiff but weak, steel is stiff and strong, nylon is flexible (low E) and strong; raspberry jelly is flexible (low E) and weak. The two properties together describe a solid about as well as you can reasonably expect two figures to do.*

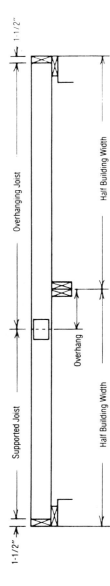

Building Width	Overhanging Joist	Supported Joist	Overhang
22' - 0"	12' - 0"	9' - 9"	1' - 1-1/2"
24' - 0"	14' - 0"	9' - 9"	2' - 1-1/2"
26' - 0"	15' - 9"	10' - 0"	2' - 10-1/2"
28' - 0"	16' - 0"	11' - 9"	2' - 1-1/2"
30' - 0"	17' - 9"	12' - 0"	2' - 10-1/2"
32' - 0"	18' - 0"	13' - 9"	2' - 1-1/2"
34' - 0"	20' - 0"	13' - 9"	3' - 1-1/2"

Fig. 10—4. Joist cutting schedule. This table lists net lumber lengths for the floor joist system in the sketch. Combinations were selected to result in a minimum of cut-off waste. *(Courtesy American Plywood Association)*

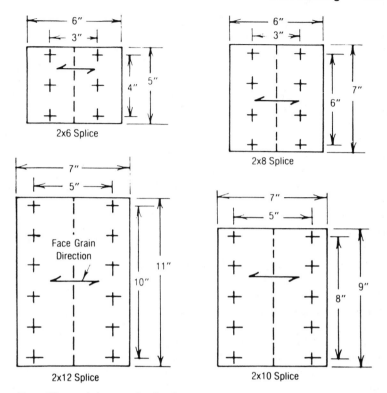

Note: *The symbol* + *on the sketches indicates nail locations.*

Fig. 10–5. Splice size and nailing schedule. These plywood splice patterns were developed through an APA test program, and require that minimum APA recommendations for subfloor or combination subfloor-underlayment be followed. One-half-inch-thick plywood splices are used on both sides of the joist. Fasteners are 10d common nails driven from one side and clinched on the other (double shear). The direction of the splice face grain parallels the joist. No glue is to be used. *(Courtesy American Plywood Association)*

By using stronger species, such as Douglas fir and Southern yellow pine, longer spans are permitted. The span of these stronger woods is determined by stiffness, which is increased by gluing the subfloor.

Gluing creates a floor *system,* as though one continuous piece of plywood was glued to the joists, and the joists and plywood are one. The floor loads are transferred to every member of this continuous system, because of the bond formed by the adhesive. There is less stress, less deflection, a stiffer floor, and a modest increase in strength.

The APA recommends that a nailed-only floor, with $19/32$-inch plywood and joists at 16 inches oc be secured with 6d annular ring nails spaced 6 inches at the edges, and 10 inches elsewhere. Gluing and nailing allows all nails to be spaced 12 inches oc. Squeaky floors and nail pops are just about eliminated.

Use $3/4$-inch T&G plywood or OSB. Follow the APA recommendation to leave $1/8$-inch spacing at all edges and end joints. This is especially important with OSB. DO NOT whack the panel with a hammer to drive the T&G tight. OSB is dimensionally stable, and the edges treated to minimize moisture absorption. Few builders cover the subfloor to protect it from getting rain soaked. A tight OSB joint, combined with a few good soakings invites trouble. Tightly butted panels provide no escape route for water. Therefore, whether the panels are roof- or floor-applied, leave the required spacings. This may not be accepted practice with most builders, but it is good practice.

Apply the adhesive correctly by moving the adhesive gun toward you, not by pushing it away. Cut the end of the plastic nozzle square, do not taper it, and position the gun straight up, perpendicular—at a right angle—to the floor joist. Use *only* 8d annular ring nails spaced 12 inches on center.

The gluing reduces air leakage and the subfloor now becomes the *air retarder,* which helps reduce one of the major sources of heat loss—infiltration. The monolithic (one piece) floor is now the air retarder that reduces air infiltration from the basement into the first floor. Oriented strand board has been considerably improved by new edge-sealing techniques which reduce moisture absorption and swelling. Because it is less expensive than $3/4$-inch plywood, the cost savings can offset the cost of additional underlayment needed for sheet vinyl, or the extra cost of using annular ring nails.

Weyerhaeuser manufactures the only OSB underlayment approved by some resilient flooring manufacturers. Weyerhaeuser installed and monitored test floors over a four-year period before putting STRUCTUREWOOD on the market. It is ¼-inch thick and carries a full one-year warranty. Should defects develop, Weyerhaeuser guarantees to assume the total cost of replacement, as long as the Structurewood was professionally installed in strict compliance with its published standards. Surface unevenness has sometimes been a problem with OSB. But Weyerhaeuser claims that there is no smoothness problem with their OSB underlayment. Structurewood should be installed over a *minimum* subfloor thickness of ⅝ inches, and in strict accordance with Weyerhaeuser's instructions. Look for the Weyerhaeuser stamp on each panel.

Which Nails?

Do not use hot-dipped galvanized, cement-coated, or cut nails to install the subfloor. It is claimed that hot-dipped galvanized nails hold better because the rough zinc deposits act like barbs. It is argued that fish hooks and porcupine quills are barbed, and painfully difficult to remove from flesh; therefore, a barbed nail will hold better in wood than a common bright (smooth shank) nail. This incorrect reasoning assumes that wood and flesh are identical substances, rather than two different materials. As we shall see, the roughness decreases the nail's withdrawal resistance.

The withdrawal resistance—how strongly the nail resists being pulled out—of a nail may actually increase with an evenly-applied coating of zinc. The rough surface of hot-dipped galvanized nails may actually *reduce* the withdrawal resistance, because the coating increases the hole size. Barbed nails were long believed (and still believed by some) to have greater holding power than plain shank nails. This is true, but only under certain conditions. The surface roughness actually decreases the tight contact between the nail shank and the surrounding wood. The purpose of the heavy zinc coating is to protect the nail from rusting. But if the hammer head chips or breaks the zinc coating, the nail will rust.

Cement-coating a nail can increase and even double its withdrawal resistance *immediately after driving.* The withdrawal resistance can be so great that the heads break while trying to remove the nail. Effective in light woods, they offer no advantage in oak, birch, maple, or other hard woods. The high resistance is short-lived, however, and in about 30 days the nail's withdrawal resistance is reduced to 50 percent or more.

The use of cut nails in subfloors is rare, but there are believers. Cut nails do have very good holding power *immediately after being driven.* But under adverse conditions of drying, shrinking, and high humidity, they lose much of their holding power. This is why old floors, nailed with cut nails, are so squeaky.

Given two nails of the same size and diameter, driven into wood with the same density, the nail with annular rings—deformations—has 40 percent more withdrawal resistance than the nail with the smooth shank.

Construction Adhesives

Although there is a bewildering variety of types and brands of adhesives for every conceivable construction project, most small builders buy whatever is on hand at the local lumberyard. Bread dough is sticky, but no one would use it to glue wood.

Adhesives should be matched to the job: Are the strength, filling properties, and bonding suitable for use in the sun? In heat and cold? With high humidity? Some adhesives will work on frozen wood, as long as there is no frost on the surface.

Adhesion takes place on or between surfaces, surfaces that are sun-baked, rain-soaked, frozen, contaminated or dirty. If the adhesives are to work, these surfaces must be free of dirt, standing water, ice, snow, and contaminants.

Although construction adhesives work over a temperature range of 10 degrees to 100 degrees Fahrenheit, they are temperature-dependent: how fast they harden and strengthen depends on the temperature. When used in freezing temperatures, they should be stored at room temperature. The usual on-site practice is to either put the cartridges in the truck, turn on the heater, and hope the core does not burn out, or put them near some other source of heat.

Many builders, frustrated by callbacks because of squeaky and less-than-stiff floors, say gluing does not work. If the conditions and the adhesives are not matched, if they are applied during long spells of 30 degree Fahrenheit temperatures, the glues cannot work. Framing crews often glue the joists, place the underlayment, and tack it here and there. Half-a-day or days later, the subfloor nailing is finished. Is the glue at fault, or the builder? Adhesives designed to work on frozen surfaces will never develop their full strength if the air temperature does not reach 40 degrees Fahrenheit in about a week or so. Glues used in attics where temperatures can be as high as 160 degrees Fahrenheit can soften and weaken.

Adhesive must be applied so the glue joint is not starved for glue, but not so much that it oozes all over. Apply a 3/16-inch-diameter bead to one side only in a continuous line. Where the butts ends of plywood meet over a joist, apply two beads to the joist. On wide faces such as top chords of trusses, or gluelam beams, apply the beads in a zigzag pattern. Do not allow the adhesive to remain exposed to air longer than 15 minutes. When exposed to air the organic solvents or water evaporates, and the adhesive begins to cure and harden. Nails and screws exert the necessary pressure to keep the glued surfaces together until the bond sets up.

We have seen that many materials under continuous pressure tend to creep, that is, sag or deform. Because construction adhesives are nonrigid when cured, they are classified as semistructural, and they too can creep. Off-the-shelf white glues creep and become brittle as they age. Do not use these glues in structural applications.

Several manufacturers such as H.B. Fuller, and Dap, Inc. sell constructions adhesives that can be used on frozen, wet, or treated lumber. DAP's DAP4000 works at lower temperatures. H.B. Fuller's STURDIBOND is usable with frozen, wet, or treated lumber. Miracle Adhesives Company's SFA/66 adhesive is designed for wet or treated lumber.

For a list of adhesives that have been both lab- and field-tested, refer to three APA publications: *Adhesives for APA Glued Floor System,* Form V450; *Caulks and Adhesives for Permanent Wood Foundation System,* Form H405; and Technical Note Y391, *Structural Adhesives For Plywood-Lumber Assemblies.*

I-Joists

Quality lumber is becoming more difficult and more expensive to obtain from available forests. Old-growth logs yield less than 20 percent of log volume as quality structural lumber. It is nearly unobtainable in second-growth logs. As a result, more reconstituted engineered lumber is being used. Engineered lumber is known as *Laminated Veneer Lumber* (LVL), *Parallel Strand Lumber* (PSL) and laminated composite lumber. One new engineered product is the *I*-joist. Dimensionally more stable, more consistent in size, stiffer, and lighter, it is available in longer lengths than common framing lumber. *I*-joists allow holes to be cut into the web material for plumbing and mechanicals. Many brands have full code acceptance. *I*-joists may be used as roofing joists—their greater depth often allows for more insulation while still allowing room above the insulation for ventilation. Because rafter/collar ties can be tricky, consult the manufacturer's literature for proper installation, or consult with the manufacturer's engineering department.

There are a number of manufacturers now producing these products and their technical literature is excellent. However, there are important differences, advantages, and characteristics, between brands.

Caution—Do not use one manufacturer's span or load tables with another manufacturer's product.

The various brands are made in different depths; some do not match common framing nominal sizes. This is done to utilize the plywood better that is sometimes allowed to be used as rim or band joists, and to discourage the mixing with common framing and prevent the resulting problems with differences in shrinking and swelling.

Trus Joist Corporation, maker of TJI brand joists, is probably the best known. Their Silent Floor residential joists come in 9½-inch, 11⅞-inch, 14-inch and 16-inch depths in several different series or strengths (Fig. 10–6). Many other sizes are available on a factory-direct basis. TJI brand joists use Laminated Veneer Lumber (LVL) flanges at the top and bottom, and either plywood or a new composite lumber web material between the flanges. The web in this brand of joists is punched with 1½-inch knockouts at 12 inches oc for wiring. These joists are not cambered or crowned and may be installed with either side up.

Georgia-Pacific's WOOD I BEAM™ joist, are made with high-strength solid sawn lumber flanges and plywood webs (Fig. 10–7). The sizes commonly available for residential construction are 9¼-inch, 11¼-inch and from 12 to 24 inches in 2-inch increments. The joist is

Fig. 10–6. TJI® joists are available in long lengths to accommodate multiple spans. *(Courtesy Trus Joint Corporation)*

cambered; there is a definite *UP,* and a definite *DOWN* when installing these joists.

Louisiana-Pacific produces the GNI™ (Fig. 10–8) and INNER-SEAL™ joists, and GLUE-LAM™ laminated veneer lumber (LVL) (Fig. 10–9). Two other firms, Alpine (ASI brand), and Boise-Cascade (BCI brand), also manufacture joists, which are pending code acceptance. Do not assume that because LVL is engineered to high performance standards, it can be used without regard to its specifications. Follow the manufacturer's recommendations.

Manufactured *I*-joists may be used as rafters (Fig. 10–9). Laminated Veneer Lumber (LVL) is produced by all three manufacturers (Figs. 10–10 and 10–11).

Advantages of Engineered Joists

Engineered joists are lightweight, easy to handle and install. They are strong and provide more load-bearing capacity per pound than

Fig. 10–7.   Georgia-Pacific Wood I Beam™. *(Courtesy Georgia-Pacific)*

Fig. 10–8. Louisiana-Pacific GNI™ joists. *(Courtesy Louisiana-Pacific)*

Fig. 10–9. Louisiana-Pacific's I-joist used as rafters, and Glue-Lam™ used as a header. *(Courtesy Louisiana-Pacific)*

Fig. 10–10. Trus Joist Micro-Lam® laminated veneer lumber. *(Courtesy Trus Joist Corporation)*

Fig. 10–11. Georgia-Pacific G-PLAM®laminated veneer lumber (LVL).
(Courtesy Georgia-Pacific)

solid sawn 2 × 10s or 2 × 12s. This means fewer beams and longer un-
supported spans. Warping and shrinkage are minimized.

Engineered joists offer a number of advantages, but at a price that
many builders might find too high. A TJI 9½-inch joist 24 inches oc
will span 15 feet and support a load of 101 pounds/LF. A 2 × 10 Doug-

las fir-Larch #1 joist at 24 inches oc will carry 100 pounds/LF, and span 15′10″. At an *average* price of $1.55 per foot, a 16-foot 9½-inch TJI costs $24.80. A 16-foot Douglas fir-Larch Select Structural 2 × 10 costs $20.80. However, less expensive SYP joists can be used.

The TJI has 1½-inch knockout holes. Holes will have to be drilled in the solid joist. However, if larger holes and different locations are necessary, they will have to be drilled in the TJI, or any of the engineered joists.

Cost should not be the only consideration when choosing between engineered products and solid sawn lumber. Just as plywood replaced boards for sheathing (even though the boards cost less), the engineered joist will soon replace solid sawn joists in quality residential use.

CHAPTER 11

Outer Wall Framing

Framing exterior walls with studs 16 inches on center (oc) is the dominant method of framing in the United States. This spacing is believed to have originated in 17th-century Europe. Wood lath was hand split from 4-foot-long cord wood. The strength and stiffness of the laths varied considerably. Plastering them on studs 24 inches oc was difficult. Another stud was added, making 4 studs in 4 feet, and thus the 16-inch oc stud spacing.

In an article in the February 1988 issue of *Progressive Architecture*, the author asked,

> *One wonders if Chicago builder George Washington Snow realized the importance of his 1832 invention: the balloon frame. It not only simplified the building of shelter in the rapidly-growing industrial cities of the East and in the sparsely-forested areas of the West, but it amounted to a whole new conception of wood as a structural material. The massive, irregular, tree-like qualities of the timber used in post-and-beam construction gave way to the lightweight, uniform, machine-made characteristics of the 2 × 4. What Snow set in motion—the development of wood as a mass-produced commodity and wood construction as an highly efficient process—continues to this day.* (Reprinted with permission of *Progressive Architecture*, Penton Publishing.)

Although the conventional wood-framed wall is basically efficient construction, in its present form it does not and cannot achieve its full potential; it is neither as cost effective nor as energy efficient as it could be. The manufacturers of insulated structural panels—incorrectly called stressed skin panels—see stick building as backward. Lightweight structural polymer concretes, such as the Swedish 3L and Finnish wood fiber concrete, have insulating value and can be sawn and nailed. They pose real challenges to conventional and nonconventional stick building. Newspaper is recycled into cellulose insulation; and now wheat straw, rye, and sugar-cane rind are being turned into 5-inch-thick panels called Envirocor panels. With an insulating value of R1.8 per inch and fire resistance, they perform better in earthquakes and high winds than stick-built houses. The panels are the walls and insulation, with no need for structural framing. Straw panels are not new—it is an old European method which is still in use there. The USDA had developed straw panels as far back as 1935, but without the structural strength of the Envirocor panels.

The reduction of labor and material costs and ways to build better for less, achieving energy efficiency at the same time, require not only planning but a willingness to consider alternatives to conventional framing. All model building codes permit the use of alternative method and materials. In this chapter we will explore some of the alternatives.

Value Engineering

Value Analysis (VA) originated at General Electric during World War II. Labor and material shortages forced engineers to seek alternatives, substitutes, or both. When GE management began to notice that often the use of substitutes resulted in lower cost and improved products, they wondered why. Was it an accident or did it work? A system called Value Analysis was developed to analyze why the alternatives worked. Eventually VA was changed to Value Engineering (VE). VE was adopted by the Department of Defense in 1954, and by the construction industry in 1963.

VE has as its objective the optimizing of cost, performance, or both, of systems, such as a house. The house's quality, value, and economy are a result of getting rid of unnecessary spending. VE concentrates on and analyzes functions: throw out that which adds cost without adding to the quality, life expectancy, appearance, comfort, energy efficiency, or maintainability of a structure. VE is opposed to the costly, unnecessary overkill, also known as redundancy, of residential construction. The National Association of Home Builders (NAHB) uses the term Optimum Value Engineering (OVE) to describe its cost-saving techniques.

Value Engineering is not just cutting corners, using less or cheaper materials. It is not a collection of unrelated cost-cutting or quality-cutting ideas. It does not reduce quality. VE analyzes functions by asking questions such as: What is a stud? What does it do? What is it supposed to do? How much does it cost? Is there something else that could do the same job? What is the cost of the other method?

As summarized by the Alberta (Canada) Innovative Housing Grants Program, Optimum Value Engineering is:

1. A procedure of comparing alternative methods and materials to determine the least costly combination.
2. An effective and systematic total approach.
3. A wide variety of cost-saving techniques compatible with wood frame construction.
4. Effective use of materials and labor skills.
5. A series of cost-saving measures integrated to work together and compliment each other.
6. The employment of techniques adaptable to on-site, shop, or pre-fabricated building methods.
7. An approach to the planning, design, and engineering of residential construction.
8. Intended to produce maximum benefit when the total OVE approach is utilized.
9. The use of individual OVE techniques compatible with conventional methods offering significant savings.
10. Application of good judgment, sound building practices, and attention to good workmanship and quality.

The Arkansas House

The originators of the Arkansas House concept were concerned with energy conservation, not obsessed with it. Their goal was to find a way to make the conventional Federal Housing Administration (FHA) Minimum Property Standards (MPS) house more energy efficient. The concept began in the late 1950s. But cheap electric rates and unwillingness of homeowners—who were complaining about high energy costs and the lack of comfort in their homes—to invest in more insulation and other measures, effectively killed the idea. The OPEC crisis in 1973 changed all that and the concept was revived. Using the data collected over a 12-year period, Frank Holtzclaw designed a house that was energy-efficient, cost less to build than its FHA MPS brother, was less expensive to operate, and was thus affordable for a greater number of people.

The Arkansas House (AH) is the brainchild of a man who was willing to consider alternatives to traditional framing. Using VE, he eliminated everything that added cost, but did not contribute to the structural integrity, safety, comfort and energy performance of the structure. Holtzclaw's design provided a safety factor in excess of 2 to 1 for load and racking, even though it had 41 percent less framing lumber in the exterior walls than the MPS house of the same size, 1200 ft^2. Even when compared to a house with 2 × 4 studs 24 inches oc, the reduction in lumber was still a healthy 35 percent.

The Arkansas House achieved its cost effectiveness and energy efficiency without the use of complicated technology or unusual framing techniques. It is sophisticated because of the utter simplicity of the methods that were used. All framing was 24 inches oc, and in line with each other: bottom chords of trusses were directly over the outer wall studs. The studs were directly over the floor joists. Roof loads were transferred through the studs to the floor joists to the foundation. In-line framing makes a double top plate unnecessary: double top plates are necessary only when rafters or bottom chords are resting between two studs. A mid-height backer and drywall clips (Fig. 11–1) were used rather than the usual three-stud trough. Two-stud corners and sheetrock clips (Fig. 11–2) eliminated 3- and 4-stud corners. The clips shown were the type originally used in the Arkansas House. Today, the PREST-ON clip (Fig. 11–3) is the most widely used sheetrock clip. Insulated plywood headers (box beams) replaced solid uninsulated headers. Partitions were framed with 2 × 3s, single top plates, and no headers over doors. One interesting technique was the drilling of holes at the bottom of outer wall studs. This allowed the romex wire to be run along the bottom plate, and eliminated drilling holes through the studs.

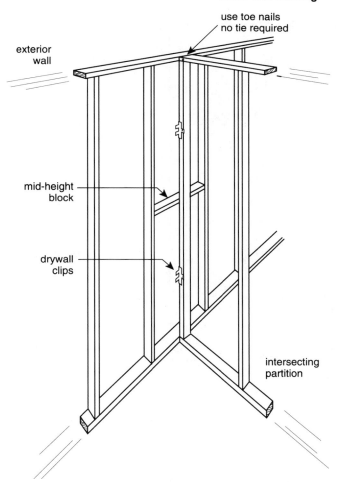

Fig. 11–1. A mid-height backer and drywall clips eliminate the 3-stud trough.

A one-inch hole can be routed at the bottom of 100 studs in about 10 minutes. The wiring no longer interferes with the insulation, and wiring costs are reduced. Ask the electrician what his labor charge is for drilling holes in the studs. Explain the new system and tell him to deduct the hole charge from his estimate.

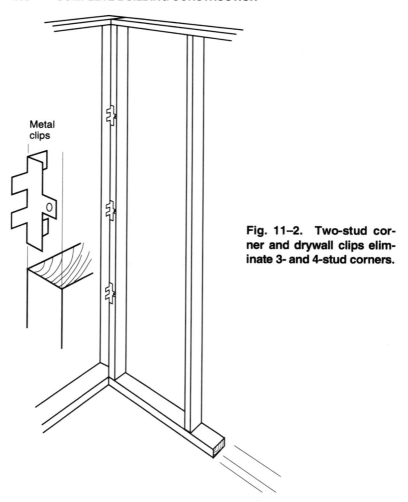

Metal
clips

Fig. 11–2. Two-stud corner and drywall clips eliminate 3- and 4-stud corners.

The Engineered Framing System

Our primary concern here is not energy efficiency, but with reducing framing costs. Yet, when cost effective framing is used, the unasked-for result is energy savings, because there is less lumber to lose heat through and more room for insulation. When framing costs

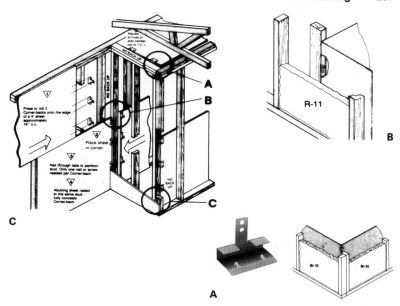

Fig. 11–3. **(A) The Corner-back eliminates the need for corner back-up studs and permits corners to be fully insulated to reduce significant thermal losses. Since the maximum load on the corner stud is one-half or less than the load on a regular stud, two 2 × 4 studs provide structural performance equivalent to three 2 × 4 studs. (B) Corner-back allows the location of partition members without the use of back-up studs. Drywall panels anchor securely to the structural studs. Tee posts are completely eliminated. In addition, cutting insulation is no longer required for a partition abutting exterior walls. The result is straighter corners. (C) The use of Corner-backs in corners and partitions.**

are reduced, the buyer's energy costs are also reduced *at no added cost to the builder.*

The Engineered 24 system is based on years of laboratory and field testing, builders' experience, structural testing, and cost studies of 24-inch wall and floor framing in two-story structures. In addition, there is the experience in Norway. For more than 40 years, floors, walls, and ceilings have been framed 24 inches oc in Norway. Generally a double wall, that is, sheathing and siding, is not used. Siding is applied directly to the studs. In a climate similar to Canada's, no problems have resulted from modular (24-inch) framing in single- or two-story houses.

The Engineered or modular system derives its structural integrity from aligning all framing members. The rafters or trusses are placed directly over the wall studs, which are located directly over the floor joists. All members are spaced 24 inches oc.

Modular Framing

Between 10 and 20 percent of residential house framing is wasted. If the depth of the house is not evenly divisible by 4, much of the usable length of floor joists is wasted. Studs, joists, and rafters are produced in 2-foot increments: 8–10–12–14 feet. The joist length is one-half the house depth, minus the thickness of the joist header, plus one-half of the required 1½-inch overlap. Thus the required joist length is one-half the house depth (Fig. 11–4).

Only when the house depth is on the 4-foot module of 24, 28 or 32 feet will the required joist length be equal to the manufactured lengths. A 25-foot deep house requires two 14-foot long joists; the same length as required in a 28-foot deep house. The joist spacing, whether the joist ends are butted or overlapped makes no difference (Table 11–1). In

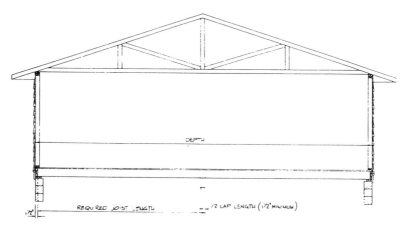

Fig. 11–4. Joist length is equal to one-half the house width. *(Courtesy National Forest Products Association)*

Table 11–1. Lineal Footage of Joists Required for Various House Depths

House Depth	Joist Length		Required Footage of Joists Per 4' of House Length Based on Standard Joist Length					
	Required	Standard	Lineal Feet		Bd. Ft. Per Sq. Ft. of Floor Area[1]			
			16" Joist Spacing	24" Joist Spacing	16" Spacing		24" Spacing	
Ft.	Ft.	Ft.			2 × 8	2 × 10	2 × 10	2 × 12
21	10½	12	72	48	1.14	1.43	.95	1.14
22	11	12	72	48	1.09	1.36	.91	1.09
23	11½	12	72	48	1.04	1.30	.87	1.04
24	**12**	**12**	**72**	**48**	**1.00**	**1.25**	**.83**	**1.00**
25	12½	14	84	56	1.12	1.40	.93	1.12
26	13	14	84	56	1.08	1.35	.89	1.08
27	13½	14	84	56	1.04	1.30	.86	1.04
28	**14**	**14**	**84**	**56**	**1.00**	**1.25**	**.83**	**1.00**
29	14½	16	96	64	1.10	1.38	.92	1.10
30	15	16	96	64	1.07	1.33	.89	1.07
31	15½	16	96	64	1.03	1.29	.86	1.03
32	**16**	**16**	**96**	**64**	**1.00**	**1.25**	**.84**	**1.00**

[1] Floor Area = House Depth Times 4'
(Courtesy National Forest Products Association)

the nonmodular 25-foot house, 3 feet of joist is wasted as overlap above the girder. A longer overlap does not increase joist strength or make for a better floor; it wastes material and money.

Increasing the house depth to 28 feet and changing the length keeps the same floor area. Changing from nonmodular, such as 22–25–26–27–29 and 30 feet, to modular can reduce joist costs without changing joist size. Total floor area can be maintained by adjusting the house length (Fig. 11–5). Subfloor costs are also reduced. Full width panels can be used, without ripping, in house widths of 24–28 and 32 feet. If spans are increased when changing over, check to be sure that the present joists will work with the new spans.

Reducing Wall Framing Costs

Figs. 11–6 and 11–7 illustrate conventional versus modular wall framing. Table 11–2 is an explanation of the differences and the potential cost savings. However, dollar savings shown do not reflect current material and labor costs.

Use utility or stud grade lumber in non-load-bearing walls. Double top plates and solid headers are not necessary. Two-by-three studs, 24 inches oc, are more than adequate for non-load-bearing partitions. Although scrap pieces of 2 × 3s can be used for mid-height blocking to stiffen the wall, this is not structurally necessary, and may not be cost effective. Gable end walls support essentially only their own weight. Because they are non-load-bearing, double top plates, headers, and double stools are unnecessary.

There is a widespread but mistaken belief that corners must be strong, solid, or both. For some builders a four-by-four post is preferred, but three- and four-stud corners are acceptable. Sheetrockers and builders alike argue that both sheets of gypsum board have to be nailed solidly in the corners or the paper will tear.

The corner studs carry half or less of the weight that load-bearing studs support. A three- or four-stud, or solid-post corner cannot be insulated. When the studs are arranged to have a small opening, it is usually left empty. Trying to cram 3½ inches of fiberglass in such a small opening does little to insulate, or stop wind wash, but a great deal to reduce the R-value of the insulation.

Case 1

Original House Floor Plan
22´ x 48´ Area—1056 sq. ft.

22´ | A = 1056 | 2 x 8 | 16˝ o.c. | 48´

Framing system—2 x 8 joists 16˝ o.c.
No. 2 Southern Pine
Joist length required—11´;
Standard—12´
No. of joist rows including end
wall joists—37
Total lineal feet of joists—37 rows
x 2 joists/row 12´ or 888´
Total board feet—1184; Bd. ft. of
joist per sq. ft. of floor area—1.121

Cost-Saving Alternative
24´ x 44´ Area—1056 sq. ft.

24´ | A = 1056 | 2 x 8 | 16˝ o.c. | 44´

8% SAVINGS

Framing system—2 x 8 joints 16˝ o.c.
No. 2 Southern Pine
Joist length required—12´;
Standard—12´
No. of joist rows including end
wall joists—34
Total lineal feet of joist—34 rows x
2 joists/row x 12´ or 816
Total board feet—1088; Bd. ft. of
joist per sq. ft. of floor area—1.030

Savings

Board feet—96

Bd. ft. per sq. ft.
of floor area—
.091 or 8.1%

Dollars—96 @
$200/M = $19.20

Case 2

Original House Floor Plan
25´ x 52´ Area—1300 sq. ft.

25´ | A = 1300 | 2 x 8 | 16˝ o.c. | 52´

Framing system: 2 x 8 joists 16˝ o.c.
No. 2 Hem-Fir
Joist length required—12½´;
Standard—14´
No. of joist rows including end
wall joists—40
Total lineal feet of joist—1120
Total board feet—1493; Bd. ft. of
joist per sq. ft. of floor area—1.148

Cost-Saving Alternative
24´ x 56´ Area—1344 sq. ft.

24´ | A = 1344 | 2 x 8 | 16˝ o.c. | 56´

11% SAVINGS

Framing system: 2 x 8 joists 16˝ o.c.
No. 2 Hem-Fir
Joist length required—12´;
Standard—12´
No. of joist rows including end
wall joists—43
Total lineal feet of joist—1032
Total board feet—1376; Bd. ft. per
sq. ft. of floor area—1.024

Savings

Board feet—117

Bd. ft. per sq. ft.
of floor area—
.124 or 10.8%

Dollars—117 @
$200/M = $23.40

Case 3

Original House Floor Plan
26´ x 60´ Area 1560 sq. ft.

26´ | A 1560 | 2 x 10 | 16˝ o.c. | 60´

Framing system: 2 x 10 joists 16˝ o.c.
No. 2 Spruce-Pine-Fir
Joist length required—13´;
Standard—14´
No. of joist rows including end
wall joists—46
Total lineal feet of joist—1288
Total board feet—2147; Bd. ft. of
joist per sq. ft. of floor area—1.376

Cost-Saving Alternative
28´ x 56´ Area—1568 sq. ft.

28´ | A 1568 | 2 x 10 | 16˝ o.c. | 56´

7% SAVINGS

Framing system: 2 x 10 joists 16˝ o.c.
No. 2 Spruce-Pine-Fir
Joist length required—14´;
Standard—14´
No. of joist rows including end
wall joists—43
Total lineal feet of joists—1204
Total board feet—2007; Bd. ft. of
joist per sq. ft. of floor area—1.280

Savings

Board feet—140

Bd. ft. per sq. ft.
of floor area—
.096 or 7.0%

Dollars—140 @
$200/M = $28.00

Case 4

Original House Floor Plan
30´ x 60´ Area—1800 sq. ft.

Framing system: 2 x 12 joists 24˝ o.c.
No. 2 Douglas Fir-Larch
Joist length required—15´;
Standard—16´
No. of joist rows including end
wall joists—31
Total lineal feet of joist—992
Total board feet—1984; Bd. ft. of
joist per sq. ft. of floor area—1.102

30´ | A 1800 | 2 x 12 | 24˝ o.c. | 60´

Cost-Saving Alternative No. 1
32´ x 56´ Area—1792 sq. ft.

Framing system: 2 x 12 joists 24˝ o.c.
No. 2 Douglas Fir-Larch
Joist length required—16´;
Standard—16´
No. of joist rows including end
wall joists—29
Total lineal feet of joist—928
Total board feet—1856; Bd. ft. of
joist per sq. ft. of floor area—1.036

32´ | A 1972 | 2 x 12 | 24˝ o.c. | 56´ 6% SAVINGS

Savings:
Board feet—128; Bd. ft. per sq. ft. of
floor area—.066 or 6.0%
Dollars—128 @ $200 = $25.60

Cost-Saving Alternative No. 2
28´ x 64´ Area—1792 sq. ft.

Framing system: 2 x 10 joists 24˝ o.c.
No. 2 Douglas Fir-Larch
Joist length required—14´;
Standard—14´
No. of joist rows including end
wall joists—33
Total lineal feet of joist—924
Total board feet—1540; Bd. ft. of
joist per sq. ft. of floor area—0.859

28´ | A 1792 | 2 x 10 | 24˝ o.c. | 64´ 17% SAVINGS

Savings:
Board feet—444; Bd. ft. per sq. ft. of
floor area—.177 or 17.1%
Dollars—444 @ $200/M = $88.80

Fig. 11–5. Examples of board foot and dollar savings achievable through use of 4-foot depth module. *(Courtesy National Forest Products Association)*

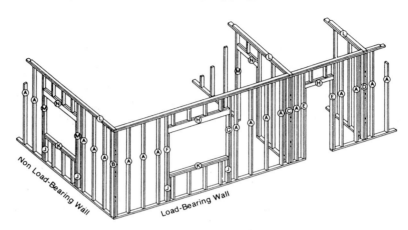

Fig. 11–6. Wall framing with cost-saving principles not applied. *(Courtesy National Forest Products Association)*

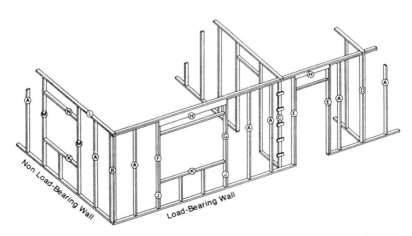

Fig. 11–7. Wall framing incorporating cost-saving principles. *(Courtesy National Forest Products Association)*

Table 11–2. Exterior Wall Framing: Potential Cost Savings for a One-Story House

Cost Saving Principle	Illustration	Potential Cost Saving
1. 24" stud spacing versus 16" spacing; wall framing costs plus application of cladding, insulation, electrical, etc.	Figs. 11–6, 11–7	$ 36.05 (Framing) 36.30 (Other)
2. One side of door and window openings located at regular 16" or 24" stud position	Figs. 11–6, 11–7, Details E, F, M	14.00
3. Modular window sizes used, with both side studs located at normal 16" or 24" stud position (Savings additional to no. 2, above)	Figs. 11–6, 11–7, Details G, M	9.70
4. Optimum three-stud arrangement at exterior corners	Figs. 11–6, 11–7, Detail B	4.70
5. Cleats instead of backup nailer studs where partitions intersect exterior walls	Figs. 11–6, 11–7, Detail C	13.15
6. Single sill member at bottom of window openings	Figs. 11–6, 11–7, Detail K	5.35
7. Support studs under window sill eliminated	Figs. 11–6, 11–7, Detail J	7.55
8. Window and door headers located at top of wall; short in-fill studs (cripples) eliminated	Figs. 11–6, 11–7, Details H, N	12.55 (16" o.c.) 9.65 (24" o.c.)
9. Single top plate for non-bearing end walls	Figs. 11–6, 11–7, Detail L	11.70
10. Single 2 × 4 header for openings in non-bearing walls	Figs. 11–6, 11–7, Detail N	5.20
11. Header support studs eliminated for openings in non-bearing walls	Figs. 11–6, 11–7	5.75
12. Mid-height wall blocking eliminated	· · ·	26.45 (16" o.c.) 25.35 (24" o.c.)
Potential cost saving if all cost saving principles are applicable in one house		**$188.45**

NOTES: 1. This example is based on a one-story house having 1660 square feet of floor area, 9 windows, 4 doors, 6 exterior corners, 8 partitions intersecting exterior walls, 2 non-bearing end walls, and a total of 196 lineal feet of exterior wall.

2. Estimated savings in labor and materials were calculated by the National Association of Home Builders Research Foundation, based on time and materials studies of actual construction. Potential cost savings are based on labor at $6.50 per hour and lumber at $175 per thousand board feet.

(Courtesy National Forest Products Association)

Mold and Mildew

The inside of exterior corners are common breeding grounds for mold and mildew because of the higher interior humidity, lack of air circulation, and cold corner wall surfaces. The corners are cold because (a) there is more wood than insulation in the corners; (b) low or zero insulation levels; (c) wind-washing, where wind enters the corner cavity, short circuits the insulation—if any—and exits through some other part of the corner cavity; (d) the corner has a greater surface area and therefore has greater heat loss. The colder corner surfaces cause the warm moist air to condense and allow mold and mildew to grow. The corners become, in addition to the windows, another *first condensing surface*. The solid wood corner is a thermal bridge or thermal short-circuit. Although the wood has an insulating value of about R-1 per inch, it will conduct heat more readily than insulation and provides an easy path for heat to *bridge* the wall (Fig. 11–8).

Floating Interior Angle Application

According to U.S. Gypsum's *Gypsum Construction Handbook*, 2nd Edition, page 175,

> *The floating interior angle method of applying gypsum board effectively reduces angle cracking and nail pops resulting from stresses at intersections of walls and ceilings. Fasteners are eliminated on at least one surface of all interior angles, both where walls and ceilings meet and where sidewalls intersect. Apply the first nails or screws approximately <u>8 inches below the ceiling at each stud</u>. At vertical angles omit corner fasteners for the first board applied at the angle. This panel edge will be overlapped and held in place by the edge of the abutting board, (Fig. 11–9). (emphasis added)*

The two-stud corner [Fig. 11–8(c)] originated with the invention of balloon framing in 1832, and was resurrected 140 years later by the designer of the Arkansas House. It is sometimes called the California corner, but because a third backer stud is added, it is not a true two-stud corner. It is also called *Advanced Framing* by those who do not know its origin.

The Arkansas House used a gypclip to replace the third corner

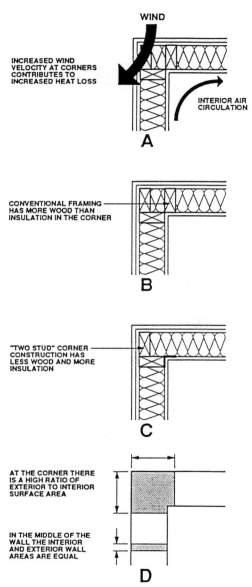

Fig. 11–8. Corner problems. *(Courtesy Joseph Lstiburek)*

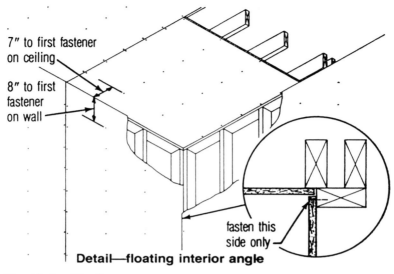

7" to first fastener on ceiling

8" to first fastener on wall

fasten this side only

Detail—floating interior angle

Fig. 11–9. Floating corners.

stud. The gypclip had to be nailed to the stud before the sheetrock was installed. In 1975, the Prest-on clip was introduced to the construction industry. The missing nailer or backer is replaced with the Prest-on clip (Fig. 11–3). The result is incorrectly called a *floating corner.* Many builders are opposed to floating corners, because they mistakenly believe the corner studs actually move and separate. The corners are secure; the only movement comes from racking forces, or the natural cyclical seasonal movements of the structure. It is the corners of the gypsum board that are said to be *floating* because they are not nailed at the corners. The first nails are 8 to 12 inches away from the corners (Fig. 11–9). When Prest-on clips are used, they are spaced 16 inches oc on the board. The tab is secured to the stud. The next sheet of gypsum board is placed at right angles to the first sheet and covers the tabs. Now both sheets are on the same stud. Accepted, but incorrect, practice nails the 4 separate ends of the sheetrock to each of the two separate studs, one in each wall. When the wood shrinks, each stud shrinks in a direction opposite the other stud. This movement in two different directions, and not the floating sheetrock, is why the corners break.

Another area where common practice fails to account for wood shrinkage is headers. The *Gypsum Construction Handbook*, page 103, cautions installers,

> *Do not anchor panel surfaces across the flat grain of wide dimensional lumber such as floor joists or headers. Float panels over these members or provide a control joint to compensate for wood shrinkage.*

Caution—The two-corner stud may not be permitted in multifamily residences in some seismic zones in California, and some regions of the midwest. The Prest-on fastener is permitted in single-family residences in California.

Window and Door Framing

Whenever possible, use windows that fit the 2-foot module, that is, windows that fit between the studs spaced 24 inches oc. This eliminates cripples, jacks, header studs, and headers. Double stools, letter *K* in Fig. 11–6, are unnecessary (Fig. 11–10). Because the header studs transfer the vertical loads downward, there is no load on the stool, let-

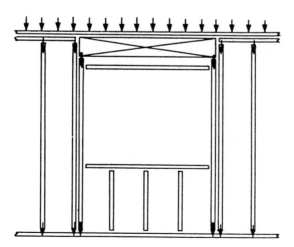

Fig. 11–10.   Load distribution through header and support studs at opening in load-bearing wall. *(Courtesy National Forest Association)*

ter *K* in Fig. 11–7, or on the two cripple studs shown supporting the stool. The ends of the stool are supported by end nailing through the header studs; the stool end cripple supports are unnecessary.

Try to locate one edge of door/window openings where a stud would be, thus using fewer studs (Fig. 11–11). If possible, door and window rough openings should be a multiple of stud openings. This reduces the number of studs and makes locating the sheathing and siding more efficient (Fig. 11–11).

Fig. 11–6 letter *H* shows the conventional method of locating the header. Figs. 11–7 and 11–10 illustrate a more cost-effective method of framing headers. The header should be moved up under the top plate. The lower top plate can be eliminated at this point. This method eliminates the need for cripples to transfer the roof load to the header. The deflection of the header, caused by the cripples, often results in binding of doors and windows. When using a 7'-6" ceiling height, a 2 × 8 header will fill the space and make the use of a horizontal head block—just below the header in Figs. 11–6 and 11–10—unnecessary.

Interior/Exterior Wall Junction

Fig. 11–6 letter *C* shows classic framing at the junction of the exterior wall and partition. This nailer, backer, stud pocket, flat, tee, or par-

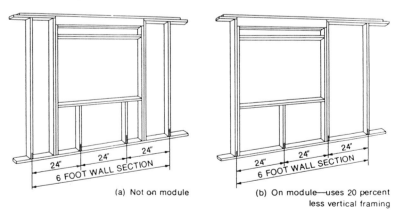

(a) Not on module

(b) On module—uses 20 percent less vertical framing

Fig. 11–11. Windows located on modules can save framing. *(Courtesy National Forest Products Association)*

tition post, as it is variously called, is usually formed from three studs in the form of a trough or U. It rarely gets insulated, and is the reason why so many closets on outside walls are always cold in the winter time. There are several alternatives, and Fig. 11–7 shows one using backup cleats. If the partition is located at a wall stud (Fig. 11–7 letter *D*), the stud becomes the nailer. A 2 × 6, placed in the wall with its wide face exposed, serves as both a nailer and a return for the sheetrock. These methods are not cost effective or necessary.

When the lead framer is laying out the walls, he should indicate that a 3½-inch wide space is to be left in the double top plate at the point where the partition is to be located. The double top plate of the partition, if load-bearing, can be moved into the open space in the double top plate, thus allowing it to be tied into the exterior wall (Fig. 11–12). Returns for the sheetrock are unnecessary as Prest-on clips can be used. Another method is to use a solid metal strap extending from the partition top plate to the exterior wall double top plate.

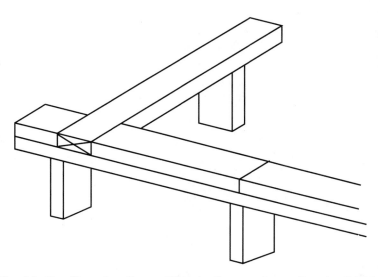

Fig. 11–12. Securing the partition to the exterior wall at the double top plates.

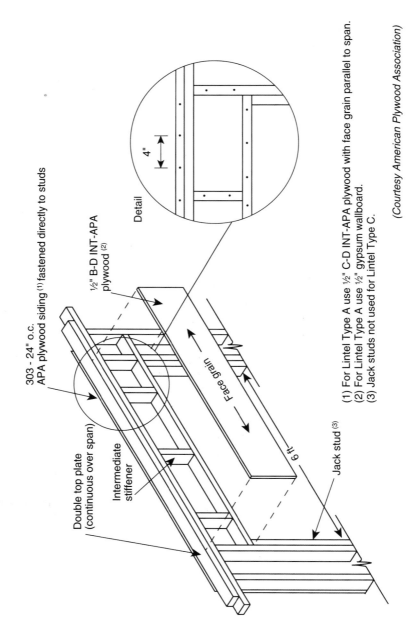

303 - 24" o.c.
APA plywood siding [1] fastened directly to studs

Double top plate
(continuous over span)

Intermediate
stiffener

½" B-D INT-APA
plywood [2]

Detail

4"

Face grain

9 ft

Jack stud [3]

(1) For Lintel Type A use ½" C-D INT-APA plywood with face grain parallel to span.
(2) For Lintel Type A use ½" gypsum wallboard.
(3) Jack studs not used for Lintel Type C.

(Courtesy American Plywood Association)

Fig. 11–13. Plywood box beams or box beams.

Headers

Spans and loads permitting, headers should be insulated for the same reasons as corners. The preferred method is to make a sandwich using rigid plastic foam insulation. In a 2 × 6 wall, this would require an insulation thickness of 2½ inches sandwiched between the inner and outer headers. The headers can be prefabricated, or put together when the wall is being framed—this is the better way. If the headers are not insulated during framing and the windows/doors are installed, it may be difficult to know if they are insulated. There is the added problem of how to get insulation into them. If the code official requires insulated headers, and cannot see the insulation, be prepared to drill holes. If box beams (Fig. 11–13) are used instead of headers, ask the code official if he wants to see them filled with insulation before the plywood web is glued and nailed.

Box beams, or plywood headers as they are called in the Arkansas House, can be prefabricated or constructed as the exterior walls are framed. They are formed by gluing and nailing a plywood web to the framing members above openings in a load-bearing wall. In Fig. 11–13, the upper flange would be the lower and upper top plates. The stiffeners would be the studs at 16 or 24 inches oc. The lower flange would be moved up or down depending on the needed depth. The plywood web must be Grade I exterior with the grain running horizontally. The glue must be an approved water-resistant structural adhesive. The plywood web is glued and nailed with 8d common nails at 6 inches oc along the edges and stiffeners. Plywood web thickness is ⅜ inch or thicker. The exterior plywood sheathing can be used as the web on the outer side. The interior plywood can be ½ inch to blend with the ½ inch sheetrock.

Pre-Framing Conferences

The success of these cost-saving measures is heavily dependent on planning and coordination between architect—if one is used—project manager, field superintendent, lead carpenter and his crew, or the framing subcontractor, and estimator. If lumberyard estimators are used to do the take-off, give them clear and specific instructions to follow; tell them not to make assumptions. Too many estimators assume everything is to be 16 inches oc even when the plans say otherwise. If they notice that double top plates are not shown they immediately say, "Nobody frames that way." Do not make assumptions that, without specific instructions, the framers will *know what to do.*

Hold a preconstruction conference with all supervisors and the framing crew. Without specific instructions, the lead framer and his crew will do their best to follow a modular framing plan. But deeply ingrained conventional framing habits are hard to overcome, and framers can and do easily revert to conventional habits. The framers may add solid headers, and double top plates in non-load-bearing walls. They may assume that 2 × 3s are a mistake and build the non-load-bearing partitions with 2 × 4s. Or add cripples under the stools, or headers in the gable-end non-load-bearing wall. Modular framing is not complicated; it is *different,* it is not the *usual* way of framing.

Include the sheetrock and insulation subcontractors in the conference. It is pointless to use two-stud corners if the insulators do not fill the corners with insulation. In spite of the growing use of Prest-on clips, many sheetrockers have never heard of or seen them. Call the Prest-on Company at 1–800–323–1813 for a free videotape that will help both contractor and sheetrock subcontractor understand the use of the clip. They must be told that nailers, backers, flats, are not used in exterior walls or backer or nailer studs in the exterior corners. Obtain a copy of the *Gypsum Handbook* from the local gypsum board distributor and have your sheetrock contractor review it and the sections cited above. Call the Extension Service of the nearest university and ask for an expert on wood who can explain how the wood shrinkage affects sheetrock. An excellent and effective way to deal with the sheetrockers is to show them what's in it for them: fewer screws, less labor, more profit.

CHAPTER 12

Roof Framing

In order to be accurate and reasonably proficient in the many phases of carpentry, particularly in roof framing and stair building, a good understanding of how to use the steel square is necessary. This tool is invaluable to the carpenter in roof framing; the reader is urged to purchase a quality square and thoroughly study the instructions for its use. A knowledge of how to use the square is assumed here.

Types of Roofs

There are many forms of roofs and a great variety of shapes. The carpenter and the student, as well as the architect, should be familiar with the names and features of each of the various types.

Shed or Lean-to Roof

This is the simplest form of roof (shown in Fig. 12–1), and is usually employed for small sheds and outbuildings. It has a single slope and is not a thing of beauty.

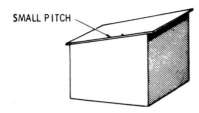

SMALL PITCH

Fig. 12–1. Shed or lean-to roof used on small sheds or buildings.

Saw-Tooth Roof

This is a development of the shed or lean-to roof, being virtually a series of lean-to roofs covering one building, as in Fig. 12–2. It is used on factories, principally because of the extra light which may be obtained through windows on the vertical sides.

Gable or Pitch Roof

This is a very common, simple, and efficient form of roof, and is used extensively on all kinds of buildings. It is of triangular section, having two slopes meeting at the center or *ridge* and forming a *gable,* as in Fig. 12–3. It is popular because of the ease of construction, economy, and efficiency.

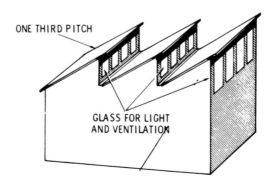

ONE THIRD PITCH

GLASS FOR LIGHT AND VENTILATION

Fig. 12–2. A saw-tooth roof used on factories for light and ventilation.

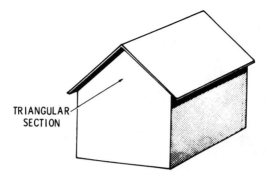

Fig. 12–3. **Gable or pitch roof that can be used on all buildings.**

Gambrel Roof

This is a modification of the gable roof, each side having two slopes, as shown in Fig. 12–4.

Hip Roof

A hip roof is formed by four straight sides, all sloping toward the center of the building, and terminating in a ridge instead of a deck, as in Fig. 12–5.

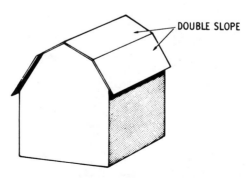

Fig. 12–4. **Gambrel roof used on barns.**

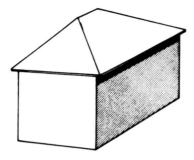

Fig. 12–5. Hip roof used on all buildings.

Pyramid Roof

A modification of the hip roof in which the four straight sides sloping toward the center terminate in a point instead of a ridge, shown in Fig. 12–6. The pitch of the roof on the sides and ends is different. This construction is not often used.

Hip-and-Valley Roof

This is a combination of a hip roof and an intersecting gable roof covering a T- or L-shaped building, as in Fig. 12–7, and so called because both hip and valley rafters are required in its construction. There are many modifications of this roof. Usually the intersection is at right angles, but it need not be; either ridge may rise above the other and the pitches may be equal or different, thus giving rise to an endless variety, as indicated in Fig. 12–7.

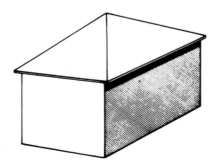

Fig. 12–6. Pyramid roof, which is not often used.

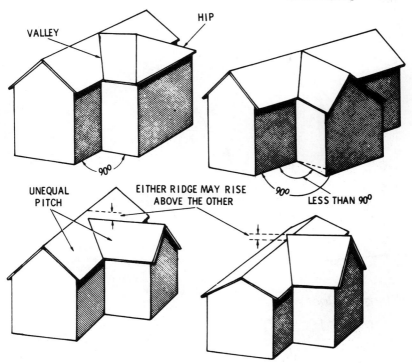

Fig. 12–7. Various hip and valley roofs.

Double-Gable Roof

This is a modification of a gable or a hip-and-valley roof in which the extension has two gables formed at its end, making an M-shape section, as in Fig. 12–8.

Ogee Roof

A pyramidal form of roof having steep sides sloping to the center, each side being ogee-shaped, lying in a compound hollow and round curve, as in Fig. 12–9.

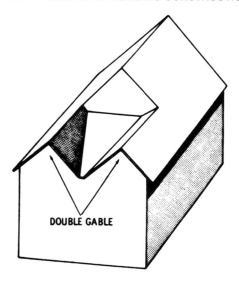

DOUBLE GABLE

Fig. 12–8. Double-gable roof.

Mansard Roof

The straight sides of this roof slope very steeply from each side of the building toward the center, and the roof has a nearly flat deck on top, as in Fig. 12–10. It was introduced by the architect whose name it bears.

French or Concave Mansard Roof

This is a modification of the Mansard roof, its sides being concave instead of straight, as in Fig. 12–11.

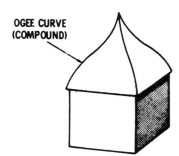

OGEE CURVE
(COMPOUND)

Fig. 12–9. Ogee roof.

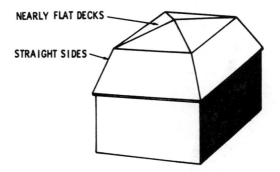

Fig. 12–10. Mansard roof.

Conical Roof or Spire

A steep roof of circular section which tapers uniformly from a circular base to a central point. It is frequently used on towers, as in Fig. 12–12.

Dome

A hemispherical form of roof (Fig. 12–13) used chiefly on observatories.

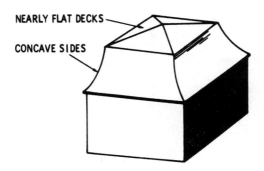

Fig. 12–11. French or concave Mansard roof.

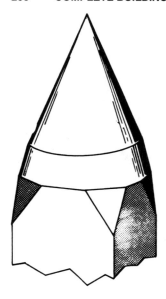

Fig. 12–12.   Conical or spire roof.

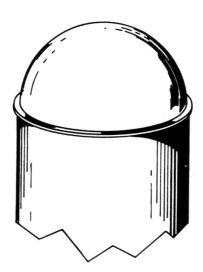

Fig. 12–13.   Dome roof.

Roof Construction

The frame of most roofs is made up of timbers called rafters. These are inclined upward in pairs, their lower ends resting on the top plate, and their upper ends being tied together with a ridge board. On large buildings, such framework is usually reinforced by interior supports to avoid using abnormally large timbers.

The primary object of a roof in any climate is to keep out the elements and the cold. The roof must be sloped to shed water. Where heavy snows cover the roof for long periods of time, it must be constructed more rigidly to bear the extra weight. Roofs must also be strong enough to withstand high winds.

The most commonly used types of roof construction include:

1. Gable,
2. Lean-to or shed,
3. Hip,
4. Gable-and-valley.

Terms used in connection with roofs are:

Span—The *span* of any roof is the shortest distance between the two opposite rafter seats. Stated in another way, it is the measurement between the outside plates, measured at right angles to the direction of the ridge of the building.

Total Rise—The *total rise* is the vertical distance from the plate to the top of the ridge.

Total Run—The term *total run* always refers to the level distance over which any rafter passes. For the ordinary rafter, this would be one-half the span distance.

Unit of Run—The unit of measurement, 1 foot or 12 inches, is the same for the roof as for any other part of the building. By the use of this common unit of measurement, the framing square is employed in laying out large roofs.

Rise in Inches—The *rise in inches* is the number of inches that a roof rises for every foot of run.

Pitch—*Pitch* is the term used to describe the amount of slope of a roof.

Cut of Roof—The *cut of a roof* is the *rise in inches* and *the unit of run* (12 inches).

Line Length—The term *line length* as applied to roof framing is the hypotenuse of a triangle whose base is the total run and whose altitude is the total rise.

Plumb and Level Lines—These terms have reference to the direction of a line on a rafter and not to any particular rafter cut. Any line that is vertical when the rafter is in its proper position is called a *plumb line*. Any line that is level when the rafter is in its proper position is called a *level line*.

Rafters

Rafters are the supports for the roof covering and serve in the same capacity as joists for the floor or studs for the walls. According to the expanse of the building, rafters vary in size from ordinary 2 × 4s to 2 × 10s. For ordinary dwellings, 2 × 6 rafters are used, spaced from 16 to 24 inches on center.

The various kinds of rafters used in roof construction are:

1. Common,
2. Hip,
3. Valley,
4. Jack (hip, valley, or cripple),
5. Octagon.

The carpenter should thoroughly know these various types of rafters, and be able to distinguish each kind as they are briefly described.

Common Rafters

A rafter extending at right angles from plate to ridge, as shown in Fig. 12–14.

Hip Rafter

A rafter extending diagonally from a corner of the plate to the ridge, as shown in Fig. 12–15.

Valley Rafter

A rafter extending diagonally from the plate to the ridge at the intersection of a gable extension and the main roof (Fig. 12–16).

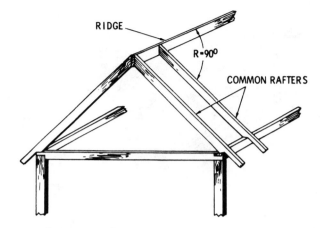

Fig. 12–14. Common rafters.

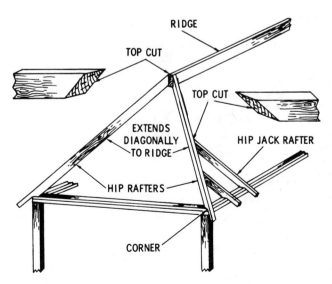

Fig. 12–15. Hip roof rafters.

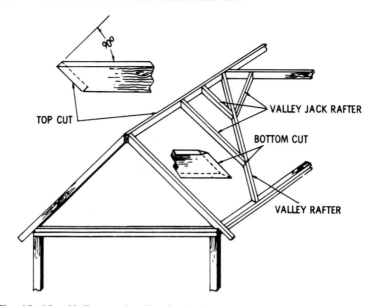

Fig. 12–16. Valley and valley jack rafters.

Jack Rafter

Any rafter which does not extend from the plate to the ridge.

Hip Jack Rafter

A rafter extending from the plate to a hip rafter, and at an angle of 90° to the plate, as shown in Fig. 12–15.

Valley Jack Rafter

A rafter extending from a valley rafter to the ridge and at an angle of 90° to the ridge, as shown in Fig. 12–16.

Cripple Jack Rafter

A rafter extending from a valley rafter to hip rafter and at an angle of 90° to the ridge, as shown in Fig. 12–17.

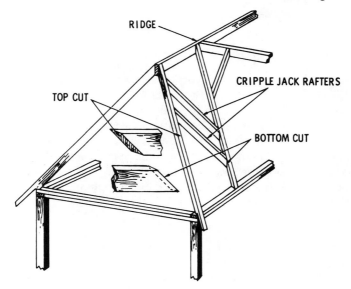

RIDGE

TOP CUT

CRIPPLE JACK RAFTERS

BOTTOM CUT

Fig. 12–17. Cripple jack rafters.

Octagon Rafter

Any rafter extending from an octagon-shaped plate to a central apex, or ridge pole.

A rafter usually consists of a main part or rafter proper, and a short length called the *tail,* which extends beyond the plate. The rafter and its tail may be all in one piece, or the tail may be a separate piece nailed on to the rafter.

Length of Rafters

The length of a rafter may be found in several ways:

1. By calculation;
2. With steel framing square—
 a. Multi-position method,
 b. By scaling,
 c. By aid of the framing table.

Example—What is the length of a common rafter having a run of 6 feet and rise of 4 inches per foot?

1. *By calculation* (Fig. 12–18).

The total rise = 6×4 = 24 inches = 2 feet

Since the edge of the rafter forms the hypotenuse of a right triangle whose other two sides are the run and rise, then the length of the rafter = $\sqrt{\text{run}^2 + \text{rise}^2}$ = $\sqrt{6^2 + 2^2}$ = $\sqrt{40}$ = 6.33 feet, as illustrated in Fig. 12–18.

Practical carpenters would not consider it economical to find rafter lengths in this way, because it takes too much time and there is the chance for error. It is to avoid both of these objections that the *framing square* has been developed.

2. *With steel framing square.*

The steel framing square considerably reduces the mental effort and chances of error in finding rafter lengths. An approved method of finding rafter lengths with the square is by the aid of the rafter table included on the square for that purpose. However, some carpenters may possess a square which does not have rafter tables. In such case, the rafter length can be found either by the *multi-position* method shown in Fig. 12–19, or by *scaling* as in Fig. 12–20. In either of these methods, the measurements should be made with care because, in the *multi-position* method, a separate measurement must be made for each foot run with a chance for error in each measurement. The following refers to Fig. 12–19.

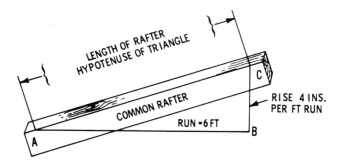

Fig. 12–18. Method of finding the length of a rafter by calculation.

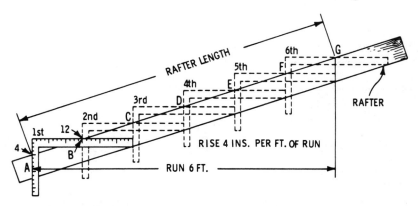

Fig. 12–19. Multi-position method of finding rafter length.

Problem—Lay off the length of a common rafter having a run of 6 feet and a rise of 4 inches per foot. Locate a point *A* on the edge of the rafter, leaving enough stock for a lookout, if any is used. Place the steel framing square so that division 4 coincides with *A*, and 12 registers with the edge of *B*. Evidently, if the run were 1 foot, distance *AB* thus obtained would be the length of the rafter *per foot run*. Apply the square six times for the 6-foot run, obtaining points *C, D, E, F,* and *G*. The distance *AG*, then, is the length of the rafter for a given run.

Fig. 12–20 shows readings of rafter tables from two well-known makes of squares for the length of the rafter in the preceding example, one giving the length per foot run, and the other the total length for the given run.

Problem 1—*Given the rise per foot in inches.* Use two squares, or a square and a straightedge scale, as shown in Fig. 12–21. Place the straightedge on the square so as to be able to read the length of the diagonal between the rise of 4 inches on the tongue and the 1-foot (12-inch) run on the body as shown. The reading is a little over 12 inches. To find the fraction, place dividers on 12 and a point *A*, as in Fig. 12–22. Transfer to the hundredths scale and read

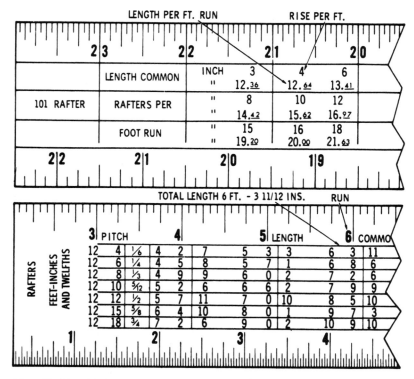

Fig. 12–20. Rafter table readings from two well-known makes of steel framing squares.

.65, as in Fig. 12–23, making the length of the rafter 12.65 inches *per foot run*, which for a 6-foot run is

$$\frac{12.65 \times 6}{12} = 6.33 \text{ feet}$$

Problem 2—*Total rise and run given in feet.* Let each inch on the tongue and body of the square = 1 foot. The straightedge should be divided into inches and 12ths of an inch so that on a scale, 1 inch = 1 foot. Each division will therefore equal 1 inch. Read the diagonal length between the numbers representing the run and rise (12 and 4), taking the whole number of inches as feet, and the fractions as inches. Transfer the fraction with dividers and apply the

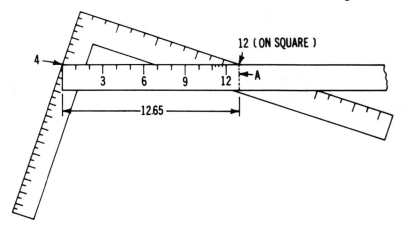

Fig. 12–21.   Method of finding rafter length by scaling.

100th scale, as was done in Problem 1, Figs. 12–22 and
12–23.

In estimating the total length of stock for a rafter having a tail, the
run of the tail or length of the lookout must of course be considered.

Rafter Cuts

All rafters must be cut to the proper angle or bevel at the points
where they are fastened and, in the case of overhanging rafters, also at
the outer end. The various cuts are known as:

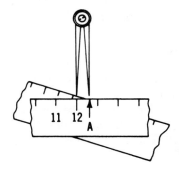

**Fig. 12–22. Reading the straight-
edge in combination with the carpenter's square.**

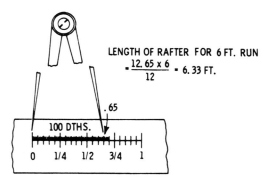

LENGTH OF RAFTER FOR 6 FT. RUN

$$= \frac{12.65 \times 6}{12} = 6.33 \text{ FT.}$$

Fig. 12–23. Method of reading hundredths scale.

1. Top or plumb;
2. Botoom, seat, or heel;
3. Tail or lookout;
4. Side, or cheek.

Common Rafter Cuts

All of the cuts for common rafters are made at right angles to the sides of the rafter; that is, not beveled as in the case of jacks. Fig. 12–24 shows a common rafter from which the nature of two of these various cuts are seen.

In laying out cuts for common rafters, one side of the square is always placed on the edge of the stock at 12, as shown in Fig. 12–24. This distance 12 corresponds to 1 foot of the run; the other side of the square is set with the edge of the stock to the rise in inches *per foot run*. This is virtually a repetition of Fig. 12–19, but it is very important to understand why one side of the square is set to 12 for common rafters—not imply to know that 12 must be used. On rafters having a full tail, as in Fig. 12–25(B), some carpenters do not cut the rafter tails, but wait until the rafters are set in place so that they may be lined and cut while in position. Certain kinds of work permit the ends to be cut at the same time the remainder of the rafter is framed.

The method of handling the square in laying out the bottom and lookout cuts is shown in Fig. 12–26. In laying out the top or plumb cut, if there is a ridge board, one-half of the thickness of the ridge must be

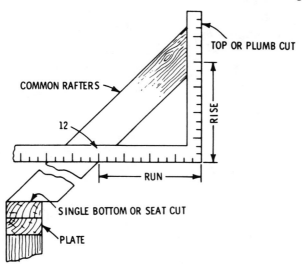

TOP OR PLUMB CUT

COMMON RAFTERS

RISE

12

RUN

SINGLE BOTTOM OR SEAT CUT

PLATE

Fig. 12–24. Placement of steel square for proper layout of plumb cut; seat cut has already been laid out and made, using the opposite leg of the square.

deducted from the rafter length. If a lookout or a tail cut is to be vertical, place the square at the end of the stock with the rise and run setting as shown in Fig. 12–26, and scribe the cut line *LF*. Lay off *FS* equal to the length of the lookout, and move the square up to *S* (with the same setting) and scribe line *MS*. On this line, lay off *MR*, the length of the vertical side of the bottom cut. Now apply the same setting to the bottom edge of the rafter, so that the edge of the square cuts *R*, and scribe *RN*, which is the horizontal side line of the bottom cut. In making the bottom cut, the wood is cut out to the lines *MR* and *RN*. The lookout and bottom cuts are shown made in Fig. 12–25(B), *RN* being the side which rests on the plate, the *RM* the side which touches the outer side of the plate.

Hip-and-Valley Rafter Cuts

The hip rafter *lies in the plane of the common rafters* and forms *the hypotenuse of a triangle,* of which one leg is the adjacent common raf-

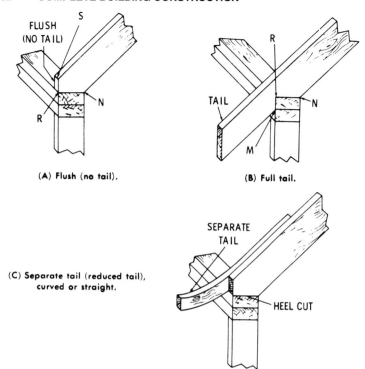

(A) Flush (no tail).

(B) Full tail.

**(C) Separate tail (reduced tail),
curved or straight.**

Fig. 12–25.   Various common rafter tails.

ter and the other leg is the portion of the plate intercepted between the feet of the hip and common rafters, as in Fig 12–27.

Problem—In Fig. 12–27, take the run of the common rafter as 12, which may be considered as 1 foot (12 inches) of the run or the total run of 12 feet (½ the span). Now for 12 feet, intercept on the plate the hip run inclined 45 degrees to the common run, as in the triangle *ABC*. Thus, $AC^2 = \sqrt{AB^2 + BC^2} = \sqrt{12^2 + 12^2} = 16.97$, or approximately 17. Therefore, the run of the hip rafter is to the run of the common rafter as 17 is to 12. Accordingly, in laying out the cuts, use figure 17 on one side of the square and the given rise in *inches per foot* on the other side. This also holds true for

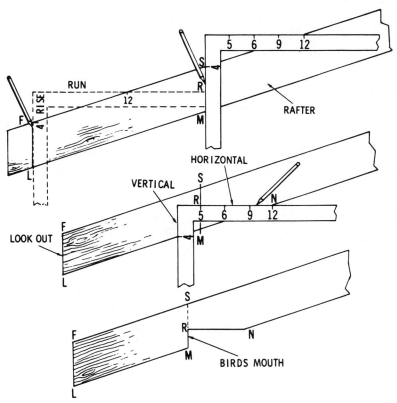

Fig. 12–26. Method of using the square in laying out the lower or end cut of the rafter.

top and bottom cuts of the valley rafter when the plate intercept *AB* equals the run *BC* of the common rafter.

The line of measurement for the length of a hip and valley rafter is along the middle of the back or top edge, as on common and jack rafters. The rise is the same as that of a common rafter, and the run of a hip rafter is the horizontal distance from the plumb line of its rise to the outside of the plate at the foot of the hip rafter, as shown in Fig. 12–28.

In applying the source for cuts of hip or valley rafters, *use the distance 17 on the body of the square the same way as 12 was used for*

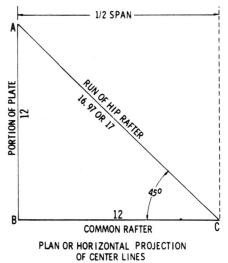

PLAN OR HORIZONTAL PROJECTION
OF CENTER LINES

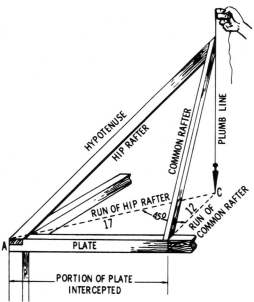

Fig. 12–27. View of hip and common rafters in respect to each other.

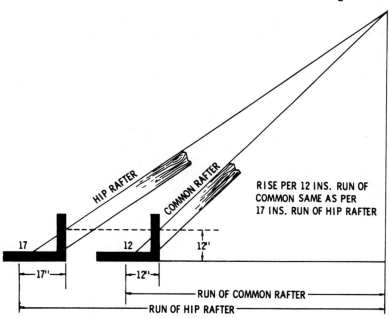

RISE PER 12 INS. RUN OF
COMMON SAME AS PER
17 INS. RUN OF HIP RAFTER

Fig. 12–28. Hip and common rafters shown in the same plane, illustrating the use of 12 for the common rafter and 17 for the hip rafter.

common rafters. When the plate distance between hip and common rafters is equal to half the span or to the run of the common rafter, the line of run of the hip will lie at 45 degrees to the line of run of the common rafter, as indicated in Fig. 12–27.

The length of a hip rafter, as given in the framing table on the square, is the distance from the ridge board to the outer edge of the plate. In practice, deduct from this length one-half the thickness of the ridge board, and add for any projection beyond the plate for the eave. Fig. 12–29(A) shows the correction for the table length of a hip rafter to allow for a ridge board, and Fig. 12–29(B) shows the correction at the plate end. Fig. 12–30 shows the correction at the plate end of a valley rafter.

The table length, as read from the square, must be reduced an

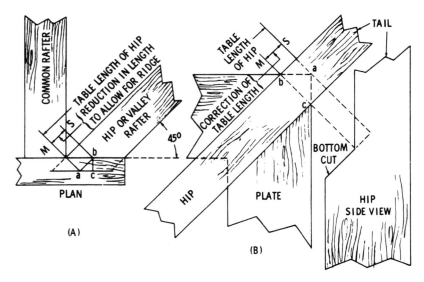

Fig. 12–29. Correction in table length of hip to allow for half-thickness of ridge board.

amount equal to *MS*. This is equal to the hypotenuse (*ab*) of the little triangle abc, which in value =

$$\sqrt{ac^2 + bc^2} = \sqrt{ac^2 + (\text{half thickness of ridge})^2}$$

In ordinary practice, take *MS* *as equal to* half the thickness of the ridge. The plan and side view of the hip rafter shows the table length and the correction *MS,* which must be deducted from the table length so that the sides of the rafter at the end of the bottom cut will intersect the outside edges of the plate. The table length of the hip rafter, as read on the framing square, will cover the span from the ridge to the outside cover *a* of the plate, but the side edges of the hip intersect the plates at *b* and *c.* The distance that *a* projects beyond a line connecting *bc* or *MS,* must be deducted; that is, measured backward toward the ridge end of the hip. In making the bottom cut of a valley rafter, it should be noted that a valley rafter differs from a hip rafter in that the correction distance for the table length must be *added* instead of subtracted, as for a hip rafter. A distance *MS* was subtracted from the table length of the

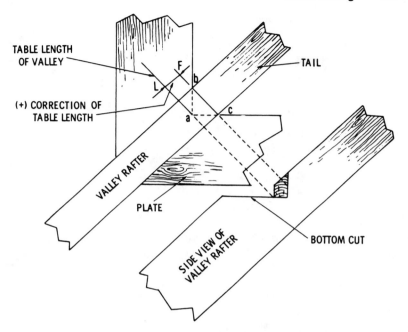

TABLE LENGTH
OF VALLEY

F

L

b

(+) CORRECTION OF
TABLE LENGTH

a

c

TAIL

VALLEY RAFTER

PLATE

SIDE VIEW OF
VALLEY RAFTER

BOTTOM CUT

Fig. 12–30. Correction in table length of valley rafter to allow for half-thickness of ridge. Correction is added, not subtracted.

hip rafter in Fig. 12–29(B), and an equal distance (*LF*) was added for the valley rafter in Fig. 12–30.

After the plumb cut is made, the end must be mitered outward for a hip, as in Fig. 12–31, and inward for a valley, as in Fig. 12–32, to receive the *facia*. A *facia* is the narrow vertical member fastened to the outside ends of the rafter tails. Other miter cuts are shown with full tails in Fig. 12–33, which also illustrates the majority of cuts applied to hip and valley rafters.

Side Cuts of Hip and Valley Rafters

These rafters have a side or cheek cut at the ridge end. In the absence of a framing square, a simple method of laying out the side cut for a 45-degree hip or valley rafter is as follows:

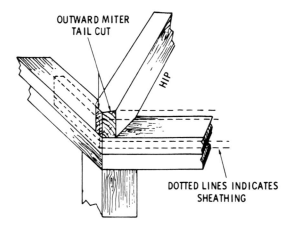

Fig. 12–31. Flush hip rafter miter cut.

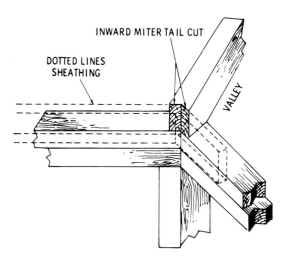

Fig. 12–32. Flush valley miter cut.

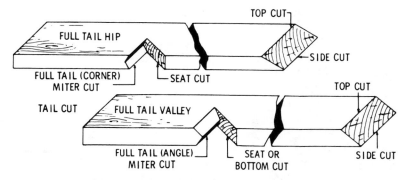

Fig. 12–33. Flush tail hip and valley rafters showing all cuts.

Measure back on the edge of the rafter from point *A* of the top cut, as shown at left in Fig. 12–34. Distance *AC* is equal to the thickness of the rafter. Square across from *C* to *B* on the opposite edge, and scribe line *AB*, which gives the side cut. At right in Fig. 12–34, *FA* is the top cut, and *AB* is the side cut. The plumb and side cuts should be made at the same time by sawing along lines *FA* and *AB*, in order to save extra labor.

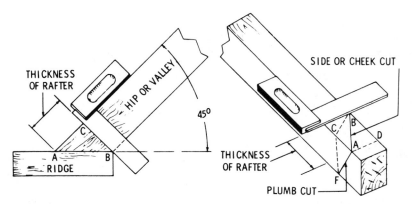

Fig. 12–34. A method of obtaining a side cut of 45-degree hip or valley rafter without the aid of a framing square.

This rule for laying out hip side cuts does not hold for any angle other than 45 degrees.

Backing of Hip Rafters

By definition, the term *backing* is the bevel upon the top side of a hip rafter which allows the roofing boards to fit the top of the rafter without leaving a triangular hole between it and the back of the roof covering. The height of the hip rafter, measured on the outside surface vertically upward from the outside corner of the plate, will be the *same as that of the common rafter measured from the same line,* whether the hip is backed or not. This is not true for an unbacked valley rafter when the measurement is made at the center of the timber.

The graphical method of finding the backing of hip rafters is shown in Fig. 12–35. Let *AB* be the span of the building, and *OD* and *OC* the plan of two unequal hips. Lay off the given rise as shown. Then *DE* and *CF* are the lengths of the two unequal hips. Take any point, such as *G* on *DE*, and erect a perpendicular cutting *DF* at *H*. Resolve *GH* to *J*, that is, make *HJ* = *GH*, draw *NO* perpendicular to *OD* and through *H*. Join *J* to *N* and *O*, giving a bevel angle *NJO*, which is the backing for rafter *DE*. Similarly, the bevel angle *NJO* is found for the backing of rafter *CF*.

Jack Rafters

As outlined in the classification, there are several kinds of jack rafters as distinguished by their relation with other rafters of the roof. These various jack rafters are known as:

1. Hip jacks,
2. Valley jacks,
3. Cripple jacks.

The distinction between these three kinds of jack rafters, as shown in Fig. 12–36, is as follows: Rafters which are framed between a hip rafter and the plate are *hip jacks;* those framed between the ridge and a valley rafter are *valley jacks;* those framed between hip and valley rafters are *cripple jacks.*

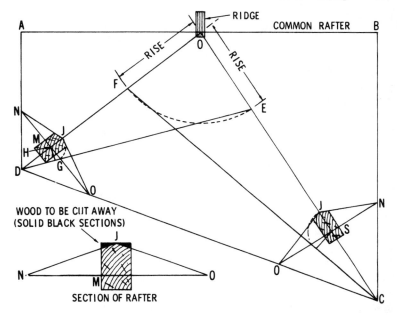

Fig. 12–35. Graphical method of finding length of rafters and backing of hip rafters.

The term cripple is applied because the ends or *feet* of the rafters are cut off—the rafter does not extend the full length from ridge to plate. From this point of view, a valley jack is sometimes erroneously called cripple; it is virtually a semi-cripple rafter, but confusion is avoided by using the term cripple for rafters framed between the hip and valley rafters, as above defined.

Jack rafters are virtually *discontinuous common rafters*. They are cut off by the intersection of a hip or valley, or both, before reaching the full length from plate to ridge. Their lengths are found in the same way as for common rafters—the number 12 being used on one side of the square and the rise in inches per foot run on the other side. This gives the length of jack rafter per foot run, and is true for all jacks—hip, valley, and cripple.

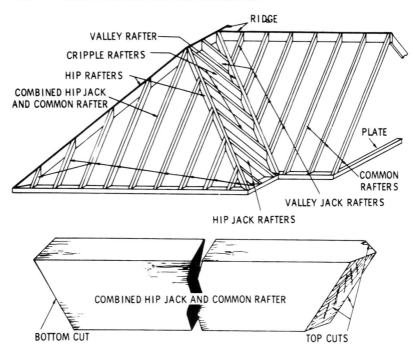

Fig. 12–36. A perspective view of hip and valley roof showing the various jack rafters, and enlarged detail of combined hip jack and common rafters showing cuts.

In actual practice, carpenters usually measure the length of hip or valley jacks from the long point to the plate or ridge, instead of along the center of the top, no reduction being made for one-half the diagonal thickness of the hip or valley rafter. Cripples are measured from long point to long point, no reduction being made for the thickness of the hip or valley rafter.

As no two jacks are of the same length, various methods of procedure are employed in framing, as:

1. Beginning with shortest jack,
2. Beginning with longest jack,
3. Using framing table.

Shortest Jack Method

Begin by finding the length of the shortest jack. Take its spacing from the corner, measured on the plates, which in the case of a 45-degree hip is equal to the jack's run. The length of this first jack will be the *common difference* which must be added to each jack to get the length of the next longer jack.

Longest Jack Method

Where the longest jack is a full-length rafter (that is, a common rafter), first find the length of the longest jack, then count the spaces between jacks and divide the length of the longest jack by number of spaces. The quotient will be the common difference. Then frame the longest jack and make each jack shorter than the preceding jack by this common difference.

Framing Table Method

On various steel squares, there are tables giving the length of the shortest jack rafters corresponding to the various spacings, such as 16, 20, and 24 inches between centers for the different pitches. This length is also the common difference and thus serves for obtaining the length of all the jacks.

Example—Find the length of the shortest jack or the common difference in the length of the jack rafters, where the rise of the roof is 10 inches per foot and the jack rafters are spaced 16 inches between centers; also, when spaced 20 inches between centers. Fig. 12–37 shows the reading of the jack table on one square for 16-inch centers, and Fig. 12–38 shows the reading on another square for 20-inch centers.

Jack-Rafter Cuts

Jack rafters have top and bottom cuts which are laid out the same as for common rafters, and also side cuts which are laid out the same as for a hip rafter. To lay off the top or plumb cut with a square, take 12 on the body and the rise in inches (of common rafter) per foot run on

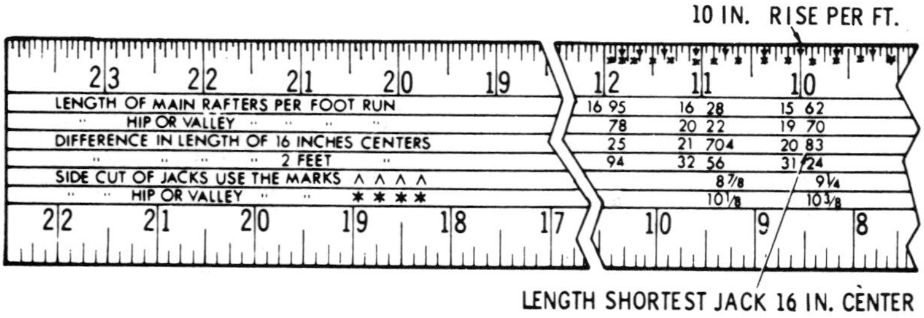

10 IN. RISE PER FT.

2	3	2	2	2	1	2	0	1	9	1	2	1	1	1	0
LENGTH OF MAIN RAFTERS PER FOOT RUN					16 95	16 28	15 62
" HIP OR VALLEY " " "					78	20 22	19 70
DIFFERENCE IN LENGTH OF 16 INCHES CENTERS					25	21 704	20 83
" " " " 2 FEET "					94	32 56	31 /24
SIDE CUT OF JACKS USE THE MARKS ∧ ∧ ∧ ∧						8 7/8	9 1/4
" " HIP OR VALLEY " " * * * *						10 1/2	10 3/8
2	2	2	1	2	0	1	9	1	8	1	7	1	0	9	8

LENGTH SHORTEST JACK 16 IN. CENTER

Fig. 12–37. Square showing table for shortest jack rafter at 16 inches oc.

the tongue, and mark along the tongue as in Fig. 12–39. The following example illustrates the use of the framing square in finding the side cut.

Example—Find the side cut of a jack rafter framed to a 45-degree hip or valley for a rise of 8 inches per foot run. Fig. 12–40 shows the reading on the jack side-cut table of the framing square, and Fig. 12–41 shows the method of placing the square on the timber to obtain the side cut. It should be noted that different makers of squares use different setting numbers, but the ratios are always the same.

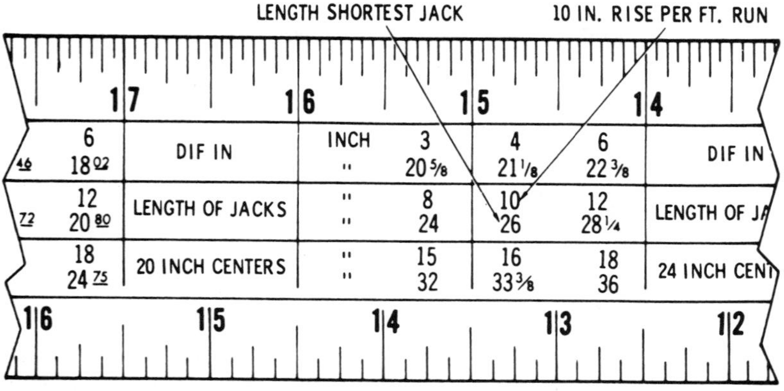

LENGTH SHORTEST JACK   10 IN. RISE PER FT. RUN

1	7	1	6	1	5	1	4
6 18 02	DIF IN	INCH 3 " 20 5/8	4 21 1/8	6 22 3/8	DIF IN
12 20 80	LENGTH OF JACKS	" 8 " 24	10 26	12 28 1/4	LENGTH OF J
18 24 75	20 INCH CENTERS	" 15 " 32	16 33 3/8	18 36	24 INCH CEN
1	6	1	5	1	4	1	3	1	2

Fig. 12–38. Square showing table for shortest jack rafter at 20 inches oc.

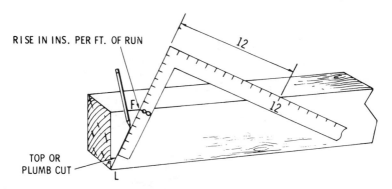

Fig. 12–39. Method of finding plumb and side cuts of jack framed to 45-degree hip or valley.

Method of Tangents

The tangent value is made use of in determining the side cuts of jack, hip, or valley rafters. By taking a circle with a radius of 12 inches, the value of the tangent can be obtained in terms of the constant of the common rafter run.

Considering rafters with zero pitch, as shown in Fig. 12–42, if the common rafter is 12 feet long, the tangent *MS* of a 45-degree hip is the

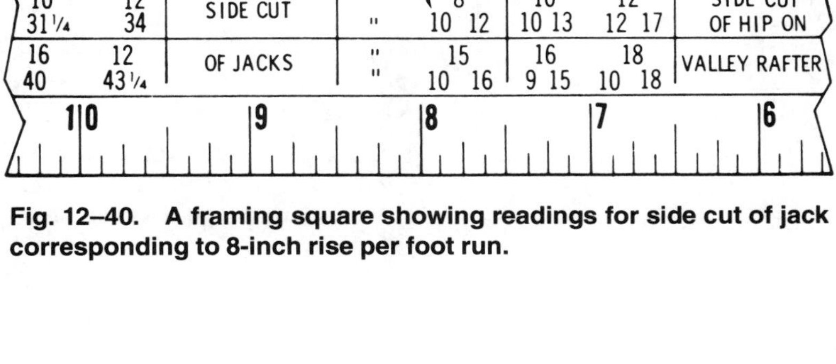

Fig. 12–40. A framing square showing readings for side cut of jack corresponding to 8-inch rise per foot run.

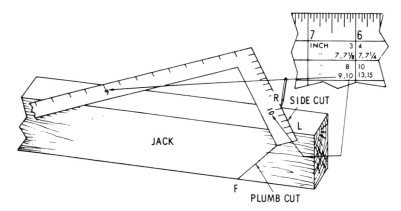

Fig. 12–41.   Method of placing framing square on jack to lay off side cut for an 8-inch rise.

same length. Placing the square on the hip, setting to 12 on the tongue and 12 on the body will give the side cut at the ridge *when there is no pitch* (at *M*) as in Fig. 12–43. Placing the square on the jack with the same setting numbers (12, 12) as at *S*, will give the face cut for the jack when framed to a 45-degree hip with zero pitch; that is, when all of the timbers lie in the same plane.

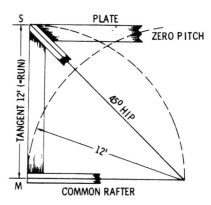

Fig. 12–42.   A roof with zero pitch showing the common rafter and the tangent as the same length.

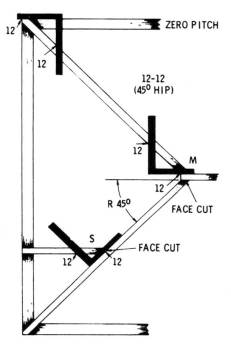

ZERO PITCH

12-12
(45° HIP)

M

FACE CUT

R 45°

FACE CUT

S

Fig. 12–43. Zero-pitch 45-degree hip roof showing application of the framing square to give side cuts at ridge.

Octagon Rafters

On an octagon or eight-sided roof, the rafters joining the corners are called octagon rafters, and are a little longer than the common rafter and shorter than the hip or valley rafters of a square building of the same span. The relation between the run of an octagon and a common rafter is shown in Fig. 12–44 as being as 13 is to 12. That is, for each foot run of a common rafter, an octagon rafter would have a run of 13 inches. Hence, to lay off the top or bottom cut of an octagon rafter, place the square on the timber with the 13 on the tongue and the rise of the common rafter per foot run on the body, as shown graphically in Fig. 12–45. The method of laying out the top and bottom cut with the 13 rise setting is shown in Fig. 12–46.

The length of an octagon rafter may be obtained by scaling the diagonal on the square for 13 on the tongue and the rise in inches per foot run of a common rafter, and multiplying by the number of feet run

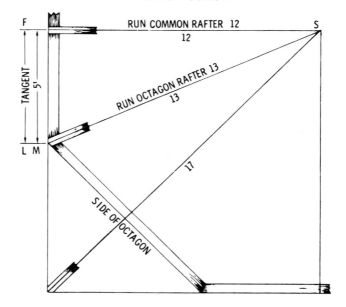

Fig. 12–44.   Details of an octagon roof showing relation in length between common and octagon rafters.

of a common rafter. The principle involved in determining the amount of backing of an octagon rafter (or rafters of any other polygon) is the same as for hip rafter. The backing is determined by the tangent of the angle whose adjacent side is one-half the rafter thickness and whose angle is equivalent to one-half the central angle.

Prefabricated Roof Trusses

Definite savings in material and labor requirements through the use of preassembled wood roof trusses make truss framing an effective means of cost reduction in small dwelling construction. In a 26′ × 32′ dwelling, for example, the use of trusses can result in a substantial cost saving and a reduction in use of lumber of almost 30 percent as compared with conventional rafter and joist construction. In addition to cost savings, roof trusses offer other advantages of increased flexibility

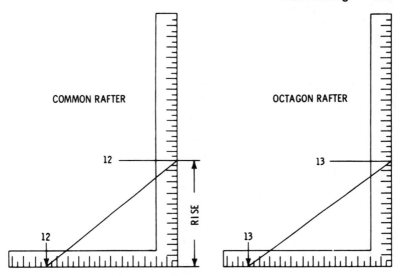

Fig. 12–45.   Diagram showing that for equal rise, the run of octagon rafters is 13 inches to 12 inches for common rafters.

for interior planning, and added speed and efficiency in site erection procedures. Today, some 70 percent of all houses built in the United States incorporate roof trusses.

For many years, trusses were extensively used in commercial and industrial buildings, and were very familiar in bridge construction. In

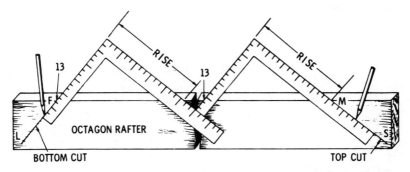

Fig. 12–46.   Method of laying off bottom and top cuts of an octagon rafter with a square using the 13 rise setting.

the case of small residential structures, truss construction took a while to catch on, largely because small-house building has not had the benefit of careful detailing and engineered design that would permit the most efficient use of materials.

During the last several years the truss has been explored and developed for small houses. One of the results of this effort has been the development of light wood trusses which permit substantial savings in the case of lumber. Not only may the framing lumber be smaller in dimension than in conventional framing, but trusses may also be spaced 24 inches on center as compared to the usual 16-inch spacing of rafter and joist construction.

The following figures show percentage savings in the 26' × 32' house through the use of trusses 24 inches on center.

Lumber Requirements for Trusses

28.4% less than for conventional framing at 16-inch spacing.

Labor Requirements for Trusses

36.8% less hours than for conventional framing at 16-inch spacing.

Total Cost of Trusses

29.1% less than conventional framing at 16-inch spacing.

The trusses consisted of 2 × 4 lumber at top and bottom chords, 1 × 6 braces, and double 1 × 6s for struts, with plywood gussets and

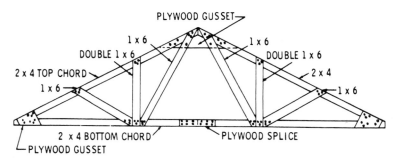

Fig. 12–47.   Wood roof truss for small dwellings.

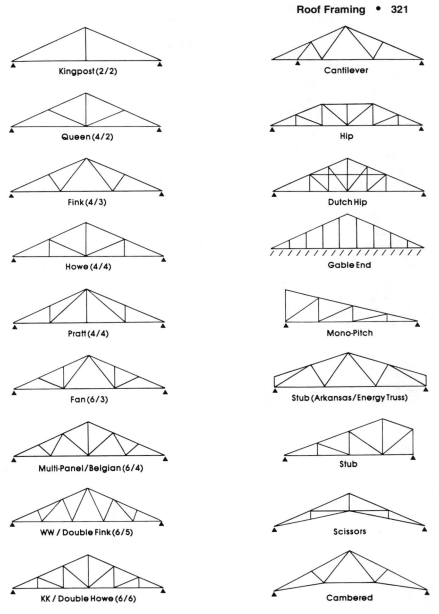

Fig. 12–48. Additional truss configurations.

splices, as shown in Fig. 12–47. The clear span of truss construction permits use of nonbearing partitions so that it is possible to eliminate the extra top plate required for bearing partitions used with conventional framing. It also permits a smaller floor girder to be used for floor construction since the floor does not have to support the bearing partition and help carry the roof load.

Aside from direct benefits of reduced cost and savings in material and labor requirements, roof trusses offer special advantages in helping to speed up site erection and overcome delays due to weather conditions. These advantages are reflected not only in improved construction methods but also in further reductions in cost. With preassembled trusses, a roof can be put over the job quickly to provide protection against the weather.

Laboratory tests and field experience show that properly designed roof trusses are definitely acceptable for dwelling construction. The type of truss shown in Fig. 12–47 is suitable for heavy roofing and plaster ceiling finish. In assembling wood trusses, special care should be taken to achieve adequate nailing since the strength of trusses is, to a large extent, dependent on the fastness of the connection between members. Care should also be exercised in selecting materials for trusses. Lumber equal in stress value to No. 2 dimension shortleaf southern pine is suitable; any lower quality is not recommended. Trusses can be assembled with nails and other fasteners, but those put out by such companies as Teco work particularly well.

Fig. 12–48 shows some additional truss configurations.

CHAPTER 13

Roofing

A roof includes the roof cover (the upper layer which protects against rain, snow, and wind) or roofing, the sheathing to which it is fastened, and the framing (rafters) which support the other components.

Because of its exposure, roofing usually has a limited life, and so is made to be readily replaceable. It may be made of many widely diversified materials, among which are the following:

1. Wood, usually in the form of shingles (which are uniform, machine-cut) or shakes (which are hand-cut) (Fig. 13–1).
2. Metal or aluminum, which simulates other kinds of roofing.
3. Slate, which may be the natural product or rigid manufactured slabs, often cement asbestos, though these are on the decline since the controversy over asbestos.
4. Tile (Fig. 13–2), a burned clay or shale product, available in several standard types.
5. Built-up covers of asphalt or tar-impregnated felts, with moppings of hot tar or asphalt between the plies and a mopping of tar or asphalt overall. With tar-felt roofs, the top is usually covered with embedded gravel or crushed slag.
6. Roll roofing, which, as the name implies, is marketed in rolls containing approximately 108 ft^2. Each roll is usually 36 inches wide

323

Fig. 13–1.   A wood shingle roof.

and may be plain or have a coating of colored mineral granules. The base is a heavy asphalt-impregnated felt.

7. Asphalt shingles (Fig. 13–3), usually in the form of strips with two, three, or four tabs per unit. These shingles are asphalt with the surface exposed to the weather heavily coated with mineral granules. Because of their fire resistance, cost, and durability, asphalt shingles are the most popular roofing material for homes. Asphalt shingles are available in a wide range of colors, including black and white.

8. Glass fiber shingles. These are made partly of a glass fiber mat, which is waterproof, and partly of asphalt. Like asphalt shingles, glass fiber shingles come with self-sealing tabs and carry a Class-A fire-resistance warranty. For the do-it-yourselfer, they may be of special interest because they are lightweight, about 220 pounds per square (100 ft^2 of roofing).

Fig. 13–2. A tile roof. This roof is popular in southwestern states.

Slope of Roofs

The slope of the roof is frequently a factor in the choice of roofing materials and method used to put them in place. The lower the pitch of the roof, the greater the chance of wind getting under the shingles and tearing them out. Interlocking cedar shingles resist this wind prying better than standard asphalt shingles. For roofs with less than a 4-inch slope per foot, do not use standard asphalt. Down to 2 inches, use self-sealing asphalt. Roll roofing can be used with pitches down to 2 inches when lapped 2 inches. For very low pitched slopes, the manufacturers of asphalt shingles recommend that the roof be planned for some other type of covering.

Aluminum strip roofing virtually eliminates the problem of wind prying, but these strips are noisy. Most homeowners object to the noise

Fig. 13–3. These asphalt shingles have a three-dimensional look. Asphalt shingles are the most popular.

during a rainstorm. Even on porches, the noise is often annoying to those inside the house.

Spaced roofing boards are sometimes used with cedar shingles as an economy measure and because the cedar shingles themselves add considerably to the strength of the roof. The spaced roofing boards reduce the insulating qualities, however, and it is advisable to use a tightly sheathed roof beneath the shingles if the need for insulation overcomes the need for economy.

For drainage, most roofs should have a certain amount of slope. Roofs covered with tar-and-gravel coverings are theoretically satisfactory when built level, but standing water may ultimately do harm. If you can avoid a flat roof, do so. Level roofs drain very slowly; slightly smaller eave troughs and downspouts are used on these roofs. They are quite common on industrial and commercial buildings.

Roll Roofing

Roll roofing (Fig. 13–4) is an economical cover especially suited for roofs with low pitches. It is also sometimes used for valley flashing instead of metal. It has a base of heavy asphalt-impregnated felt with additional coatings of asphalt that are dusted to prevent adhesion in the roll. The weather surface may be plain or covered with fine mineral granules. Many different colors are available. One edge of the sheet is left plain (no granules) where the lap cement is applied. For best results, the sheathing must be tight, preferably 1 × 6 tongue-and-groove, or plywood. If the sheathing is smooth, with no cupped boards or other protuberance, the slate-surfaced roll roofings will withstand a surprising amount of abrasion from foot traffic, although it is not generally recommended for that purpose. Windstorms are the most relentless enemy of roll roofings. If the wind gets under a loose edge, almost certainly a section will be blown off.

Built-Up Roof (BUR)

A built-up roof is constructed of sheathing paper, a bonded base sheet, perforated felt, asphalt, and surface aggregates (Fig. 13–5). The

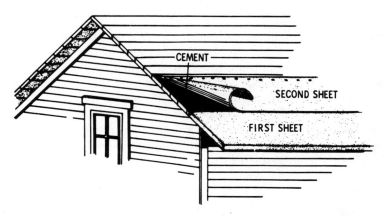

Fig. 13–4. Method of cementing and lapping the first and second strips of roll roofing.

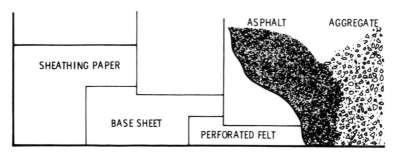

Fig. 13–5. Sectional plan of a built-up roof.

sheathing paper comes in 36-inch-wide rolls and has approximately 500 ft^2 per roll. It is a rosin-size paper and is used to prevent asphalt leakage to the wood deck. The base sheet is a heavy asphalt-saturated felt that is placed over the sheathing paper. It is available in 1, 1½, and 2 square rolls. The perforated felt is one of the primary parts of a built-up roof. It is saturated with asphalt and has tiny perforations throughout the sheet. The perforations prevent air entrapment between the layers of felt. The perforated felt is 36 inches wide and weighs approximately 15 lbs. per square. Asphalt is also one of the basic ingredients of a built-up roof. There are many different grades of asphalt, but the most common are low-melt, medium-melt, high-melt, and extra-high-melt.

Prior to the application of the built-up roof, the deck should be inspected for soundness. Wood board decks should be constructed of ¾-inch seasoned lumber or plywood. Any knotholes larger than one inch should be covered with sheet metal. If plywood is used as a roof deck it should be placed with the length at right angles to the rafters and be at least ½ inch in thickness.

The first step in the application of a built-up roof is the placing of sheathing paper and base sheet. The sheathing paper should be lapped in 2 inches and secured with just enough nails to hold it in place. The base sheet is then placed with 2-inch side laps and 6-inch end laps. The base sheet should be secured with ½-inch diameter head galvanized roofing nails placed 12 inches on center on the exposed lap. Nails should also be placed down the center of the base sheet. The nails should be placed in two parallel rows 12 inches apart.

The base sheet is then coated with a uniform layer of hot asphalt.

While the asphalt is still hot, a layer of roofing felt is placed and mopped with the hot asphalt. Each succeeding layer of roofing felt is placed and mopped in a similar manner with asphalt. Each sheet should be lapped 19 inches, leaving 17 inches exposed.

Once the roofing felt is placed, a gravel stop is installed around the deck perimeter, Fig. 13–6. Two coated layers of felt should extend 6 inches past the roof decking where the gravel stop is to be installed. When the other plies are placed, the first two layers are folded over the other layers and mopped in place. The gravel stop is then placed in a 1/8-inch-thick bed of flashing cement and securely nailed every 6 inches. The ends of the gravel stop should be lapped 6 inches and packed in flashing cement.

After the gravel stop is placed, the roof is flooded with hot asphalt and the surface aggregate is embedded in the flood coat. The aggregates should be hard, dry, opaque, and free of any dust or foreign matter. The size of the aggregates should range from 1/4 inch to 5/8 inch. When the aggregate is piled on the roof, it should be placed on a spot that has been mopped with asphalt. This technique assures proper adhesion in all areas of the roof.

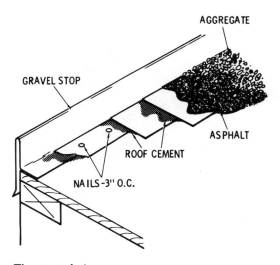

Fig. 13–6. The gravel stop.

Wood Shingles

The better grades of shingles are made of cypress, cedar, and red-wood and are available in lengths of 16 and 18 inches and thicknesses at the butt of 5⁄16 and 7⁄16 inches, respectively. They are packaged in bundles of approximately 200 shingles in random widths from 3 to 12 inches.

An important requirement in applying wood shingles is that each shingle should lap over the two courses below it, so that there will always be at least three layers of shingles at every point on the roof. This requires that the amount of shingle exposed to the weather (the spacing of the courses) should be less than one-third the length of the shingle. Thus in Fig. 13–7, 5½ inches is the maximum amount that 18-inch shingles can be laid to the weather and have an adequate amount of lap. This is further shown in Fig. 13–8.

In case the shingles are laid more than one-third of their length to the weather, there will be a space, as shown by *MS* in Fig. 13–8(B), where only two layers of shingles will cover the roof. This is objection-able, because if the top shingle splits above the edge of the shingle below, water will leak through. The maximum spacing to the weather for 16-inch shingles should be 4⅞ inches and for 18-inch shingles

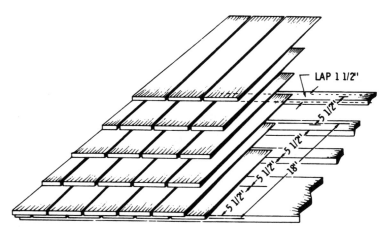

Fig. 13–7. Section of a shingle roof showing the amount of shingle that may be exposed to the weather as governed by the lap.

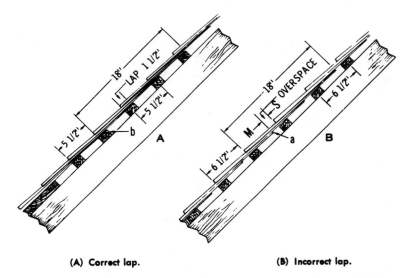

(A) Correct lap.   **(B) Incorrect lap.**

Fig. 13–8. The amount of lap is an important factor in applying wood shingles.

should be 5½ inches. Strictly speaking, the amount of lap should be governed by the pitch of the roof. The maximum spacing may be followed for roofs of moderate pitch, but for roofs of small pitch, more lap should be allowed, and for a steep pitch the lap may be reduced somewhat, but it is not advisable to do so. Wood shingles should not be used on pitches less than 4 inches per foot.

Table 13–1 shows the number of square feet that 1,000 (five bundles) shingles will cover for various exposures. This table does not allow for waste on hip and valley roofs.

Shingles should not be laid too close together, for they will swell when wet, causing them to bulge and split. Seasoned shingles should not be laid with their edges nearer than 3⁄16 inch when laid by the

Table 13–1.   Space Covered by 1,000 Shingles

Exposure to weather (inches)	4¼	4½	4¾	5	5½	6
Area covered (ft^2)	118	125	131	138	152	166

American method. It is advisable to thoroughly soak the bundles before opening.

Great care must be used in nailing wide shingles. When they are over 8 inches in width, they should be split and laid as two shingles. The nails should be spaced such that the space between them is as small as is practical, thus directing the contraction and expansion of the shingle toward the edges. This lessens the danger of wide shingles splitting in or near the center and over joints beneath. Shingling is always started from the bottom and laid from the eaves or cornice up.

There are various methods of laying shingles, the most common known as:

1. The straightedge,
2. The chalkline,
3. The gage-and-hatchet.

The straightedge method is one of the oldest. A straightedge having a width equal to the spacing to the weather or the distance between courses is used. This eliminates measuring, it being necessary only to keep the lower edge flush with the lower edge of the course of shingles just laid; the upper edge of the straightedge is then in line for the next course. This is considered to be the slowest of the three methods.

The chalkline method consists of snapping a chalkline for each course. To save time, two or three lines may be snapped at the same time, making it possible to carry two or three courses at once. This method is still extensively used. It is faster than the straightedge method, but not as fast as the gage-and-hatchet method.

The gage-and-hatchet method is extensively used in the western states. The hatchet used is either a lathing or a boxmaker's hatchet, as shown in Fig. 13–9. Hatchet gages to measure the space between courses are shown in Fig. 13–10. The gage is set on the blade at a distance from the hatchet poll equal to the exposure desired for the shingles.

Nail as close to the butts as possible, if the nails will be well covered by the next course. Only galvanized shingle nails should be used. The 3d shingle nail is slightly larger in diameter than the 3d common nail, and has a slightly larger head.

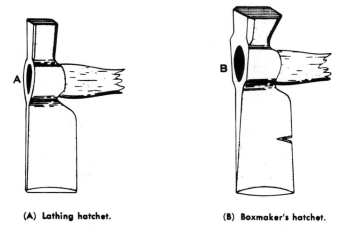

(A) Lathing hatchet.   **(B) Boxmaker's hatchet.**

Fig. 13–9.   Hatchets used for shingling.

Hips

The hip is less liable to leak than any other part of the roof as the water runs away from it. However, since it is so prominent, the work should be well done. Fig. 13–11 shows the method of cutting shingle butts for a hip roof. After the courses *1* and *2* are laid, the top corners

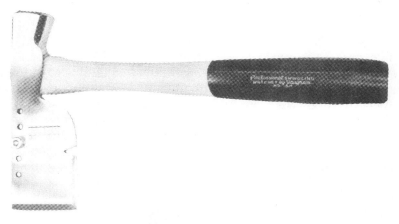

Fig. 13–10.   Shingling hatchet.

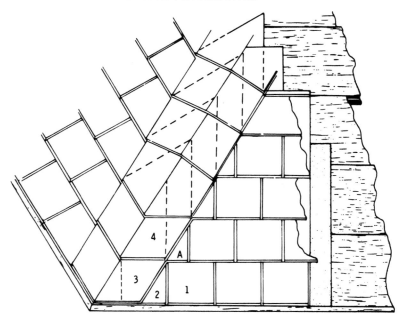

Fig. 13–11. Hip roof shingling.

over the hip are trimmed off with a sharp shingling hatchet kept keen for that purpose. Shingle 3 is trimmed with the butt cut so as to continue the straight line of courses and again on the dotted line 4, so that shingle A of the second course squares against it. This process continues from side to side, each shingle alternately lapping the other at the hip joint. When gables are shingled, this same method may be used up the rake of the roof if the pitch is moderate to steep. It cannot be effectively used with flat pitches. The shingles used should be ripped to uniform width.

For best construction, tin shingles should be laid under the hip shingles, as shown in Fig. 13–12. These tin shingles should correspond in shape to the hip shingles. They should be at least 7 inches wide and large enough to reach well under the tin shingles of the course above, as at W. At A, the tin shingles are laid so that the lower end will just be covered by the hip shingle of the course above.

A variation on the wood shingle is the recently introduced shingle that is part asphalt and part wood composition (Fig. 13–13).

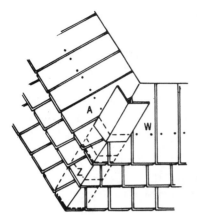

Fig. 13–12. Method of installing metal shingles under wood shingles.

Fig. 13–13. Wood fiber roofing from Masonite is a relatively new product. It is available in fire-rated versions, is bigger than standard shingle.

Valleys

In shingling a valley, first a strip of tin, lead, zinc, or copper, ordinarily 20 inches wide, is laid in the valley. Fig. 13–14 illustrates an open valley. Here the dotted lines show the tin or other material used as flashing under the shingles. If the pitch is above 30 degrees, then a width of 16 inches is sufficient; if flatter, the width should be more. In a long valley, its width between shingles should increase in width from top to bottom about 1 inch, and at the top 2 inches is ample width. This is to prevent ice or other objects from wedging when slipping down. The shingles taper to the butt, the reverse of the hip, and need no reinforcing, as the thin edge is held and protected from splitting off by the shingle above it. Care must always be taken to nail the shingle nearest the valley as far from it as practical by placing the nail higher up.

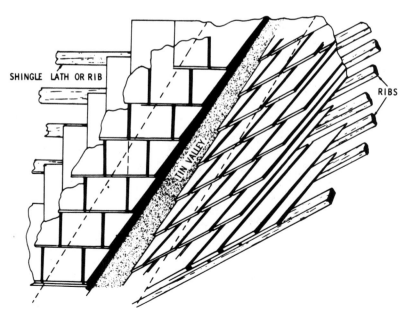

Fig. 13–14.   Method of shingling a valley.

Asphalt Shingles

Asphalt shingles are made in strips of two, three, or four units or tabs joined together, as well as in the form of individual shingles. When laid, strip shingles furnish practically the same pattern as undivided shingles. Both strip and individual types are available in different shapes, sizes, and colors to suit various requirements.

Underlayment

Impregnated paper felts have been in use for more than 130 years. However, underlayment in residential construction has been the subject of controversy for years: some builders swear by it, most swear at it. Those who argue against its use claim it is: (a) a vapor barrier; (b) a giant sponge that soaks up water and causes the shingles to wrinkle; (c) unnecessary. Some builders have claimed that its use voided the shingle warranty. But this is in direct contradiction to the shingle manufacturer's warranty which *requires* the use of felt underlayment. When questioned, shingle manufacturers claim that only when felt underlayment was improperly installed, and shingles damaged, was the warranty voided. The NAHB has been questioning the need for underlayment, and is attempting to get the underlayment requirement removed from the Codes.

The 1990 *BOCA* code requires Type 15 asphalt-saturated felt underlayment be installed on roof slopes from 2:12 and up. On slopes below 4:12, a double layer of underlayment is required. On slopes from 4:12 and up, a single-layer underlayment is required on all slopes. The 1988 *UBC* code, Table No. 32-B-1, requires nonperforated Type 15 felt on roofs with slopes from 2:12 up.

Traditionally, three reasons have been given for the use of underlayment:

1. It protects the roof sheathing until the shingles are installed;
2. It continues to protect the roof, after the shingles have been installed, from rain that may be wind-driven up under the shingles;
3. The shingles are protected from the turpentine in the plywood resins which can dissolve the asphalt.

The last item is rarely seen with plywood sheathing and OSB, but is common with board sheathing.

The director of Technology and Research of the National Roofing Contractors Association (NRCA) claims that the ability of felt to absorb moisture causes it to swell up and wrinkle after the shingles have been applied. It is this swelling-up between the shingle nails that gives the roof its wavy appearance. Spokesmen for the Asphalt Roofing Manufacturers Association (ARMA) argue that it is the use of non-standard felt that causes the problem. Only felt that complies with ASTM D4869 or D226 should be used. The ARMA representative further added that leaving the felt on the roof exposed to weather and drenching rains for a month or two does not help.

A rash of roof shingle ridging (buckling) across the United States led the APA, ARMA, and NAHB Research Foundation to form a committee to seek the cause of the problem and its solution. The researchers concluded that improper storage of plywood sheathing, improper installation of roof sheathing, improper installation of shingles, and inadequate attic ventilation were the causes.

There are two seeming contradictions in APA-published statements. APA has for many years said, "Cover sheathing as soon as possible with roofing felt for extra protection against excessive moisture prior to roofing application." Compare this with the following:

Exposure 1 panels have a fully waterproof bond and are designed for applications where long construction delays may be expected prior to providing protection, or where high moisture conditions may be encountered in service.

"Long construction delays" is not defined. According to an APA spokesperson, "it would be substantially in excess of the normal two- or three-week period that one would expect in today's construction practice." For years and years, most builders have completely ignored APA's warning not to leave CDX panels exposed to rain for weeks on end as they were not designed for that purpose.

There is no conflict between the two statements. The Exposure 1 panels will retain their structural integrity during prolonged periods of exposure. Whey then should they be covered as soon as possible? The direct rain on the panels raises the grain, causes checking, and other surface problems. Although these surface problems are not structural, they could affect the appearance of the panel by telegraphing through the thin fiberglass shingles.

The common widespread practice of not leaving the required

⅟₁₆-inch spacing at the panel ends, and ⅛ inch at the edges—with H clips—denies the rain an escape route off the panels. Continued exposure to rain causes edge swelling; with no room to move horizontally, the panels buckle. APA cautions: *"Where wet or humid conditions prevail, double these spacings."*

Commercial roofers have seen the quality of felt decline over the years. The elimination of rags in the manufacture of felt around 1945 reduced the tensile strength of the felt and increased its ability to absorb moisture. Non-standard felt, combined with poor or indifferent installation spells trouble. So does improperly applied, ASTM-specified felt.

Is underlayment necessary? No, as long as certain conditions are met:

1. Use the correct plywood, Exposure 1, or Exterior, if necessary.
2. Leave the required ⅟₁₆-inch spacing at ends and ⅛ inch at the edges. Use H clips at the edges.
3. In rain-prone areas or humid climates, double these spacings.
4. Cover the sheathing with shingles as soon as possible.
5. Install the shingles correctly.

When selecting a roofer, ask to watch his crew installing shingles; notice where they place the nails. If using your own crew, watch as they install shingles. If they *nail into or above the glueline,* stop them immediately. Unfortunately, this widespread *accepted practice* is incorrect, and contrary to the instructions printed on the wrapper that holds the shingles.

Where the nails are located is important because it affects the shingle's wind resistance. As the nails are moved up, less of the shingle is held down, exposing a greater area to the wind. Never assume roofers or your own crews know this. Make certain they do; otherwise, instruct them in proper nailing.

If underlayment is used, use Type 15 felt that meets ASTM standards: D4869 or D226. Do not allow it to sit in the rain for weeks because a schedule says when it should be covered. The weather should determine this, not CPM charts. The last measure may make little or no difference because the sheathing is gaining and losing moisture daily. But it can make a difference in the appearance of the roof shingles.

Ice Dam Shields

All three model codes, *BOCA*, *UBC*, and *SBCCI*, require ice dam protection. *BOCA* calls for it when the average daily temperature in January is 25 degrees F or less, or if there is a possibility of ice dams forming on the eaves and causing water back-up. The *UBC* requires protection in areas subject to wind-driven snow or roof ice buildup. Both require a double layer of Type 15 felt. However, a number of commercial products other than Type 15 felt are in common use. The most common is BITUTHENE®, which is a polyethylene-coated rubberized asphalt. W. R. Grace sells it under the name ICE & WATER SHIELD™, in 225 ft² rolls, 3 feet wide and 75 feet long. The codes are very specific as to how far up the roof from the eaves this flashing, or ice shield must extend.

Grace recommends it be installed at a temperature above 40 degrees Fahrenheit, and up the roof to the highest expected dam height. Common, but incorrect practice, is to extend it from the eaves up the roof 3 feet—the width of the material. The codes require that the ice shield extend from the eave edge up the roof to a line or point *no less than 24 inches inside the exterior wall line of the building.*

Caution— Ice shields, whether Type 15 felt, Bituthene, or metal, DO NOT PREVENT OR STOP ice dams. That is not their function. Their purpose is to provide some protection from the water forced up the roof by ice dams.

Ice Dams

Several inches of snow on a roof, and below-freezing temperatures for a week or so are two of the conditions necessary for ice dams. Inadequate levels of attic insulation and poor attic ventilation are the other two factors.

Heat escaping through the ceiling and into the attic warms the roof and melts the snow at or near the ridge. The snow melts at the point where the snow and shingle meet. The melted snow runs down the roof, under the snow, as snow-water. As the snow-water moves down the roof, it reaches the coldest part at the edge, and freezes. An ice dam forms and increases in size as more snow melts. The continuing melting results in the water accumulating and eventually backing up under the shingles. Correctly installed bituthene does not stop the water from backing up the roof. Because it is waterproof, it prevents water penetration. Most ice dams are found on houses with gable-end louvered vents.

Drip-Edge Flashing

Most builders install drip-edge only on the eaves. Some builders install it on both eaves and rakes. Other builders install no drip-edge; they extend the shingles slightly beyond the edge of the roof sheathing to act as drip-edge.

Ninety-nine percent of all residential roof failures start at the unprotected rake—gable—ends. If there were no wind-driven rain, moving the shingle out beyond the roof sheathing edge would be fine. But the rain goes where the wind goes: up under the projecting shingle and at the end of the sheathing. The five conditions necessary for rot are present and thus begins the eventual destruction of the roof.

There is a definite order to the installation of drip-edge, bituthene, and underlayment:

1. Install the bituthene over the eave's drip-edge and up the roof 24 inches inside the exterior wall line of the building.
2. Install the eave end drip-edge first.
3. Attach the Type 15 asphalt-saturated felt underlayment.
4. Install the drip-edge on the rake ends and over the felt.
5. Use galvanized nails with galvanized drip-edge; aluminum nails with aluminum drip-edge.

Common practice, when drip-edge is used on eaves and rakes, is to first install the drip-edge, and then the underlayment, if it is used. This is another example of incorrect common practice. If the underlayment is installed over the rake-end drip-edge, wind-driven rain gets up under the shingles, under the underlayment, and onto the roof sheathing. Installed in proper order, any wind-driven rain penetrating under the shingle gets on the paper, not on the wood sheathing.

To shed water efficiently at the roof's edge, a drip-edge is usually installed. A drip-edge is constructed of corrosion-resistant sheet metal, and extends 3 inches back from the roof edge. To form the drip-edge, the sheet metal is bent down over the roof edges.

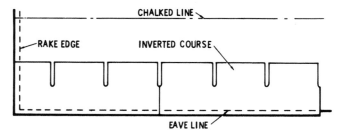

Fig. 13–15. The starter course.

Installing Asphalt Shingles

The nails used to apply asphalt singles should be hot-dipped galvanized nails, with large heads, sharp points, and barbed shanks. The nails should be long enough to penetrate the roof decking at least ¾ inch.

To ensure proper shingle alignment, horizontal and vertical chalk lines should be placed on the underlayment. It is usually recommended that the lines be placed 10 or 20 inches apart. The first course of shingles placed is the starter course. This is used to back up the first regular course of shingles and to fill in the spaces between the tabs. The starter course is placed with the tabs facing up the roof and is allowed to project one inch over the rake and eave, Fig. 13–15. To ensure that all cutouts are covered, 3 inches should be cut off the first starter shingle.

Once the starter course has been placed, the different courses of shingles can be laid. The first regular course of shingles should be started with a full shingle; the second course with a full shingle, minus one-half tab; the third course is started with a full shingle (Fig. 13–16)

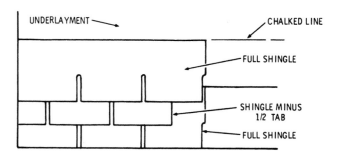

Fig. 13–16. Application of the starter shingles.

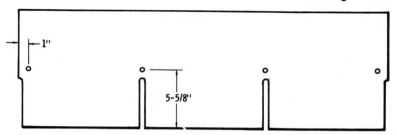

Fig. 13–17. The proper placement of nails.

and the process is repeated. As the shingles are placed, they should be properly nailed (Fig. 13–17). If a three-tab shingle is used, a minimum of 4 nails per strip should be used. The nails should be placed 5 ⅝ inch from the bottom of the shingle and should be located over the cutouts. The nails on each end of the shingle should be 1 inch from the end. The nails should be driven straight and flush with the surface of the shingle.

If there is a valley in the roof, it must be properly flashed. The two materials that are most often used for valley flashing are 90 lb. mineral surfaced asphalt roll roofing or galvanized sheet metal. The flashing is 18 inches in width and should extend the full length of the valley. Before the shingles are laid to the valley, chalklines are placed along the valley. The chalklines should be 6 inches apart at the top of the valley and should widen ⅛ inch per foot as they approach the eave line. The shingles are laid up to the chalk lines and trimmed to fit.

Hips and ridges are finished by using manufactured hip-and-ridge units, or hip-and-ridge units cut from a strip shingle. If the unit is cut from a strip shingle, the two cut lines should be cut at an angle (Fig. 13–18). This will prevent the projection of the shingle past the overlaid

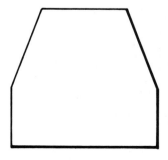

Fig. 13–18. Hip shingle.

shingle. Each shingle should be bent down the center so that there is an equal distance on each side. In cold weather the shingles should be warmed before they are bent. Starting at the bottom of the hip or at the end of a ridge the shingles are placed with a 5-inch exposure. To secure the shingles, a nail is placed on each side of the shingle. The nails should be placed 5½ inches back from the exposed edge and one inch up from the side.

If the roof slope is particularly steep, specifically if it exceeds 60 degrees or 21 inches per foot, then special procedures are required for securing the shingles. This is shown in Fig. 13–19.

Other details: for neatness when installing asphalt shingles, the courses should meet in a line above any dormer (Fig. 13–20).

Slate

Slate is an ideal roofing material and is used on permanent buildings with pitched roofs. The process of manufacture is to split the quarried slate blocks horizontally to a suitable thickness, and to cut vertically to the approximate sizes required. The slates are then passed through planers, and after the operation are ready to be reduced to the

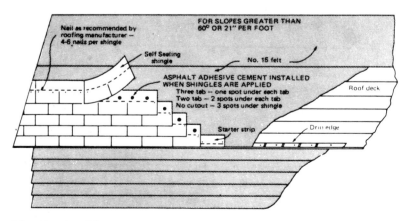

Fig. 13–19. When roof slope exceeds 60 degrees, you have to take special steps in application.

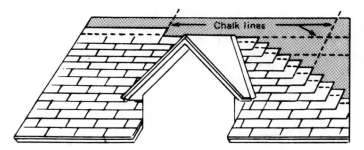

Fig. 13–20. For neatness, shingle courses should meet in a line above dormer.

exact dimensions on rubbing beds or through the use of air tools and other special machinery.

Roofing slate is usually available in various colors and in standard sizes suitable for the most exacting requirements. On all boarding to be covered with slate, asphalt-saturated rag felt of certain specified thickness is required. This felt should be laid in a horizontal layer with joints lapped toward the eaves and at the ends at least 2 inches. A well-secured lap at the end is necessary to hold the felt in place properly, and to protect the structure until covered by the slate. In laying the slate, the entire surface of all main and porch roofs should be covered with slate in a proper and watertight manner.

The slate should project 2 inches at the eaves and 1 inch at all gable ends, and must be laid in horizontal courses with the standard 3-inch headlap. Each course breaks joints with the preceding one. Slates at the eaves or cornice line are doubled and canted $\frac{1}{4}$ inch by a wooden cant strip. Slates overlapping sheet-metal work should have the nails so placed as to avoid puncturing the sheet metal. Exposed nails are permissible only in courses where unavoidable. Neatly fit the slate around any pipes, ventilators, or other rooftop protuberances.

Nails should not be driven in so far as to produce a strain on the slate. Cover all exposed nail heads with elastic cement. Hip slates and ridge slates are to be laid in elastic cement spread thickly over unexposed surfaces. Build in and place all flashing pieces furnished by the sheeting contractor and cooperate with him in doing the work of flashing. On completion, all slate must be sound, whole, and clean, and the roof left in every respect tight and a neat example of workmanship.

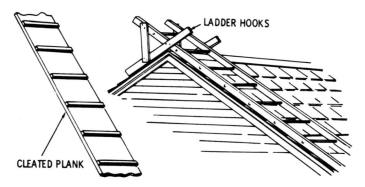

Fig. 13–21. Two types of supports used in repairs of roof.

The most frequently needed repair of slate roofs is the replacement of broken slates. When such replacements are necessary, supports similar to those shown in Fig. 13–21 should be placed on the roof to distribute the weight of the roofers while they are working. Broken slates should be removed by cutting or drawing out the nails with a ripper tool. A new slate shingle of the same color and size as the old should be inserted and fastened by nailing through the vertical joint of the slates in the overlying course approximately 2 inches below the butt of the slate in the second course, as shown in Fig. 13–22.

A piece of sheet copper or terneplate about 3″ × 8″ should be inserted over the nail head to extend about 2 inches under the second course above the replaced shingle. The metal strip should be bent slightly before being inserted so that it will stay securely in place. Very old slate roofs sometimes fail because the nails used to fasten the slates have rusted. In such cases, the entire roof covering should be removed

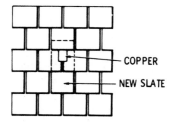

Fig. 13–22. Method of inserting new pieces of slate shingles.

and replaced, including the felt underlay materials. The sheathing and rafters should be examined and any broken boards replaced with new material. All loose boards should be nailed in place and, before laying the felt, the sheathing should be swept clean, protruding nails driven in, and any rough edges trimmed smooth.

If the former roof was slate, all slates that are still in good condition may be salvaged and relaid. New slates should be the same size as the old ones and should match the original slates as nearly as possible in color and texture. The area to be covered should govern the size of slates to be used and whatever the size, the slates may be of random widths, but they should be of uniform length and punched for a head lap of not less than 3 inches. The roof slates should be laid with a 3-inch headlap and fastened with two large-head slating nails. Nails should not be driven too tightly, for the nail heads should barely touch the slate. All slates within 1 foot of the top and along the gable rakes of the roof should be bedded in flashing cement.

Gutters and Downspouts

Most roofs require gutters and downspouts (Fig. 13–23) in order to convey the water to the sewer or outlet. They are usually built of metal. In regions of heavy snowfall, the outer edge of the gutter should be ½ inch below the extended slope of the roof to prevent snow banking on the edge of the roof and causing leaks. The hanging gutter is best adapted to such construction.

Downspouts should be large enough to remove the water from the gutters. A common fault is to make the gutter outlet the same size as the downspout. At 18 inches below the gutter, a downspout has nearly four times the water-carrying capacity of the inlet at the gutter. Therefore, a good-sized ending spout should be provided. Wire baskets or guards should be placed at gutter outlets to prevent leaves and trash from collecting in the downspouts and causing damage during freezing weather.

Gutters come in a variety of materials including wood, metal, and vinyl. Most people favor metal gutters. You can get them in enameled steel or aluminum, with the latter the favorite.

Though it comes in sections, the so-called seamless gutter is easiest

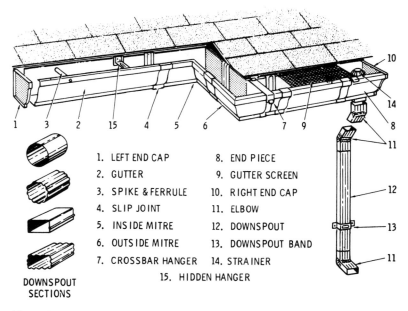

1. LEFT END CAP
2. GUTTER
3. SPIKE & FERRULE
4. SLIP JOINT
5. INSIDE MITRE
6. OUTSIDE MITRE
7. CROSSBAR HANGER

8. END PIECE
9. GUTTER SCREEN
10. RIGHT END CAP
11. ELBOW
12. DOWNSPOUT
13. DOWNSPOUT BAND
14. STRAINER
15. HIDDEN HANGER

DOWNSPOUT
SECTIONS

Fig. 13–23. Various downspouts and fittings. *(Courtesy Billy Penn Gutters)*

to install. A specialist cuts the gutter to the exact lengths needed; no joining of lengths is necessary, and therefore there are no possible leaks.

It should be noted that aluminum gutter is available in various gauges. The .027 size is standard, but .032 is standard in seamless types; .014 is also available, but this should be avoided because it is too flimsy.

Selecting Roofing Materials

Roofing materials are commonly sold by dealers or manufacturers on the basis of quantities to cover 100 ft². This quantity is commonly termed "one square" by roofers and in trade literature. When ordering roofing material, make allowance for waste such as in hips, valleys, and starter courses. This applies in general to all types of roofing.

The slope of the roof and the strength of the framing are the first determining factors in choosing a suitable covering. If the slope is slight, there will be a danger of leaks with a wrong kind of covering, and excessive weight may cause sagging that is unsightly and adds to the difficulty of keeping the roof in repair. The cost of roofing depends to a great extent on the type of roof to be covered. A roof having ridges, valleys, dormers, or chimneys will cost considerably more to cover than one having a plain surface. Very steep roofs are also more expensive than those with a flatter slope, but most roofing materials last longer on steep grades than on low-pitched roofs. Frequently, nearness to supply centers permits the use, at lower cost, of the more durable materials instead of the commonly lower-priced, shorter-lived ones.

In considering cost, one should keep in mind maintenance and repair and the length of service expected from the building. A permanent structure warrants a good roof, even though the first cost is somewhat high. When the cost of applying the covering is high in comparison with the cost of the material, or when access to the roof is hazardous, the use of long-lived material is warranted. Unless insulation is required, semipermanent buildings and sheds are often covered with low-grade roofing.

Frequently, the importance of fire resistance is not recognized, and sometimes it is wrongly stressed. It is essential to have a covering that will not readily ignite from glowing embers. The building regulations of many cities prohibit the use of certain types of roofings in congested areas where fires may spread rapidly. The Underwriters Laboratories has grouped many of the different kinds and brands of roofing in classes from A to C according to the protection afforded against the spread of fire. Class A is best.

The appearance of a building can be changed materially by using the various coverings in different ways. Wood shingles and slate are often used to produce architectural effects. The roofs of buildings in a farm group should harmonize in color, even though similarity in contour is not always feasible.

The action of the atmosphere in localities where the air is polluted with fumes from industrial works or saturated with salt (as along the seacoast) shortens the life of roofing made from certain metals. Sheet aluminum is particularly vulnerable to acid fumes.

All coal-tar pitch roofs should be covered with slag or a mineral coating, because when fully exposed to the sun, they deteriorate. Observation has shown that, in general, roofings with light-colored surfaces absorb less heat then those with dark surfaces. Considerable attention should be given to the comfort derived from a properly insulated roof. A thin, uninsulated roof gives the interior little protection from heat in summer and cold in winter. Discomfort from summer heat can be lessened to some extent by ventilating the space under the roof. None of the usual roof coverings have any appreciable insulating value. Installing insulation and providing for ventilation are other things considered elsewhere in this book.

Detection of Roof Leaks

A well-constructed roof should be properly maintained. Periodic inspections should be made to detect breaks, missing shingles, choked gutters, damaged flashings, and also defective mortar joints of chimneys, parapets, coping, and such. At the first appearance of damp spots on the ceilings or walls, a careful examination of the roof should be made to determine the cause, and the defect should be promptly repaired. When repairs are delayed, small defects extend rapidly and involve not only the roof covering, but also the sheathing, framing, and interior.

Many of these defects can be readily repaired to keep water from the interior and to extend the life of the roof. Large defects or failures should be repaired by people familiar with the work. On many types of roofs, an inexperienced person can do more damage than good. Leaks are sometimes difficult to find, but an examination of the wet spots on a ceiling furnishes a clue to the probable location. In some cases, the actual leak may be some distance up the slope. If near a chimney or exterior wall, the leaks are probably caused by a defective or narrow flashing, loose mortar joints, or dislodged coping. On flat roofs, the trouble may be the result of choked downspouts or an accumulation of water or snow on the roof higher than the flashing. Defective and loose flashing is not uncommon around scuttles, cupolas, and plumbing vent pipes. Roofing deteriorates more rapidly on a south exposure than on a north exposure, which is especially noticeable when wood or composition shingles are used.

Wet spots under plain roof areas are generally caused by holes in the covering. Frequently, the drip may occur much lower down the slope than the hole. Where attics are unsealed and roofing strips have been used, holes can be detected from the inside by light shining through. If a piece of wire is stuck through the hole it can be located from the outside.

Sometimes gutters are so arranged that when choked, they overflow into the house, or ice accumulating on the eaves will form a ridge that backs up melting snow under the shingles. This is a common trouble if roofs are flat and the eaves wide. Leaky downspouts permit water to splash against the wall and the wind-driven water may find its way through a defect into the interior. The exact method to use in repairing depends on the kind of roofing and the nature and extent of the defect.

CHAPTER 14

Cornice Construction

The cornice is the projection of the roof at the eaves that forms a connection between the roof and the side walls. The three general types of cornice construction are the *box,* the *closed,* and the *open.*

Box Cornices

The typical box cornice shown in Fig. 14–1 utilizes the rafter projection for nailing surfaces for the facia and soffit boards. The soffit provides a desirable area for inlet ventilators. A frieze board is often used at the wall to receive the siding. In climates where snow and ice dams may occur on overhanging eaves, the soffit of the cornice may be sloped outward and left open ¼ inch at the facia board for drainage.

Closed Cornices

The closed cornice shown in Fig. 14–2 has no rafter projection. The overhang consists only of a frieze board and a shingle or crown molding. This type is not so desirable as a cornice with a projection, because it gives less protection to the side walls.

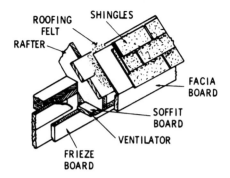

Fig. 14–1. Box cornice construction.

Wide Box Cornices

The wide box cornice in Fig. 14–3 requires forming members called lookouts, which serve as nailing surfaces and supports for the soffit board. The lookouts are nailed at the rafter ends and are also toenailed to the wall sheathing and directly to the stud. The soffit can be of various materials, such as beaded ceiling boards, plywood, or aluminum, either ventilated or plain. A bed molding may be used at the juncture of the soffit and frieze. This type of cornice is often used in hip roof houses, and the facia board usually carries around the entire perimeter of the house.

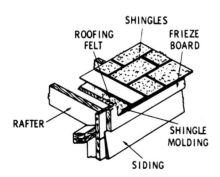

Fig. 14–2. Closed cornice construction.

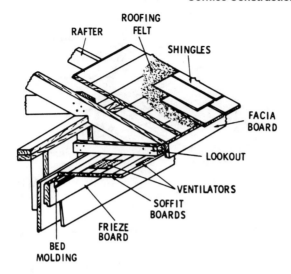

RAFTER

ROOFING
FELT

SHINGLES

FACIA
BOARD

LOOKOUT

VENTILATORS

SOFFIT
BOARDS

FRIEZE
BOARD

BED
MOLDING

Fig. 14–3. Wide cornice construction.

Open Cornices

The open cornice shown in Fig. 14–4 may consist of a facia board nailed to the rafter ends. The frieze is either notched or cut out to fit between the rafters and is then nailed to the wall. The open cornice is often used for a garage. When it is used on a house, the roof boards are visible from below from the rafter ends to the wall line, and should consist of finished material. Dressed or matched V-beaded boards are often used.

Cornice Returns

The cornice return is the end finish of the cornice on a gable roof. The design of the cornice return depends to a large degree on the rake or gable projection, and on the type of cornice used. In a close rake (a gable end with very little projection), it is necessary to use a frieze or rake board as a finish for siding ends, as shown in Fig. 14–5. This board is usually 1⅛ inch thick and follows the roof slope to meet the return of

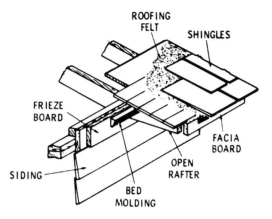

Fig. 14–4. Open cornice construction.

the cornice facia. Crown molding or other type of finish is used at the edge of the shingles.

When the gable end and the cornice have some projection as shown in Fig. 14–6, a box return may be used. Trim on the rake projection is finished at the cornice return. A wide cornice with a small gable projection may be finished as shown in Fig. 14–7. Many variations of this trim detail are possible. For example, the frieze board at the gable end might be carried to the rake line and mitered with a facia board of the cornice. This siding is then carried across the cornice end to form a return.

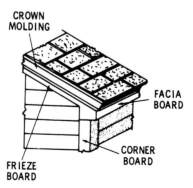

Fig. 14–5. The closed cornice return.

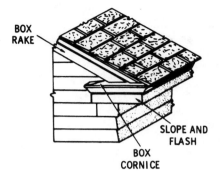

BOX
RAKE

SLOPE AND
FLASH

BOX
CORNICE

Fig. 14–6.   The box cornice return.

Rake or Gable-End Finish

The rake section is that trim used along the gable end of a house. There are three general types commonly used: the *closed*, the *box with a projection*, and the *open*. The closed rake, as shown in Fig. 14–8, often consists of a frieze or rake board with a crown molding as the finish. A 1″ × 2″ square edge molding is sometimes used instead of the crown molding. When fiberboard sheathing is used, it is necessary to use a narrow frieze board that will leave a surface for nailing the siding into the end rafters.

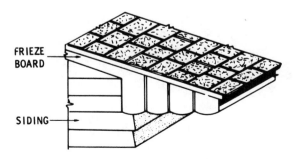

FRIEZE
BOARD

SIDING

Fig. 14–7.   The wide cornice return.

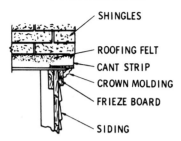

SHINGLES

ROOFING FELT

CANT STRIP

CROWN MOLDING

FRIEZE BOARD

SIDING

Fig. 14–8. The closed end finish.

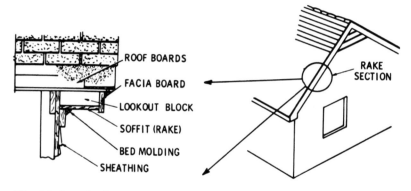

ROOF BOARDS

FACIA BOARD

LOOKOUT BLOCK

SOFFIT (RAKE)

BED MOLDING

SHEATHING

RAKE SECTION

Fig. 14–9. The box end finish.

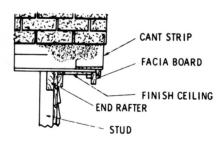

CANT STRIP

FACIA BOARD

FINISH CEILING

END RAFTER

STUD

Fig. 14–10. The open end finish.

If a wide frieze is used, nailing blocks must be provided between the studs. Wood sheathing does not require nailing blocks. The trim used for a box rake section requires the support of the projected roof boards, as shown in Fig. 14–9. In addition, lookouts for nailing blocks are fastened to the side wall and to the roof sheathing. These lookouts serve as a nailing surface for both the soffit and the facia boards. The ends of the roof boards are nailed to the facia. The frieze board is nailed to the side wall studs, and the crown and bed moldings complete the trim. The underside of the roof sheathing of the open projected rake as shown in Fig. 14–10, is generally covered with liner boards such as 5/8-inch beaded ceiling. The facia is held in place by nails through the roof sheathing.

CHAPTER 15

Sheathing and Siding

Sheathing is nailed directly to the framework of the building. Its purpose is to strengthen the building, to provide a base wall to which the finish siding can be nailed, to act as insulation, and in some cases to be a base for further insulation. Some of the common types of sheathing include *fiberboard, wood,* and *plywood.*

Fiberboard Sheathing

Fiberboard usually comes in $2' \times 8'$ or $4' \times 8'$ sheets which are tongue-and-grooved, and generally coated or impregnated with an asphalt material which increases water resistance. Thickness is normally $\frac{1}{2}$ and $\frac{25}{32}$ inch. Fiberboard sheathing may be used where the stud spacing does not exceed 16 inches, and it should be nailed with 2-inch galvanized roofing nails or other type noncorrosive nails. If the fiberboard is used as sheathing, most builders will use plywood at all corners (the thickness of the sheathing), to strengthen the walls, as shown in Fig. 15–1.

PLYWOOD

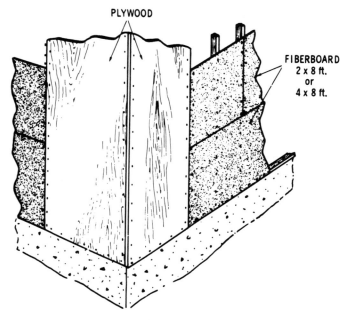

FIBERBOARD
2 x 8 ft.
or
4 x 8 ft.

Fig. 15–1. Method of using plywood on all corners as bracing when using fiberboard as exterior sheathing.

Solid Wood Sheathing

Wood wall sheathing can be obtained in almost all widths, lengths, and grades. Generally, widths are from 6 to 12 inches, with lengths selected for economical use. Almost all solid wood wall sheathing used is $25/32$ to 1 inch in thickness. This material may be nailed on horizontally or diagonally, as shown in Fig. 15–2. Wood sheathing is laid on tight, with all joints made over the studs. If the sheathing is to be put on horizontally, it should be started at the foundation and worked toward the top. If the sheathing is installed diagonally, it should be started at the corners of the building and worked toward the center or middle.

Diagonal sheathing should be applied at a 45-degree angle. This method of sheathing adds greatly to the rigidity of the wall and eliminates the need for the corner bracing. It also provides an excellent tie to the sill plate when it is installed diagonally. There is more lumber

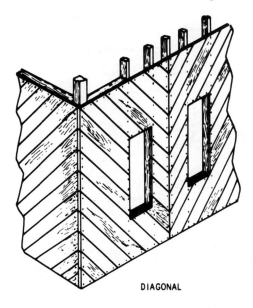

DIAGONAL

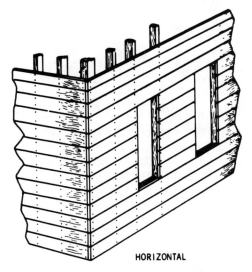

HORIZONTAL

Fig. 15–2. Two methods of nailing on wood sheathing.

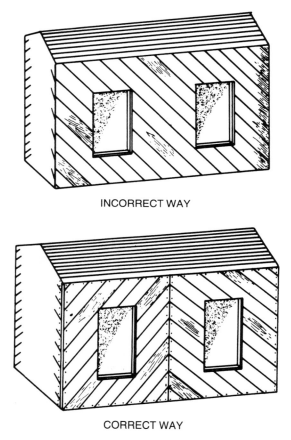

INCORRECT WAY

CORRECT WAY

Fig. 15–3. The incorrect and correct ways of laying sheathing.

waste than with horizontal sheathing because of the angle cut, and the application is somewhat more difficult. Fig. 15–3 shows the wrong way and the correct way of laying diagonal sheathing.

Plywood Sheathing

Plywood as a wall sheathing is highly recommended because of its size, weight, and stability, plus the ease and rapidity of installation (Fig.

15–4). It adds considerably more strength to the frame structure than the conventional horizontal or diagonal sheathing.

Plywood sheathing is easy and fast to install, adds great strength to the structure, and eliminates the need for corner let-in wood bracing. The sheathing may be plywood, OSB, or waferboard. Although plywood can be installed vertically or horizontally, it should be installed horizontally with studs 24 inches oc. Three-ply ⅜-inch plywood, applied horizontally, is an acceptable thickness for studs spaced 24 inches oc.

Rigid Exterior Foam Sheathing

The OPEC crisis in 1973 saw a sudden increase in the use of nonstructural insulating sheathing materials. The common form of the material today is XPS, such as Dow STYROFOAM, FORMULAR®,

Fig. 15–4. Plywood is a popular sheathing. Here it is used at corners with fiberboard.

AMOFOAM®, and polyisocyanurate aluminum covered boards such as THERMAX®, HI-R SHEATHING®, ENERGY SHIELD®, and a relative newcomer, a phenolic foam board called KOPPERS RX. Originally manufactured by Koppers, the product line was sold to Johns Mansville in 1989. Mansville has renamed it WEATHERTIGHT PREMIER.

Because these boards are nonstructural, the building must be braced against racking forces. The traditional form of bracing has been wood let-in corner bracing (Fig. 15–5). For many years let-in corner bracing was the standard of construction. The introduction of more labor-efficient structural materials such as plywood made the use of let-in bracing unnecessary. It is rarely seen today. Some builders and code officials still believe it must be used even with plywood sheathing; other building officials require its use with waferboard, but not plywood.

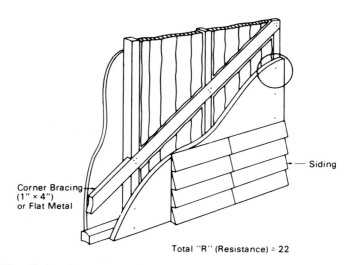

Corner Bracing
(1" × 4")
or Flat Metal

Siding

Total "R" (Resistance) = 22

Fig. 15–5. Corner wood let-in bracing.

The *BOCA, UBC* and *SBCCI* codes do not require let-in corner bracing when diagonal wood sheathing, plywood, or waferboard such as OSB and ASPENITE® panels are used. Other sheathings, such as THERMO-PLY®, are also permitted when used vertically. Thermo-Ply sheathing is manufactured of specially treated water- and weather-resistant Kraft long-fibered plies. Using water-resistant adhesives, the plies are pressure-laminated. The surface finish is reflective aluminum foil continuously pressure-laminated to the multi-ply substrate. Thermo-Ply is available in 3 grades: RED, for 16-inch oc d framing, BLUE for 24-inch oc m, and GREEN which is the Utility grade. Both the Red and Green are called Stormbrace grades and can be used without let-in corner bracing. The Red utility grade requires let-in corner bracing.

Technical Circular No. 12, "A standard for testing sheathing materials for resistance to racking," was released in 1949 by the Federal Housing Administration (FHA). It was intended as an interim standard until a new performance standard was introduced. As with crawlspace ventilation standards, *temporary* standards have a habit of becoming permanent. This is still the standard used by the *BOCA* code. It requires that the wall withstand a racking load of 5200 pounds. The existing standards were based on the racking strength of walls with let-in corner bracing and horizontal board sheathing. Nonstructural sheathing and let-in corner bracing is allowed by some codes. But little testing of wall panels with let-in bracing has been done for nearly 30 years. Workmanship and lumber quality was not specified. Engineering laboratories performing the same racking test often came up with different results.

Because of the increasing use of let-in corner bracing, the lack of engineering standards, and little information as to how let-in corner bracing actually performed when installed according to the 1949 standard, engineers decided to so some research to get answers and develop engineering standards for racking tests. This research showed that let-in bracing without horizontal board sheathing failed well below the 5200 pound level. The quality of the lumber and workmanship affects the strength of the brace. Although the strength and stiffness of the brace are important, it is the stud frame that controls the ultimate strength. Unfortunately, most let-in bracing is butchered. A better method of reinforcing the wall (less butchering) is to use a heavy metal brace, such as the Simpson STRONG-TIE® TWB T-Type Wall Brace (Fig. 15–6). A kerf is made with the saw, the brace is then tapped in and nailed.

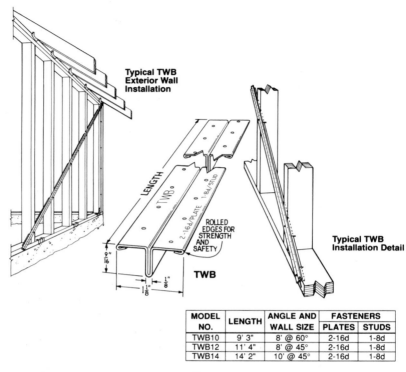

MODEL NO.	LENGTH	ANGLE AND WALL SIZE	FASTENERS	
			PLATES	STUDS
TWB10	9' 3"	8' @ 60°	2-16d	1-8d
TWB12	11' 4"	8' @ 45°	2-16d	1-8d
TWB14	14' 2"	10' @ 45°	2-16d	1-8d

Fig. 15–6. Simpson Strong-Tie® TWB T-type wall brace. *(Courtesy Simpson Strong-Tie Company, Inc.)*

Sheathing Paper

Both Type 15 felt and rosin paper were the traditional sheathing papers. In the 1950s, Dupont developed a spun-bonded olefin material, which was first used as a covering over bedsprings. Another intended use was to protect attic insulation from the effects of wind blowing over, under, and through it as the wind would reduce the insulation's effectiveness. Experts argued pro and con over this, but Canadian researchers found that if enough outside air gets into the insulation, its effectiveness will be reduced. The state of Minnesota has revised its building code to require an air retarder to prevent or reduce air infiltration into the attic floor insulation. This might be done in several ways, such as extending the sheathing up between the rafters/trusses to within 2 inches of the underside of the roof sheathing. This is the first time any state building code has included such details to prevent the degradation of the insulation. The air retarder also prevents loose fill insulations from entering and blocking the soffit vent.

Dupont's spun-bonded olefin, TYVEK®, is made from polyethylene. The sheets are formed by spinning polyethylene into short threads which are then sprayed into a mat and bonded together. The mat is spun tightly so that it becomes an air retarder, but readily allows water vapor to get through it.

Tyvek is only one of several *air retarders,* or housewraps, on the market. Parsec AIRTIGHT WHITE® is actually Tyvek. Other available housewraps are Parsec AIRTIGHT WRAP®, RUFCO-WRAP®, VER-SAWRAP®, AIR SEAL®, BARRICADE®, ENERGY SEAL®, and TYPAR®. Parsec Airtight Wrap, Rufco-wrap, Tu-Tuf Air Seal, and VersaWrap are polyethylene films that have holes in them, and are called perforated films. Barricade and Typar are spun-bonded polypropylene, whereas Tyvek is made from polyethylene. These housewraps have just about replaced felt and rosin papers as exterior sheathing papers. Dupont has made a number of changes in Tyvek to improve its performance, tear resistance, and give it greater protection against ultraviolet radiation from the sun.

Housewraps are not vapor diffusion retarders (VDR), as some believe. Housewraps have perm ratings that range from 10 to 80, much too high to be a VDR. According to the American Society of Heating, Refrigerating and Air-Conditioning Engineers (ASHRAE) *Handbook of Fundamentals,* the generally accepted rating of a VDR is 1 perm. Permeance, which is given in perms, is a measure of how readily or easily water vapor flows through a material. Most building materials such as plaster, gypsum board, paper, wood, concrete, and insulation are porous enough to allow water vapor to pass through. How easily the water vapor moves through the material depends on the material. Loose-fill insulation readily allows water vapor to pass through. Such materials are said to have high *permeance.* Concrete and wood offer greater resistance to the flow of water vapor, and have a lower permeance.

The purpose of a VDR is to control, retard, slow down, the movement of water vapor. Polyethylene, the most common VDR, must have a very low permeance; it must be highly impermeable to moisture. A VDR must have a perm rating of 1 or less. There are 7000 grains of water in one pound of water.

A VDR is a substance that will allow no more than one grain of water vapor to pass through one ft² of that material in one hour, with a vapor pressure difference, from one side of the material to the other side, of one inch of mercury (Perm = gr/hr/ft²/in.Hg).

This is approximately one-half pound per square inch. A vapor diffusion retarder then, is any material that has a perm rating of 1 or less. Table 15–1 is a listing of the perm ratings of some common building materials.

Although not VDRs, housewraps too are designed to control the passage of moisture by readily allowing water vapor to pass through them. For example, Tyvek will allow 77 times more moisture vapor to pass through it than would pass through the same size piece of felt underlayment. When wrapped around the outside of a building, housewraps serve as air retarders to resist the flow of wind into the walls. They are highly, but not totally, impervious (have a low permeance) to air movement through them. For example, Tyvek has a measured permeance of 0.035 cubic feet per square foot. Thus they serve a dual purpose: keep wind and rain out of the wall, but allow water vapor to escape.

Whatever brand of housewrap is used, all seams should be taped, and the bottom of the wrap at the sill plate should also be sealed. Merely covering the outside of the house with a housewrap does not stop the air leaks. The Canadian Construction Materials Centre, a division of the Canadian National Research Council, has conducted extensive research to housewrap tapes. The following brands are recommended:

> Tyvek Housewrap Tape—sold in Canada under the Tuck brand name;
> 3M Contractors' Sheathing Tape No. 8086;
> Venture HouseWrap Contractors' Sheathing Tape no. 1585 CW-2 and 1586 CW.

The wrap that covers window and door openings is cut in an X pattern, the flaps folded in and around the rough opening, and doors and windows installed.

Table 15–1. Permeance Values of Some Common Building Materials

Material	Permeance (Perm)
Materials Used in Construction	
Brick Masonry (4" thick)	0.8
Concrete block (8" cored, limestone aggregate)	2.4
Tile masonry, glazed (4" thick)	0.12
Plaster on metal lath (¾")	15
Plaster on wood lath	11
Gypsum wall board (⅜" plain)	50
Hardboard (⅛" standard)	11
Plywood (douglas-fir, exterior glue, ¼" thick)	0.7
Thermal Insulations	
Air (still) (1")	120
Cellular glass	0.0
Corkboard (1")	2.1–2.6
Expanded polystyrene—extruded (1")	1.2
Plastic and Metal Foils and Films	
Aluminum foil (1 mil)	0.0
Aluminum foil (0.35mil)	0.05
Polyethylene (4 mil)	0.08
Polyethylene (6 mil)	0.06
Building Paper, Felts, Roofing Papers	
Duplex sheet, asphalt laminated, aluminum foil one side	0.002
Saturated and coated roll roofing	0.05
Kraft paper and asphalt laminated, reinforced 30–120–30	0.03
Blanket thermal insulation back up paper, asphalt coated	0.04
Asphalt saturated and coated vapor barrier paper	0.2–0.3
15 lb asphalt felt	1.0
15 lb tar felt	4.0
Single kraft, double	31
Liquid-Applied Coating Materials	
Paint-2 Coats	
Aluminum varnish on wood	0.3–0.5
Enamels on smooth plaster	0.5–1.5
Various primers plus 1 coat flat oil paint on plaster	1.6–3.0
Paint-3 Coats	
Exterior paint, white lead and oil on wood siding	0.3–1.0
Exterior paint, white lead-zinc oxide and oil on wood	0.9

Caution—To seal the space between the window and door frames and the wall rough opening properly requires the use of spray-in polyurethane foam (PUR). DO NOT USE FIBERGLASS. Fiberglass will not stop wind from entering the house. When PUR foam is used, **DO NOT** fold the flaps in and around the rough opening. If the flaps are bunched-up in the cavity, a complete sealing with PUR is not possible. Make certain that the flaps are flat against the inside of the rough opening studs. These are the sort of details that should appear on the drawings, and be discussed at the pre-construction conference.

Although housewraps have some ultraviolet protection, they should not be allowed to remain uncovered for weeks or months. Cover the south walls with housewrap first, and install siding as soon as possible.

Are Housewraps Necessary?

Housewraps do reduce air infiltration both into the exterior walls and into the house. Air infiltration into the exterior walls causes windwash, helps promote mold and mildew on interior outside corners, and degrades the performance of the insulation. There is little real difference in performance between the brands; however, there is also little documented evidence to prove their energy savings claims. There have been one or two studies, one by the NAHB which showed a 5 percent reduction in the natural infiltration rate of a house after it was wrapped with Tyvek. But studies also show that just taping or caulking the sheathing joints works about as well as wrapping the entire house.

Housewraps should not be seen as *the* sealing technique. It is just one element in a system that includes the sheathing. A tightly sealed house will show little improvement with housewraps. A leaky house will show considerable improvement. All housewraps will stop most of the air leaks, regardless of advertised claims. Not all building scientists believe housewraps are necessary. Other sealing techniques, they believe, may be just as effective. Any builder using Airtight Drywall Approach (ADA), which uses caulks, gaskets or tapes, will gain little from the use of housewraps. Unfortunately, housewraps have been so effectively marketed that many buyers will not purchase a new house that does not have housewrap.

Wood Siding

One of the materials most characteristic of the exteriors of American houses is wood siding. The essential properties required for wood siding are good painting characteristics, easy working qualities, and freedom from warp. These properties are present to a high degree in the cedars, eastern white pine, sugar pine, western white pine, cypress, and redwood.

Material used for exterior siding should preferably be of a select grade, and should be free from knots, pitch pockets, and wavy edges. The moisture content at the time of application should be that which it would attain in service. This would be approximately 12 percent, except in the dry southwestern states, where the moisture content should average about 9 percent.

Bevel Siding

Plain bevel siding, as shown in Fig. 15–7, is made in nominal 4-, 5-, and 6-inch widths from $\frac{7}{16}$-inch butts, 6-, 8-, and 10-inch widths with $\frac{9}{16}$- and $1\frac{1}{16}$-inch butts. Bevel siding is generally furnished in random lengths varying from 4 to 20 feet. Illustration details are shown in Fig. 15–8.

Drop Siding

Drop siding is generally $\frac{3}{4}$-inch thick, and is made in a variety of patterns with either matched or shiplap edges. Fig. 15–9 shows three common patterns of drop siding that are applied horizontally. Fig. 15–9(A) may be applied vertically, for example, at the gable ends of a house. Drop siding was designed to be applied directly to the studs, and it thereby serves as sheathing and exterior wall covering. It is widely used in this manner in farm structures, such as sheds, and garages in all parts of the country. When used over or when in contact with other material, such as sheathing or sheathing paper, water may work through the joints and be held between the sheathing and the siding.

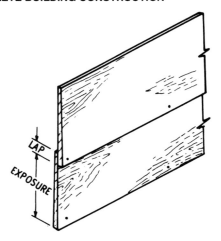

Fig. 15–7. Bevel siding.

This sets up a condition conducive to paint failure and decay. Such problems can be avoided when the side walls are protected by a good roof overhang.

Square-Edge Siding

Square-edge or clapboard siding made of $^{25}/_{32}$-inch board is occasionally selected for architectural effects. In this case, wide boards are generally used. Some of this siding is also beveled on the back at the top to allow the boards to lie rather close to the sheathing, thus providing a solid nailing surface.

Vertical Siding

Vertical siding is commonly used on the gable ends of a house, over entrances, and sometimes for large wall areas. The type used may be plain-surfaced matched boards, patterned matched boards or square-edge boards covered at the joint with a batten strip. Matched vertical siding should preferably not be more than 8 inches wide and should have two 8d nails not more than 4 feet apart. Backer blocks

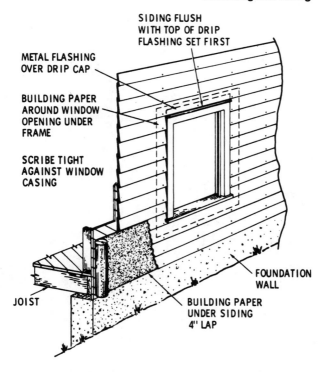

SIDING FLUSH
WITH TOP OF DRIP
FLASHING SET FIRST

METAL FLASHING
OVER DRIP CAP

BUILDING PAPER
AROUND WINDOW
OPENING UNDER
FRAME

SCRIBE TIGHT
AGAINST WINDOW
CASING

FOUNDATION
WALL

JOIST

BUILDING PAPER
UNDER SIDING
4" LAP

Fig. 15–8.   Installation of bevel siding.

should be placed between studs to provide a good nailing base. The bottom of the boards should be undercut to form a water drip.

Batten-type siding is often used with wide square-edged boards which, because of their width, are subject to considerable expansion and contraction. The batten strips used to cover the joints should be nailed to only one siding board so the adjacent board can swell and shrink without splitting the boards or the batten strip.

Plywood Siding

Plywood is often used in gable ends, sometimes around windows and porches, and occasionally as an overall exterior wall covering. The sheets are made either plain or with irregularly cut striations. It can be

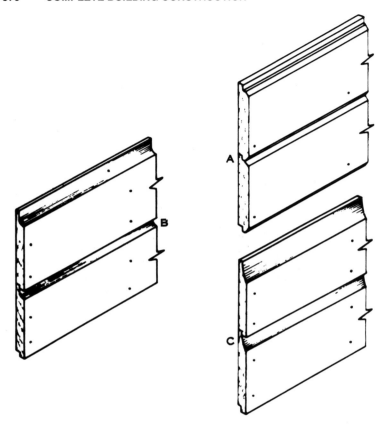

Fig. 15–9. Types of drop siding: (A) V-rustic, (B) drop, (C) rustic drop.

applied horizontally or vertically. The joints can be molded batten, V-grooves, or flush. Sometimes it is installed as lap siding. Plywood siding should be of exterior grade, since houses are often built with little overhang of the roof, particularly on the gable end. This permits rainwater to run down freely over the face of the siding. For unsheathed walls, the following thicknesses are suggested:

Minimum thickness	Maximum stud space
3⁄8 inch	16 inches on center
1⁄2 inch	20 inches on center
5⁄8 inch	24 inches on center

Treated Siding

In a construction situation which allows rainwater to flow down freely over the face of the siding (such as when there is little roof overhang along the sides or gable ends of the house), water may work up under the laps in bevel siding or through joints in drop siding by capillary action, and provide a source of moisture that may cause paint blisters or peeling.

A generous application of a water-repellent preservative to the back of the siding will be quite effective in reducing capillary action with bevel siding. In drop siding, the treatment would be applied to the matching edges. Dipping the siding in the water repellent would be still more effective. The water repellent should be applied to all end cuts, at butt points, and where the siding meets door and window trim.

Hardboard Siding

The manufacturing, basic types, sizes, and finishing of hardboard siding was covered in Chapter 7, **Wood.** The successful performance of hardboard siding is dependent on the quality of installation and moisture.

Moisture—Although hardboard is *real* wood, it is an homogenized wood, and behaves differently than solid natural unprocessed wood. It is vulnerable to moisture. The manufacturing process produces a very dry hardboard. It must be allowed to absorb moisture and stabilize before it is installed. Because it is homogenized wood, it expands more in length than a piece of solid wood siding. Nailed down dry, the hardboard siding will expand and buckle. Most hardboard failures are caused by moisture absorption. The hardboard industry has established a standard of 2.4 inches of expansion for every 50 feet of siding. That much expansion will cause severe buckling, bowing of the siding away from the wall, and pulling the nails right through the siding. The expansion/buckling can exert enough force on the studs to cause cracks in the sheetrock.

The bowing/buckling becomes a vicious cycle. Nails pulled deeper into the siding, unpainted butts, and uncaulked seams leave unprotected wood open to more moisture absorption. More moisture means more expansion, more nail pulling and so on, and on. Often the builder, siding contractor, or both are at fault for butting the ends too tightly, painting improperly, and failing to caulk. Yet failures have resulted even when all the precautions were religiously followed.

Hardboard performs best in the dry regions of the country, and poorly in the hot humid southeastern regions. There have been a lot of problems with the siding in New England. As with any product, there are builders who say they have never had any trouble. Therefore, if hardboard siding is the choice, the manufacturer's instructions must be rigidly followed.

Here are some recommendations for using hardboard siding:

1. Check to see if an instruction brochure came with the siding.
2. Store the siding in a dry unheated building, or under a tarpaulin. This allows the siding to stabilize and minimizes expansion or shrinkage.
3. Break the bundles apart and separate the layers with sticks to allow air to circulate completely around each piece.
4. Keep the bottom of the siding 8 inches above grade.
5. Follow the manufacturer's instructions for spacing between the boards. The spacings can vary from $\frac{1}{16}$ inch to $\frac{3}{16}$ inch. If H-strips are used, an even wider gap may be necessary.
6. Seal the gaps or use H-strips manufactured specifically to seal the gaps.
7. Whether to caulk or use H-strip may be a matter of preference and *looks:* for some, the caulking looks smoother than the H-strips. Caulking can lead to moisture absorption problems if it should fall out when the board expands or contracts. Silicone caulks will not take paint; use urethane or acrylic caulks.
8. Both the inside and the outside corner boards must be thick enough to cover the ends of the siding. Leave an $\frac{1}{8}$-inch gap between the board and trim, and caulk.
9. Follow manufacturer's instructions as to nail type and size. Drive the nails carefully and flush with the surface. Overdriving breaks the surface of the siding and opens it to moisture absorption. Because hardboard is solid material, the nail pushes the siding mate-

rial aside. This material may then be forced to the surface of the board where it forms a little mound. If the nail is overdriven, caulk it. If the mound shows up, caulk it.

10. Do not let the siding age; paint or stain it within the time limit set by the manufacturer. Use only the paint/stain recommended for use on hardboard.

11. Assign someone to inspect the painting/priming for bare spots, called *holidays* by painters.

One solution to some of these problems is to use prefinished siding with hidden nails rather than face-nailed board. This eliminates one moisture entry path.

Wood Shingles and Shakes

Cedar shingles and shakes (Fig. 15–10) are also available. They come in a variety of grades and may be applied in several ways. You can get them in random widths 18 to 24 inches long or in a uniform 18

Fig. 15–10. Wood shingles blend well with stone veneer on this home.

inches. The shingles may be installed on regular sheathing or on an undercourse of shingles, which produces a shadowed effect. Cedar, of course, stands up to weather well and does not have to be painted.

Installation of Siding

The spacing for siding should be carefully laid out before the first board is applied. The bottom of the board that passes over the top of the first-floor windows should coincide with the top of the window cap, as shown in Fig. 15–11. To determine the maximum board spacing or exposure, deduct the minimum lap from the overall width of the siding. The number of board spaces between the top of the window and the bottom of the first course at the foundation wall should be such that the maximum exposure will not be exceeded. This may mean that the boards will have less than the maximum exposure.

Siding starts with the bottom course of boards at the foundation, as shown in Fig. 15–12. Sometimes the siding is started on a water table, which is a projecting member at the top of the foundation to throw off water, as shown in Fig. 15–13. Each succeeding course overlaps the upper edge of the lower course. The minimum head lap is 1 inch for 4- and 6-inch widths, and 1¼-inch lap for widths over 6 inches. The joints between boards in adjacent courses should be staggered as much as possible. Butt joints should always be made on a stud, or where boards butt against window and door casings and corner boards. The siding should be carefully fitted and be in close contact with the member or

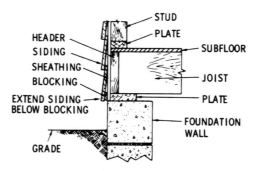

Fig. 15–11.   Installation of the first or bottom course.

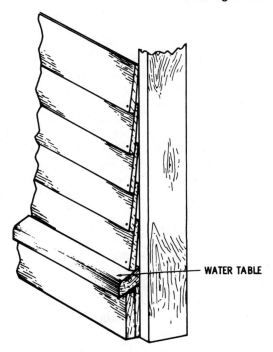

Fig. 15–12. Water table is sometimes used.

adjacent pieces. Some carpenters fit the boards so tightly that they have to spring the boards in place, which assures a tight joint. Loose-fitting joints allow water to get behind the siding and thereby cause paint deterioration around the joints, which also sets up conditions conducive to decay at the ends of the siding.

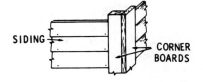

Fig. 15–13. Corner treatment for bevel siding using for corner board.

Types of Nails

Nails cost very little compared to the cost of siding and labor, but the use of good nails is important. It is poor economy to buy siding that will last for years and then use nails that will rust badly within a few years. Rust-resistant nails will hold the siding permanently and will not disfigure light-colored paint surfaces.

There are two types of nails commonly used with siding, one having a small head and the other a slightly larger head. The small-head casing nail is set (driven with a nailset) about $\frac{1}{16}$ inch below the surface of the siding. The hole is filled with putty after the prime coat of paint is applied. The large-head nail is driven flush with the face of the siding, with the head being later covered with paint. Ordinary steel wire nails tend to rust in a short time and cause a disfiguring stain on the face of the siding. In some cases, the small-head nail will show rust spots through the putty and paint. Noncorrosive-type nails (galvanized, aluminum, and stainless steel) that will not cause rust stains are readily available.

Bevel siding should be face-nailed to each stud with noncorrosive nails, the size depending upon the thickness of the siding and the type of sheathing used. The nails are generally placed about $\frac{1}{2}$ inch above the butt edge, in which case it passes through the upper edge of the lower course of siding. Another method recommended for bevel siding by most associations representing siding manufacturers, is to drive the nails through the siding just above the lap so that the nail misses the thin edge of the piece of siding underneath. The latter method permits expansion and contraction of the siding board with seasonal changes in moisture content.

Corner Treatment

The method of finishing the wood siding at the exterior corners is influenced somewhat by the overall house design. Corner boards are appropriate to some designs, and mitered joints to others. Wood siding is commonly joined at the exterior corners by corner boards, mitered corners, or by metal corners.

Corner Boards—Corner boards, as shown in Fig. 15–13, are used with bevel or drop siding and are generally made of nominal 1- or 1¼-inch material, depending upon the thickness of the siding. Corner boards may be either plain or molded, depending on the architectural treatment of the house. The corner boards may be applied vertically against the sheathing, with the siding fitting tightly against the narrow edge of the corner board. The joints between the siding and the corner boards and trim should be caulked or treated with a water repellent. Corner boards, and trim around windows and doors, are sometimes applied over the siding, a method that minimizes the entrance of water into the ends of the siding.

Mitered Corners—Mitered corners, such as shown in Fig. 15–14, must fit tightly and smoothly for the full depth of the miter. To maintain a tight fit at the miter, it is important that the siding is properly seasoned before delivery, and is stored at the site so as to be protected from rain. The ends should be set in white lead when the siding is applied, and the exposed faces should be primed immediately after it is applied. At interior corners, shown in Fig. 15–15, the siding is butted against a corner strip of nominal 1- or 1¼-inch material, depending upon the thickness of the siding.

Metal Corners—Metal corners, as shown in Fig. 15–16, are made of 8-gauge metals, such as aluminum and galvanized iron. They are used with bevel siding as a substitute for mitered corners, and can be purchased at most lumber yards. The application of metal corners takes less skill than is required to make good mitered corners, or to fit the siding to a corner board. Metal corners should always be set in white lead paint.

Other Siding

Aluminum Siding

The most popular metal siding is aluminum. It is installed over most types of sheathing with an aluminum building paper (for insulation) nailed on between the sheathing and siding. Its most attractive

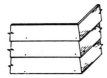

Fig. 15–14. The mitered corner treatment.

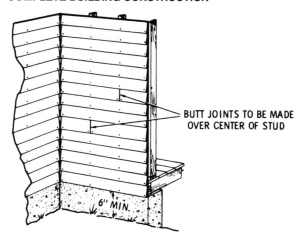

BUTT JOINTS TO BE MADE
OVER CENTER OF STUD

6" MIN.

Fig. 15–15.   Construction of an interior corner using bevel siding.

characteristic is the long-lasting finish obtained on the prefinished product. The cost of painting and maintenance has made this type of siding doubly attractive. Aluminum siding can be installed over old siding that has cracked and weathered, or where paint will not hold up. Installation instructions are furnished with the siding, which is available with insulation built on and in various gauges.

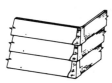

Fig. 15–16.   Corner treatment for bevel siding using the corner metal caps.

Fig. 15–17. **Vinyl siding comes in a variety of colors and is easy to keep clean.** *(Courtesy Vinyl Siding Institute)*

Vinyl Siding

Also popular is vinyl siding (Fig. 15–17). This comes in a wide variety of colors, textures, and styles. As with aluminum siding, the big advantage of vinyl siding is that it does not need to be painted and will not corrode, dent, or pit. When it is very cold, it is relatively susceptible to cracking if hit.

CHAPTER 16

Windows

Basic Considerations and Recent Developments

Windows make up between 10 to 45 percent of the total wall area of a house. Views and architectural considerations determine glass areas, but the building codes dictate the minimum allowable amount of glass area. The *BOCA* code requires that any space intended for human occupancy must have a glass area equal to 8 percent of the floor area. State energy codes, on the other hand, dictate the maximum amount of glass area allows.

The casement window, commonly called a *contemporary window,* originated in medieval Europe about 500 AD. The sash or double-hung window, also called guillotine, was invented by the Dutch in the seventeenth century. In 1860, a Canadian engineer named Henry Rutton mentions his idea of double glazing glass in his book, *Ventilation and Warming of Buildings.* In 1865 the first multiple-pane window was patented by Thomas Sutton. In the 1950s, sealed double-paned insulating windows saw increasing sales.

Prior to the 1972 OPEC embargo, most of the glass in American homes was single-glazed. Sky-high energy costs, however, soon forced the changeover to insulated glass (Fig. 16–1). By 1988 more than 80 percent of the residential windows sold in the USA were double-

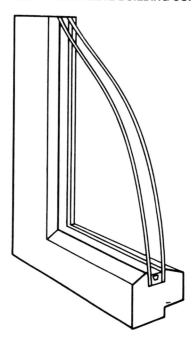

Fig. 16–1. A double-insulated window. The dead air space between the sandwich of glass helps save on fuel.

glazed. Triple-glazed windows amounted to 15 percent of the windows sold, and quadruple-pane windows made an appearance.

Perhaps no area of research into building products has been more successful than window research. The research continues even though the cost of fossil fuels has dropped considerably. Concerns over global warming, renewed interest in energy conservation, and tougher energy codes all help to keep the research going.

Multiple Glazings

A single pane of glass admits nearly all the light rays that directly strike the glass. About 85 percent of the solar energy is passed through. Some of the heat is reflected, some is absorbed, but most of the energy passes through and is absorbed by furniture or other thermal mass. The R-value of a single-glazed window is about R-1, and is due largely to the air films on the surface of the inner and outer panes.

Heat Transfer

Heat is energy, not a substance. Therefore, it can flow or be transferred from one place to another. There are three basic ways heat can be transferred: convection, conduction, radiation.

Convection. A forced hot-air system is an example of convection. The hot air is bodily moved from the furnace to the various rooms of a house. Convective loops are caused by temperature differences. A sea breeze is an example of a convective loop. As the land is warmed by the sun, the warm air, which is not as heavy as the cold air, rises and leaves an empty space, a vacuum. The colder air on the water moves in to fill up the empty space. As long as the warm air rises, there will be a continuous flow, or convective loop.

Conduction is the transfer of heat through solid objects. Anyone who has grabbed the handle of a hot metal pan knows what conduction is. It is the transfer of heat between bodies in direct contact.

Radiation. The sun radiates its heat energy through the vacuum of space. The wood stove radiates its heat energy throughout a room. Heat is a form of electromagnetic energy, like radio waves, which travels through space and fluids until it is absorbed by a solid, or reflected by a radiant barrier such as silver or aluminum foil.

According to the second law of thermodynamics, heat is cold-seeking. The radiation is always from a warm object to a cold object. You feel cold standing next to a window because your warm body is radiating heat toward a cold body—the window. Very often convection, conduction, and radiation are working at the same time. Multiple glazings increase the thermal resistance by trapping air between two or more panes of glass. The R-value ranges from R-0.7 to R-1.0 in each space. Heat is transferred across the air spaces by infrared radiation and by conduction. The heat from the warm inner pane is radiated across the air space, absorbed by each pane of glass and reradiated outward and inward. The greatest heat loss is to the exterior cold panes. Although the air is insulation, it also conducts the heat across the air space.

Because the air between the panes is the insulation, increasing the air space increases the R-value. But beyond a certain thickness, ¾ inch, no increase in R value is possible. This is because of convection which carries the heat from the inner pane to the outer pane. A window with a ½-inch air space has an R-value of R-2. Triple-glazed glass has an R-value between R-2.5 and R-3.5, depending on the thickness. A triple-glazed unit with only a ¼-inch air space on each side carries a premium price without a premium R-value. Triple glazing is not automatically better—it depends on the thickness of the air space.

Emissivity

Emissivity is the relative amount of radiant energy a surface gives off, at some temperature, compared to an *ideal* black surface at the same temperature. The emission is a coefficient—a number—that ranges from zero (no emission) to one (emitting as well as a blackbody at the same temperature). A blackbody—so-called because of its color—is an object that absorbs and reradiates all of the radiation that strikes it. It is a perfect absorber and perfect emitter of all radiation. A perfect blackbody is an *ideal*, a *concept;* no perfect blackbody exists. It has a surface emissivity of 1.0. We must deal with real objects. Many real objects such as building materials have high emissivities: anodized aluminum has a thermal emissivity of 0.65, pure aluminum 0.1, ordinary float glass, 0.88. Shiny surfaces such as polished silver (0.02) have low emissivities.

Any real object that has low emissivity will also have low radiation heat loss. Because ordinary float glass has a high emissivity (0.88), a large amount of heat energy will be transferred across a thermopane unit because of the long wave heat-carrying infrared energy from the inner pane to the outer pane.

Low-E Glazing

The OPEC crisis had a far greater impact on Europe than on the United States. Glass manufacturers there spent considerable time and money researching the heat-insulating characteristics of low-emissivity coatings. By the late 1970s, they were producing reasonably clear coatings on clear glass for residential windows. The research into low-emissivity/high-light transmittance began in the United States in the early 1980s.

Low-E is actually a microscopically thin metal coating applied to the exterior surface of the inner pane of a double glazed window or door. There are two methods of applying low-emissivity coatings to glass: vacuum sputtering and pyrolytic deposition. Pyrolytic coatings are produced by applying hot metal oxides to hot glass. Vacuum-sputtered coatings are deposited as thin films, from 20 to 200 atoms thick, in a vacuum chamber. The sputtered low-E is referred to as *soft-coat low-E,* and the pyrolytic coating as *hard-coat low-E.*

Low-emissivity coatings are filters that allow visible light to pass through, but block, by reflection and absorption, the ultraviolet and near-infrared solar radiation. The short wave radiation is converted into long wave radiation (heat) when it is absorbed by furniture or other solid objects. Because they are good reflectors and poor radiators of long-wave heat-carrying energy, much of the heat is kept in the interior of the building. It is also effective at reducing cooling loads by reflecting the heat radiated from concrete sidewalks or asphalt roads. The low-E has a slight edge over even triple-glazed windows: Triple has an R-value of R-2.9; Low-E, an R-3.2. Because the low-E coating reflects ultraviolet, it reduces color fading in carpets, curtains, and other fabrics. Even though any window is the first condensing surface, low-E windows have a warmer surface, and condensation is less likely.

Gas-Filled Windows

For some years European window manufacturers have offered gas-filled windows. American window manufacturers were slow to offer them, but they are now a standard item in this country.

The space between panes is filled with a gas rather than air. Argon, sulfur hexafloride, and carbon dioxide are the three commonly used gases; Hurd uses Krypton and Argon for its INSOL-8® window. Some window manufacturers use carbon dioxide gas, but these are specialty windows and the R-value is no better than double-glazed glass. The gases are denser and lower in thermal conductance (higher R-value) than air. Because the gas is heavier than air it is less likely to move within the air space. The performance of low-E windows is considerably improved when gas is added. The gas-filled low-E window has an R-value of R-4.

Gas Leakage—American manufacturers hesitated to use gas because of leakage concerns. Swedish manufacturers admit that within 10 to 15 years most of the gas will seep out as the seals degrade. There are no industry standards in the United States for gas leakage from windows. Based on testing in this country and European researchers' experience, a leakage rate of 1 percent per year can be expected. The rate would decrease with time because the concentration of argon decreases. In 20 years, the R-4 value would be R-3.9, a loss of 2.5 percent. Although there are no gas leakage standards, there are accelerated aging tests of the edge sealant materials: ASTM E773 and E774. Glass units are rated A, B, C. When shopping for gas-filled windows, purchase the A-rated window.

Heat Mirror

HEAT MIRROR® is the registered trademark of Southwall Technologies, Inc. Heat Mirror is a plastic film which has a low-emissivity metal coating applied to it in a vacuum. The film is not applied directly to the glass, but is suspended between the two panes of glass (Fig. 16–2). The long-wave infrared radiation from the inner pane of glass is reflected back by the Heat Mirror. Heat loss is reduced and the effective R-value increased (Fig. 16–3). Windows with Heat Mirror have an R-value, with a ½-inch air space, of R-4.3. Hurd's new InSol-8 achieves an R-8 by using two sheets of Heat Mirror. Hurd claims its InSol-8 has the highest R-value of any residential window.

Visionwall

VISIONWALL® is a product of a Swiss company named Geilinger. The window seems to be a step backward because of its aluminum frame. Although it does use two layers of Heat Mirror, the air spaces between layers is plain air. It has an *overall* R-value of R-6.3 (Fig. 16–4). In order to deal with a seeming contradiction—the Hurd R8 is higher than the Visionwall R-6.3—it is important to understand how R-values are measured.

Calculating R Values

The glass center area makes up between 65 to 85 percent of the area of the window. The traditional method of measuring the R-value of a window was at the center of the glass. An adjustment was made depending on whether a wood frame or an aluminum frame was used. Wood frames raised the overall R-value, because wood has a higher R-value than glass. Aluminum frames, with a lower R-value than glass, lowered the overall value.

Fig. 16–2. Heat Mirror® film suspended between two panes of glass. *(Courtesy of Hurd Millwork Company, Inc.)*

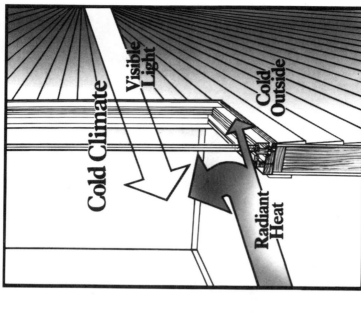

Figure A. Hurd Heat Mirror 66 Sunbelter clearly shades interior.

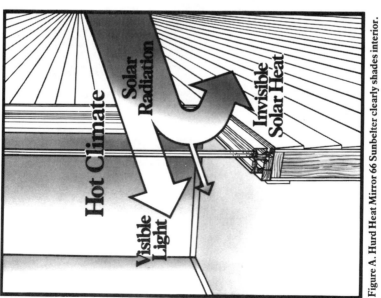

Figure B. Hurd Heat Mirror 88 reflects back escaping heat.

Fig. 16–3. Two Heat Mirror® options. *(Courtesy of Hurd Millwork Company, Inc.)*

Fig. 16–4. Visionwall® windows with two Heat Mirror films suspended between two panes of glass. *(Courtesy Visionwall Technologies)*

The spacer that keeps the panes apart is a hollow, desiccant-filled, aluminum-edge spacer. The edge of the glass is considered to be the exterior 2½-inch area around the frame.

According to Hurd's fact sheet on the InSol-8,

All calculations based on center-of-glass values for 1-inch Hurd InSol-8 windows within the Superglass System™ with Heat Mirror Film. All data were calculated using Windows 3.1 Computer Program and standard ASHRAE winter conditions of 0°F outdoor and 70°F indoor temperatures with an 15 mph outside wind.

Published R-values for manufactured windows are usually an *average* of the frame areas and the glass, calculated, as we have seen, according to ASHRAE guidelines. The glass edge area, window frame, and glass are the total unit. The R-value quoted by manufacturers is the center-of-glass R-value. The R-value of the total unit is usually lower.

Hurd's published R-value for the InSol-8 center of glass is R-8. The total unit R-value is R-4.6. The Visionwall aluminum frame is thermally broken with structurally-reinforced nylon spacers. Rockwool or polystyrene insulation is sandwiched in between the spacers. The center-of-glass area, frame area, and edge area all have about the same R-value. As of this writing, the Visionwall window is the most energy-efficient window on the market.

Some European window manufacturers are putting the glass edge deeper into the window sash as a way to reduce the edge effect. In the United States the Alaska Window Company provides a one-inch-deep channel for the window edges. Some manufacturers are replacing the aluminum spacer with lower conductivity materials, such as fiberglass spacers; others are using the Tremco Swiggle Strip. Swiggle Strip is a thin, corrugated aluminum spacer embedded in a black butyl tape. The greatly reduced amount of aluminum results in the Swiggle Strip conducting less heat. The interior of the glass surface now has a higher temperature during cold weather. This reduces condensation at the glass edge and results in a slight overall increase in the window R-value. All Peachtree windows will be available with Swiggle Strip and low-E, argon-filled glass. Alcoa's MAGNA-FRAME® series of vinyl windows will be equipped with the Swiggle Strip which Alcoa calls WARM-EDGE®, and WARM-EDGE PLUS®. Hurd uses a nonmetallic spacer in its InSol-8 window. For a flat $3.00 charge, Weathershield will install the insulated edge spacer in any window as a special order.

Another solution to the window edge problem is to use Owens/Corning fiberglass windows. As of this writing, these are the only thermally improved window frames available. The hollow fiberglass frame is insulated with high-density fiberglass. The fiberglass frame is dimensionally more stable than vinyl frames, and unlike the vinyl frame, the fiberglass frame can be painted.

Aerogels

Although aerogels are not new, their use in windows is. They are transparent, and researchers are trying to suspend aerogels between two panes of glass, create a vacuum inside, and produce an R-20 window. Aerogel windows may be available for residential use within the next 5 years.

Switchable Glazings

Switchable glazings are already in use at some airports and in GE's Living Environments House (better known as the Plastic house) in Pittsfield, Massachusetts. Simply by flipping a switch, the glazing turns from transparent (clear) to opaque, as though a shade were pulled down. Curtains, blinds, louvers are no longer necessary. Called an electrochromic glazing, it is either a five-layer laminate or a solid state design.

Transparent metal oxide films are sandwiched between two panes of glass. When voltage is applied, the metal oxides change color. When the switch is turned off, the glazing becomes clear again. The idea behind these *smart glazings* is to allow one to tune the windows to reflect or transmit different wavelengths. Winter light and heat are admitted into the interior, but heat attempting to get out is reflected back into the room. In the summer, 95 percent of the light is allowed to enter, but the unwanted heat is reflected back into the atmosphere. Car windows could be made to do the same thing by keeping the hot sun out as the car sits parked in the sun for 8 hours.

Installing Windows

It makes absolutely no sense to buy high-performance windows and degrade their performance by stuffing fiberglass between the window jamb and the wall rough opening. Fiberglass is not an air retarder; if it were, it would not be used in furnace filters. Under continuous pressure differential, air will leak through it no matter how tightly it is stuffed into the opening. Skeptics who believe to the contrary should have a blower-door test performed. Otherwise, stand next to the window on a windy day and feel the air entering through the fiberglass around the frame and exiting around the molding.

Use care when sealing the openings with sprayed-in PUR foam. Do not put the nozzle in the cavity and allow the foam to flow until the cavity is filled. The expansion of the foam can exert enough pressure against the jambs to make opening them difficult. Overfilling will leave foam all over the window frame. Use a non-expanding foam such as HANDI-FOAM™ manufactured by Fomo Products, Norton, Ohio. Their 10-pound Professional Unit will cover 1700 lineal feet. A one-inch thickness has an R-value of R-5, and is completely cured and expanded in 24 hours.

The window/door rough openings should be made larger by ½ to ¾ inch to allow the foam to enter readily and make a good seal. Mark the drawings to indicate a wider than normal opening, and discuss this at the pre-construction conference.

Purchasing Windows

Low-E and Heat Mirror make great sense in cold climates. However, only an argon-gas-filled Heat Mirror window should be used on the north side; otherwise it may actually gain heat. This allows the use of more glass on the north side without paying a large heating bill penalty. The higher R-value means a warmer window and less chance of condensation on the glass. When looking at the manufacturer's published R-values, look for a statement such as the one quoted from Hurd's literature. The R-values should be based on *Windows* 3.1 software. The new 4.1 software has been released. Most manufacturers now list two R-values: one for the center-of glass, and one for the overall unit performance. The overall R-value is the important value.

Air Leakage

Fixed windows are the tightest. Next is the casement, followed by the sliders, and the least tight of the group are the double-hungs. Do not place too much value on the listed air infiltration values. The number may refer to only a single window that was tested. Andersen seems to be the only manufacturer that *randomly* takes a window from the production line and has it tested. It is not a selected window. But even that window will not be quite the same when it is received on site. The value of the published infiltration data is that it allows a comparison of the air leakage differences between window types of the same manufacturer. Look at the seals and the general condition of the window.

In northern climates, select windows that allow the maximum solar gain. One such brand is LOF's Energy Advantage glass. In southern climates look for *selective* glazings that reduce solar gain without significantly reducing the amount of light.

Look for the *shading coefficient* (SC) which tells how much solar heat the glass transmits. The higher the SC, the more solar heat the glass allows to pass through. In northern areas the SC should be at least 0.80. In southern climates anything below 0.60 is desirable, as long as there is no substantial reduction in the amount of light allowed in. Look for a window that has the highest Luminous Efficacy Constant, preferably higher than 1.0.

The three main window types are gliding, double-hung, and casement, but there are also awning, bow, and bay windows (Fig. 16–5). Basic windows consist essentially of two parts, the frame and the sash. The frame is made up of four basic parts: the head, two jambs, and the sill. Good construction around the window frame is essential to good building. Where openings are to be provided, studding must be cut away and its equivalent in strength replaced by doubling the studs on each side of the opening to form trimmers and inserting a header at the top. If the opening is wide, the header should be doubled and trussed. At the bottom of the opening, a header or rough sill is inserted.

Window Framing

The frame into which the window sash fits is set into a rough opening in the wall framing, and is intended to hold the sash in place (Fig. 16–6).

Double-Hung Windows

The double-hung window is the most common kind of window. It is made up of two parts—an upper and lower sash, which slide vertically past each other. An illustration of this type of window, made of wood, is shown in Fig. 16–7. It has some advantages and some disadvantages. Screens can be installed on the outside of the window with-

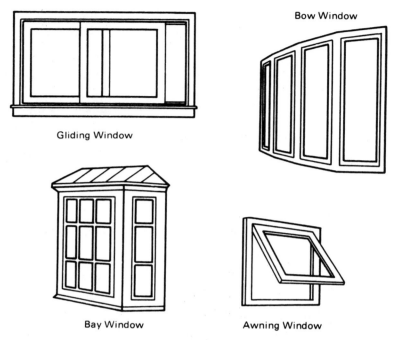

Fig. 16–5. Various kinds of windows.

out interfering with its operation. For full ventilation of a room, only one-half the area of the window can be utilized, and any current of air passing across its face is, to some extent, lost in the room. Double-hung windows are sometimes more involved in their frame construction and operation than the casement window. Ventilation fans and air conditioners can be placed in the window when it is partly closed.

Hinged or Casement Windows

There are basically two types of casement windows—the out-swinging and the inswinging. These windows may be hinged at the side, top, or bottom. The casement window which opens out requires the screen to be located on the inside. This type of window, when closed, is most efficient as far as waterproofing. The inswinging, like double-

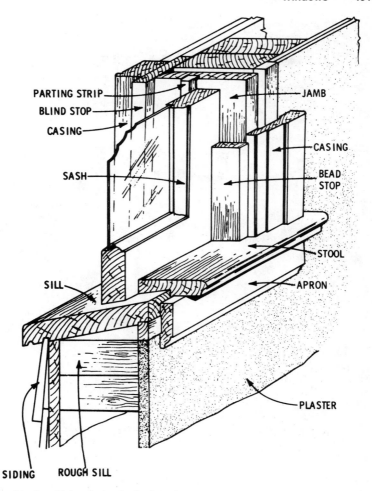

PARTING STRIP

BLIND STOP

CASING

SASH

SILL

JAMB

CASING

BEAD STOP

STOOL

APRON

PLASTER

SIDING   ROUGH SILL

Fig. 16–6.   Side view of window frame.

hung windows, are clear of screens, but they are extremely difficult to make watertight. Casement windows have the advantage of their entire area being opened to air currents, thus catching a parallel breeze and slanting it into a room. Casement windows are considerably less complicated in their construction than double-hung units. Sill construction is very much like that for a double-hung window, however, but with the

Fig. 16–7. The popular double-hung window.

stool much wider and forming a stop for the bottom rail. When there are two casement windows in a row in one frame, they are separated by a vertical double jamb called a mullion, or the stiles may come together in pairs like a french door. The edges of the stiles may be a reverse rabbet, a beveled reverse rabbet with battens, or beveled astragals. The

Fig. 16–8. A casement window.

battens and astragals ensure better weathertightness. Fig. 16–8 shows a typical casement window with a mullion.

Gliding, Bow, Bay, and Awning Windows

Gliding windows consist of two sashes that slide horizontally right or left. They are often installed high up in a home to provide light and ventilation without sacrificing privacy.

Awning windows have a single sash hinged at the top and open outward from the bottom. They are often used at the bottom of a fixed picture window to provide ventilation without obstructing the view. They are popular in ranch homes.

Bow and bay windows add architectural interest to a home. Bow windows curve gracefully, while bay windows are straight across the middle and angled at the ends. They are particularly popular in Georgian and Colonial homes.

CHAPTER 17

Insulation

The 1972 OPEC crisis resulted in an increasing emphasis on energy conservation in general, and insulation in particular. An incredible number of schemes were proposed to free us of OPEC. At every corner Americans were confronted with *the* solution to the energy crisis: woodstoves, active solar systems, Trombe walls, passive solar, mass and glass, urea formaldehyde, and so on, without end.

Both active and passive solar enthusiasts stressed active solar systems, or passive solar systems incorporating large amounts of glass and thermal mass, such as the Trombe wall, named after a Frenchman who did not invent it. Both of these approaches emphasize the active or passive solar system; the structure was ignored. In early 1970s, researchers in Saskatchewan, Canada, and Pepperell, Massachusetts, were working independently on a *systems* approach that stressed the importance of the structure: so-called superinsulation. The essence of this approach was that the house was a system of elements, of which insulation was a part. The active/passive solar philosophy saw the house as just something to which one added solar collectors, or mass and glass.

In solar houses only *the solar system* matters. *In a Micro-Energy System House (MESH), the* house *is the* only *thing that matters.*

Unfortunately, the term superinsulation should never have been used. While it is true that both the Saskatchewan Conservation House and the Leger House are superinsulated houses, *insulation was only one of the many techniques used* to give these houses their unbelievable energy and comfort performance. The term *superinsulation* forces one to concentrate on insulation to the exclusion of everything else. Insulation is important, but so are all the many other things that go into making a house comfortable, attractive, and energy efficient.

Types of Insulation

There are seven generic types of insulation:

1. Mineral fiber (glass, rock, and slag),
2. Cellulose,
3. Cellular plastics,
4. Vermiculite,
5. Perlite,
6. Reflective insulations,
7. Insulating concrete.

Mineral Fiber

Mineral wool is a generic term which includes fiberglass and rock and slag wool. Rock and slag wools were the first insulations manufactured on a large scale. Natural rocks or industrial slags were melted in a furnace, fired with coke, and the molten material was spun into fibers and made into felts or blankets. The fibers are usually coated with plastic so they do not touch. The fibers are packed or arranged in such a way so that small air pockets are formed. This composite is resistant to heat flow and is now called an insulator. By the middle of the 20th century, mineral wool was being used in houses and industrial/commercial buildings.

Fiberglass was first commercially developed in the United States in the 1930s by Owens-Illinois and Corning Glass Company. Owens and Corning later formed Owens Corning Fiberglass Company, which developed fiberglass insulation in the 1940s and 1950s. Until 1950, Owens Corning was the only manufacturer of fiberglass insulation in the United States.

Batts—Mineral fibers are available as batts, as loose fill, or as a spray-in insulation. Batts are sold in standard 15- and 23-inch widths, for 16-

and 24-inch oc framing. The widths for steel studs are 16 and 24 inches. The widths of attic batts are 16 and 24 inches for 16- and 24-inch oc framing. Why? Bottom chords and the tops of ceiling joists left uncovered are exposed to attic air, which is close to outside air temperature. This increases the heat loss through the framing members. Research by Owens Corning shows that this can result in a 20 percent reduction in the overall R-value. The narrow batts do not cover the framing, but the full-width batts do. Discuss this with your insulation contractor at the pre-framing conference. Of course, blown-in cellulose would not only cover these framing members—except for the vertical support members of the truss—but provide a better seal and thus a higher R-value.

High Density Batts—Fiberglass batts are usually manufactured at a density of 1 lb/ft^3. The R-value at this density is about R-3.1 per inch. It varies with the product and the manufacturer. The standard 3½-inch batt has an R-value of R-11, but its density is only 0.5 lb/ft^3. The R-13 batt, 3⅝ inches, has a density of 1.0 lb/ft^3. The R-value of fiberglass depends primarily on density: the higher the density, the higher the R-value. Warm-N-Dri at one inch has an R-value of 4.2, but its density is 6.9 lb/ft^3. The densities can be increased to produce higher R-values, but at a considerable increase in the amount of glass, and cost.

High density, low cost fiberglass insulation is now available for residential use: R-15 at 3½ inches; R-21 at 5½ inches; and R-30 at 8¼ inches. The R-19, 6½ inch insulation used in 2 × 6 stud walls, had to be compressed to 5½ inches. This reduced its R-value to R-18.

Caution—Compressing a fiberglass batt *increases* its R-value per inch. But the overall R-value is less because the thickness has been decreased. Recent tests at Oak Ridge National Laboratory showed that when a high density R-21, 5½-inch batt is shoved into a 2 × 4 cavity 3½ inches deep its R-value decreases to R-15.

The new R-21, 5½-inch batt does not have to be compressed to fit the 2 × 6 cavity. The 5½-inch R-19 batts were intended for attic insulation. The high density batts are offered by Owens Corning, Certainteed, and Johns Manville.

Spray-in Mineral Fiber—Ark-Seal International of Denver, Colorado, sells equipment for the installation of Blown-in-Blanket (BIB) insulation. The first BIB machine was sold in 1979. Loose-fill fiberglass, rockwool, or cellulose insulation is mixed with a moist latex binder—actually a glue—and installed into the wall cavity, behind a previously-installed nylon netting (Fig. 17–1). The insulation *moves* around pipes, work boxes, and wiring. When dry, it forms a continuous one-piece blanket of insulation. BIB is similar to wet spray, but it bridges the gap between batts and wet spray. Wet spray insulations are applied to a substrate such as plywood or OSB. The BIB system uses less moisture because it is not glued to a substrate, it just assumes the shape of the cavity. The binder serves to give consistency to the insulation. Unlike wet spray, the BIB adhesive is applied inside the nozzle. Third-party

Fig. 17–1. BIB insulation being installed behind the nylon netting.
(Courtesy Ark-Seal International)

testing by a number of private and government laboratories have established an R-value per inch of 3.8 to 3.9, which is just below their claimed R-4 per inch. Although the R-value varies with density, the densities used by BIB installers keep the R-value at 3.9 per inch.

A new Ark-Seal product called Fiberiffic is designed to eliminate the nylon netting. A special nozzle is used to foam the adhesive as it mixes with the insulation. The foamed fiberglass has a consistency between bread dough and shaving cream. Because it does not stick to walls, it is trowelled in. A special trowel the width of the studs and 12 inches high, is held against the bottom of the studs. The foam is injected behind the trowel, which is slowly moved up as the wall is filled with foam. This foam is fast-drying and remains in place as the trowel is moved up. When completely dried, the foam disappears and the *batt* has the same consistency as the BIB insulation.

Spray-Applied Rockwool—American Rockwool of North Carolina, has introduced a spray-applied rockwool insulation system called F.A.T.S: Fire Acoustical Thermal System.

Although it is just plain rockwool mixed with a wetted adhesive, it does not require special machinery as does the BIB system. Conventional blowing machines and nozzles are used. American Rockwool recommends the use of Ultra-Lok 40–0871 adhesive which gives the wall an Underwriter Laboratory (UL) Class A fire rating. Otherwise any latex adhesive can be used. The liquid adhesive is mixed at the rate of 1 gallon per 30 pound bag which gives it a moisture content of about 28 percent on a dry weight basis. Netting is not required.

When installed to the recommended 4 lb/ft^3 density, the R-value is approximately 3.8. R-value varies with density and ranges between 3.5 and 3.9.

One of the problems with any wet-spray insulation is the question of moisture. How long will it take to dry out the insulation? Will moisture problems result if a VDR is used? The F.A.T.S uses less water than cellulose, but more than the BIB system. A spokesperson at American Rockwool says the rockwool should dry out in 24 to 48 hours.

Loose-Fill Insulations—Owens Corning, Certainteed, and Johns Manville manufacture loose-fill, or blowing insulation. All loose-fill insulations settle, some more than others. Fiberglass settles between 0 and 8 percent when it is installed at or above the label density. The loss of R-value due to the settling is small, even if the settling were 8 percent. Fiberglass loses 0.5 percent in R-value for every 1 percent loss in thickness due to settling. Therefore, an 8 percent loss would reduce the R-value by only 4 percent.

Oak Ridge National Laboratory (ORNL) Oak Ridge, Tennessee, has a Large Scale Climate Simulator (LSCS) used for the dynamic testing of whole roof systems (Fig. 17–2). The LSCS can be used as an environmental chamber or as a guarded hot box for testing insulations. It operates over a temperature range of −40 degrees Fahrenheit to 200 degrees Fahrenheit. The range of climatic conditions found in the United States can be created in the upper chamber, and a wide range of indoor conditions in the lower chamber. Although originally designed to test low slope roofs, higher pitched roofs can be tested. It is also used to test the performance of whole attic insulation systems.

One such testing of Owens Corning's THERMACUBE® found that at very low temperatures, −18 degrees Fahrenheit, the insulation lost 50 percent of its apparent R-value. However, even when the attic temperature was much higher—20 degrees Fahrenheit—the effective R-value was reduced by 40 percent from its nominal value. Therma-

Fig. 17–2. ORNL large scale climate simulator. *(Courtesy Oak Ridge National Laboratory)*

cube was blown into the attic space to a nominal R-20. The chamber's temperature was varied from –18 degrees to 145 degrees Fahrenheit as the R-value of the ceiling system was measured.

As the air in the attic space cools down, it reaches a critical temperature—about 50 degrees Fahrenheit—at which point attic air convects into and within the insulation. Infrared scans of the top of the insulation and sheetrock verify this. The loss of R-value is not brand specific. Tests at the University of Illinois revealed that in one test attic blown with 14 inches of Certainteed's INSUL-SPRAY III®, the R-value dropped over 50 percent at an attic air temperature of 10 degrees Fahrenheit. The cause was convective air loops. All loose-fill fiberglass will suffer substantial R-value loss due to low temperatures.

Is This a Problem with Cellulose?—Cellulose, installed to R-19 at a density of about 2.4 lbs/ft^2, was subjected to the same test. The lower chamber temperature was maintained at 70 degrees Fahrenheit, while the upper chamber temperature was varied from 40 degrees to –18 degrees Fahrenheit. The R-value of the cellulose did not drop even at 18 degrees below zero. Even at different densities, there was no indication of convection within the cellulose.

Caution—Only when the attic temperature drops below 30 degrees Fahrenheit does the R-value of loose-fill fiberglass insulation start dropping. The large drop in R-values happens at 10 degrees and lower. In areas where 0 degrees Fahrenheit happens one or two days a year, the loss is not important. Therefore, in milder regions these findings can be ignored. In the more northerly regions where daily average winter temperature is 0 degrees or lower, loose-fill fiberglass may not be the choice.

Solutions—There are a number of ways of dealing with this—which method is used should be based on Value Engineering. Choose the solution with the lowest labor and material cost.

1. Choose fiberglass batts instead of loose fill. There is no evidence of convection in batts even at low temperatures.
2. Cover the loose-fill fiberglass with a convective blanket. This was one of the intended uses of Tyvek back in the late 1960s.
3. Cover the loose fill with a 1.0 lb/ft^3 high density fiberglass blanket.
4. Blow in loose-fill cellulose.

ORNL tried some of these solutions. All of the perforated convection blankets, such as ATTIC SEAL® do not work very well. While the Tyvek stopped attic air from penetrating the insulation, there was still air movement within the insulation. Another blanket of high density fiberglass between perforated polyethylene film worked very well, as did a high density fiberglass blanket without the poly. Loose-fill fiberglass will work well in cold climates, *if it is covered* with an effective convective blanket. The negative to the 4 ORNL solutions is the material and labor costs: how easy is it to install any kind of blanket over 15 inches of loose-fill insulation? For more information see ASTM publication STP 1116.

Finally there is the issue of the safety of fiberglass insulation. Richard Munson through his Victims of Fiberglass nonprofit organization claims that it is carcinogenic. His critics, the fiberglass manufacturers, say that one should be suspicious of Munson because he is in the cellulose manufacturing business. Naturally, the fiberglass interests claim that there is no evidence, or that the evidence is inconclusive or misinterpreted, or not significant. However, remember the manufacturers' caution: wear gloves, protective glasses, and face masks when handling fiberglass.

There is a public relations battle going on between fiberglass and cellulose manufacturers: the alleged fire hazards of cellulose on the one hand, and the claimed cancer-causing fiberglass fibers, on the other. Cellulose is not, of course, without its potential for causing health problems. It is not the homeowner who may be exposed to possible dangers from the borax in cellulose; it is the installer who faces continuous exposure.

Cellulose

Cellulose insulation has one of the highest R-values of common insulations. The patent for cellulosic fiber was issued in the 1800s, but not until the 1950s was cellulose established in the marketplace. Cellulose is made by converting old newspapers, other paper products, virgin wood, or cotton textiles into fibers. The newsprint is fed to a hammer mill or cutter/shredder mill where it is shredded and pulverized into a fibrous, homogeneous material. Fire-retardant chemicals such as borax, or boric acid are added, blended with the cellulose fibers, and the product is then bagged. Virgin wood may also be used. This is the dry process and the most common method of producing cellulose insulation.

There are two wet processes. In the first, liquid fire retardant is sprayed, misted, or sprinkled onto the raw cellulose fibers just before the mixture is fed to the finish mill. The airstream and the short duration heat buildup during the final process evaporate the excess moisture.

Wood-derived insulations are processed two ways. In the first, a rotating-disk pressurized refiner produces chips. The chips are heated with 320-degree-Fahrenheit saturated steam and turned into fibers. The 65 percent moisture content is reduced in a forced-air oven operating at 220 degrees Fahrenheit. The dried insulation is placed in a rotating drum sprayer and wet borax/boric acid is sprayed on before bagging.

In the second system, papermaking machinery is used. The wood chips are made into a pulp slurry. Compression is used to remove 50 percent of the water, the pulp is then dried, fluffed, and bagged. The papermaking process makes the best of the two wet processed insulations. The fire retardants are not only more thoroughly and uniformly mixed with the fibers, but bonded to the fibers. Conwed manufactures its FIBERFLUF™ cellulose using this process. Accepted thermal resistance values range from R-3.2 to R-3.7 per inch.

Cellulose can be blown in dry, or sprayed in wet. As with all loose-fill insulation it too settles, but more so than fiberglass. Testing and research into real attics over a number of years has clearly established that cellulose settles on an average of between 15 and 25 percent over time. For every 1 percent loss in thickness due to settling, the R-value is reduced by 1 percent.

As the only insulation regulated by the federal government, cellulose has acquired an undeserved reputation. At about the time of the OPEC crisis, there were about 50 cellulose manufacturers; by 1978 the number had grown to 700. Many of these were fly-by-night garage operations turning out untreated ground-up paper. One nationally known magazine encouraged its readers to get in on the ground floor: Buy a grinding machine, hitch it to your car, drive up to the house, grind up newspapers and pump them into the house wall cavities. Naturally, if houses with this plain paper burned down, *all cellulose* insulation was blamed.

Settling—Attic cellulose will settle because it cannot be compressed to a so-called design density, or settled density. It will continue to settle until it reaches its settled density, at which point it stops settling.

Properly installed cellulose in sidewalls will not settle. Dr. George Tsongas of Portland State University has argued that loose-fill materials do not settle to any appreciable degree, *if properly installed.* He adds that "What appears to be settling is probably incomplete filling of the wall cavity." How must cellulose be installed to prevent settling?

Cellulose has to be installed at a high enough density to prevent settling. Sidewalls have to be filled at a *design density* which is so many pounds per cubic foot. Although it can vary from one manufacturer to the next and between different batches of the same cellulose, most studies have shown that at 3.5 to 4 lbs/ft^3, settling will not occur. It also depends on the number of holes—the two-hole system is common—nozzle diameter, skill of the installer, and other variables.

Two Holes versus One Hole—Traditionally, 2 holes have been used when installing cellulose: one 8 inches below the top plate, and the other about 1 foot above the bottom plate. The one-hole method was retried and improved upon in a 200-house study in Minnesota. A 1- to 1½-inch hose was attached to blowing hose, and inserted into a 2-inch hole drilled about 1 foot above the bottom plate. The hose is moved up into the cavity until it hits blocking or the top plate. Cellulose is blown in and it fills the cavity from *bottom up* at a low density. Blowing continues until compaction begins, as indicated by the strain on the blowing machine. The hose is slowly withdrawn, which causes the cellulose to compact from *top to bottom.* At the bottom of the hole, the hose is turned downward and the insulation is compacted into the bottom of the wall cavity.

One of the advantages of this method is obvious: the wall cavity is probed for blocking, which cannot be missed. The cellulose is blown in at a higher density over the entire wall cavity. Monitoring showed that top and bottom of the cavities were well compacted, and that 10 to 20 percent more insulation is installed with this technique. Blower door tests show a 40 percent reduction in a house's air leakage with this method. Other tests on retrofitted houses, using the one hole and cellulose, confirm the validity of the one-hole technique, and the ability of cellulose to largely reduce air movement into and within the wall cavity.

Problem 1—Given a wall framed with 2 × 6s, 24 inches oc and a wall height of 7'–6", how many pounds of cellulose will each cavity require at a density of 3.5 lb/ft^3?

The problem can be solved in one of two ways: A, convert 5.5 inches to a decimal fraction of 1 foot: 5.5 inches / 12 inches = 0.4583.

$(7'\text{-}6'' \times 2' \times 0.4583'') \times 3.5 \text{ lb/ft}^3 = 24$ pounds.

B, convert all measurements to inches.

$[(90'' \times 24'' \times 5.5'') \times 3.5 \text{ lb/ft}^3]/1728 \text{ in}^3/\text{ft}^3 = 24$ pounds.

Problem 2—A wall framed with 2 × 6s has a net area of 1080 ft². How many 30 pound bags will be needed to fill the wall at 3.5 lb/ft³?

$1080 \text{ ft}^2 \times 0.4583 \times 3.5 \text{ lb/ft}^3 /30 \text{ lb bag} = 57.7$ or 58 bags.

Cellulose in Attics—Not only does attic cellulose settle, but its R-value drops as well. As the material settles, the density increases and the overall thickness decreases. For example, when 15 inches of blown cellulose at a density of 2.5 lb/ft³ settles to 12.5 inches, its R-value drops from R-55.5 down to R-43.8, and its density increases to 3.0 lb/ft³.

A contractor deciding to insulate looks at the label on a bag of cellulose. The label says (Fig. 17–3) for an R-40 in the attic, blow in 10.8 inches minimum. Two years or so later, the homeowner decides to measure the thickness and finds it is only 8 inches. Checking a cellulose label at the lumberyard shows that 8 inches is only R-30.

Unfortunately, the 10.8 inches refers *not* to the installed thickness, but to the final settled thickness for the rating of R-40. What the label does not tell the do-it-yourselfer is that for an R-40, enough cellulose must be blown in so it settles down to 10.8 inches. How much, then, has to be blown-in? Assuming a 25-percent settling, overblow by 25 percent: 10.8 × 1.25 = 13.5 or 14 inches. The initial R-value will be much higher but will drop down to R-40 at the end of the settling. Several manufacturers have now changed their labels to show both the *installed* and the *settled* thickness (Fig. 17–4).

Spray-in Cellulose—Spray-in cellulose is gaining in popularity and controversy. The major controversy over wet-spray cellulose involves the possibility of moisture damage. Properly installed cellulose at a water content of 50 percent maximum on a wet weight basis, 3 gallons of water per 30 pound bag, will not cause problems, even with double vapor diffusion retarders (VDR): a VDR on the warm interior side, XPS or PIR on the outside studs.

Wet-sprayed cellulose is cellulose to which either a dry adhesive is

R-Value at 75° Mean Temp.	Minimum Thickness	Max. Net Coverage		Max. Gross Coverage 2 × 6 Framing on 16″ Center		Min. Weight per Sq. Ft.
To obtain insulation Resistance (R) of:	Installed Insulation should not be less than (inches)	Min. Sq. Ft. coverage per bag	Bags per 1000 Sq. Ft.	Maximum Sq. Ft. Coverage per bag	Bags per 1000 Sq. Ft.	Weight per Sq. Ft. of installed insulation should be no less than (lbs.)
Attic:						
R-40	10.8	17.22	58.08	18.08	55.31	1.80
R-38	10.3	18.13	55.17	19.08	52.40	1.71
R-32	8.6	21.52	46.46	22.89	43.69	1.44
R-30	8.1	22.96	43.56	24.52	40.79	1.35
R-24	6.5	28.70	34.85	31.18	32.08	1.08
R-19	5.1	36.25	27.59	40.00	25.00	.86
R-13	3.5	52.98	18.87	58.46	17.11	.59
R-11	3.0	62.61	15.97	69.09	14.47	.50
Sidewalls:				**2 × 4 Studs on 16″ Center**		
R-13	3.5			33.90	29.59	1.02

Fig. 17–3.   Typical coverage label on a bag of cellulose insulation.

added, or a liquid is introduced at the nozzle at the time of spraying (Fig. 17–5). Not all wet-sprayed cellulose has an added adhesive—this is not necessary when spraying into stud cavities 5.5 inches deep maximum. The adhesive slightly reduces the R-value.

Spray-applied cellulose has a number of advantages:

1. It *flows* into every nook and cranny.
2. Because of the open cavity, the cellulose can be inspected as it is installed.
3. It does not settle.
4. It forms a crust which gives it added air infiltration resistance.
5. Air movement within the wall and within the dry insulation is substantially reduced.

The R-value is between 3.5 and 3.7 per inch.

Caution—The cellulose used for wet spraying must be manufactured for that purpose. The last operation before it is bagged is

R VALUE 75° MEAN TEMP	INITIALLY INSTALLED THICKNESS (Approx.)	OPEN ATTIC MINIMUM THICKNESS (Inches)	MAXIMUM COVERAGE PER BAG (Square Feet)			MINIMUM BAGS PER THOUSAND SQ. FT.			MINIMUM WEIGHT (lbs./sq. ft.) At 1.7 PCF Settled Density
			NET	2 × 6 Framing on 16″ Centers	2 × 4 Framing on 24″ Centers	NET	2 × 6 Framing on 16″ Centers	2 × 4 Framing on 24″ Centers	
R-11	4 ″	3.0	70.3	77.4	74.9	14.2	12.9	13.4	.427
R-13	4.75″	3.6	59.5	65.5	63.4	16.8	15.3	15.8	.505
R-19	6.75″	5.2	40.7	44.8	42.5	24.6	22.3	23.5	.737
R-22	8 ″	6.0	35.1	38.4	36.5	28.5	26.0	27.4	.854
R-24	8.75″	6.6	32.2	34.9	33.3	31.1	28.6	30.0	.932
R-30	10.75″	8.2	25.8	27.5	26.5	38.8	36.4	37.8	1.164
R-32	11.5 ″	8.8	24.2	25.7	24.8	41.4	39.0	40.4	1.242
R-38	13.5 ″	10.4	20.3	21.4	20.8	49.2	46.7	48.1	1.475
R-40	14.25″	11.0	19.3	20.3	19.7	51.8	49.3	50.7	1.553
R-44	15.75″	12.1	17.6	18.4	17.9	56.9	54.5	55.9	1.708
R-50	18 ″	13.7	15.5	16.1	15.7	64.7	62.3	63.7	1.941

Fig. 17–4. New labels now being used, show the initially installed thickness and the minimum settled thickness.

Fig. 17–5. Installation of wet-spray cellulose. *(Courtesy Advanced Fiber Technology)*

filtering to reduce dust and fines. If unfiltered cellulose is used, its R-value will be reduced to about R-2 per inch.

Spray-applied cellulose will not support itself in a staggered stud wall such as 2 × 3s, 24 inches oc on 2 × 6 plates. It will not support itself in a double wall, such as a staggered 2 × 4 double wall. The outer wall can be sprayed, but the remaining cavity will have to be filled with blown-in cellulose.

Story Jig—It is possible to install loose cellulose in a stud wall without having to either sheetrock it first or use wet spray. The Story Jig is manufactured by Insul-Tray, Redmond, Washington. It is a clear plexiglass panel, with four gasketed fill holes, that is placed over the wall studs. The insulation is blown through the holes. After the cavity is filled, the panel is moved to the next bay and the next until the walls are filled. When installed to 3.5 lb/ft^3 density, the cellulose remains in place, and settling should not be a problem. The jig is available in a number of sizes, but it cannot be purchased, only rented.

Caution—The success of this method depends on no less than a 3.5 lb/ft^3 installed density. The calculated number of bags must be installed. Reputable insulation contractors will keep the top of the bag as a way of tallying the number used.

Cellular Plastics

The several different plastics insulations are used in two basic forms: sprayed-in, poured-in, or frothed, and rigid foam boards.

Polystyrene is probably the most common and widely-used plastic insulation because it is not confined to building insulation. Coffee cups, packing material, wrappers for soda bottles, children's surfboards are just a few of the uses of this material. There are two methods of producing polystyrene: expanded and extruded.

Expanded, or molded, foam is made by putting polystyrene beads containing a blowing agent into a mold and heating it. The vapor pressure of the blowing agent, usually pentane, expands the beads, and produces a largely closed-cell foam. This foam is also known as *beadboard.*

Extruded polystyrene is manufactured by flowing a hot mixture of polystyrene, solvent, and a blowing agent, usually a chlorofluorocarbon, through a slit into the atmosphere. The gas expands and the result is a very fine, closed-cell foam.

There are 5 types of extruded polystyrene: IV, V, VI, VII, and X. The R-value is R-5.0 per inch at all densities.

Note—All expanded polystyrene (EPS) whether expanded or extruded, is EPS. MEPS is molded, expanded polystyrene insulation board, also known as beadboard. XPS is extruded, expanded polystyrene insulation board such as the blue STYROFOAM®, pink FORMULAR®, green AMOFOAM®, and yellow DIVERSIFOAM. For more information see ASTM standard E631–89.

PURs and PIRs age, that is, their R-values decrease with time. The chlorofluorocarbon (CFC) blowing agent in PURs and PIRs, has a lower conductivity—higher R-value—than air. Some of the CFC leaks out as the foam ages, and the foam's R-value drops. PUR may have a manufactured value of R-9, but an aged value of R-7 or lower.

Beadboard is CFC-free; the blowing agent is pentane gas, which diffuses out of the foam shortly after manufacture. The R-value of EPS is not based on pentane gas. Although the R-value depends on the board's density, it remains stable as long as the board is kept dry. The ASTM C578–85 standard listed 9 types of EPS. The C578–87A standard lists 10 and is currently under revision. The 4 types of EPS listed here are from the ASTM C578–85 standard. Density is in lb/ft^3. R-value is per inch.

Type	I	II	VIII	IX
Density	0.9	1.35	1.15	1.8
R-Value	3.6	4.0	3.8	4.2

When the term *aged* is used with EPS, it refers not to R-value age, but to dimensional stability aging. The EPS used in Exterior Insulation and Finish Systems (EIFS) is aged to stabilize its dimensions and minimize movement of the board to prevent cracking of the finish.

Pentane is not trouble-free: it is a pollutant because of VOCs, which were discussed in Chapter 3, **Foundations.** The VOCs combine with other compounds to produce smog. Pentane is regulated by the federal Clean Air Act. BASF Corporation has produced a new resin, BFL, which contains about 4 percent pentane rather than the usual 7 percent.

Polyurethane (PUR) and **Polyisocyanurate (PIR)** use a chlorofluorocarbon called Freon. It is a low conductivity (high R-value) gas inside the closed-cell foam. PUR and PIR are manufactured in three different forms: bun stock, to provide cut boards, continuous bun stock, which is cut to length depending on the ultimate uses, with or without outer coverings such as aluminum foil, and sprayed, poured, or frothed in place.

PURs and PIRs are closely related; the same basic ingredients are used, but they are put together differently. PIR foam has a lower thermal conductivity (higher R-value), more thermal stability, and lower flammability—higher fire retardant—than PURs. They are more expensive and more friable (break easily). PURs are tougher, less expensive, have a lower R-value, and must have fire retardants added. Although the two are sometimes confused, the commonly used sheathing boards with aluminum foil facings, such as THERMAX®, ENERGY-SHIELD®, R-MAX®, and CELOTEX®, are PIRs. The R-value is 7.04 per inch (Fig. 17–6).

Phenolic is a phenol-formaldehyde resin produced by mixing phenol and formaldehyde. The resin is mixed with a catalyst, a blowing agent, and a foaming agent. Either open-celled or closed-celled foam can be produced. It is produced in a continuous slabstock and cut to length as

Fig. 17–6. Celotex® Tuff-R® rigid board insulated sheathing. *(Courtesy the Celotex Corporation)*

necessary. It is presently available only as board stock, although the Japanese have a spray-in-place phenolic foam. Probably the best-known phenolic insulation board was Kopper's Exeltherm Xtra used in commercial roofing. A version for residential use with an R-value of R-8.33 per inch was produced, but the product line was sold to Johns Manville. Phenolic's R-8.33 remained the highest R-value foam insulation available until Celotex introduced its new TUFF-R BLACKORE® in July 1991. It has an aged R-value of R-8.7 per inch; up 13 percent from Tuff-R's R-7.7 per inch. The black color comes from carbon black which decreases the radiant heat transfer within the foam cells (Fig. 17–7).

The chlorofluorocarbons (CFCs) used as the blowing agents in PURs, PIRs, XPs, and phenolic foams have been identified as harmful to the ozone layer in the earth's stratosphere. As a result of an international treaty, signed by the US in 1988, there will be a gradual production phaseout. In 1989 Dow Chemical ceased using CFC in their Styrofoam XEPS foam board.

Urea-Formaldehyde (UFFI) is manufactured from urea and formaldehyde. It is generated on-site using portable equipment. A urea for-

Fig. 17–7. Celotex's New Blackore™ with an R-value of R-8.7 per inch. *(Courtesy of Celotex Corporation)*

maldehyde resin, a foaming agent, and air, are mechanically foamed with a compressed air pump, and a mixing gun. R-values range from 3.5 to 5.5, with an accepted value of R-4.2 per inch.

Urea formaldehyde has been the subject of considerable controversy, lawsuits, a government ban, and subsequent reinstatement by the federal courts. The insulation releases formaldehyde after installation and for many months later. The formaldehyde levels in houses without the foam are lower than those with the foam. There are, of course, houses insulated with UFFI that have relatively low formaldehyde levels.

Shrinkage has been a serious shortcoming of the foam, ranging from 1 to 10 percent and averaging 6 percent or more. HUD studies indicated that with a 6 percent linear shrinkage, the thermal efficiency is reduced by 28 percent. If the claimed R-value was R-4, a 28 percent reduction means that an R-4 was paid for but an R-2.88 is the end result of the shrinkage. The shrinkage does not affect the thermal conductivity of the foam itself, but does affect its installed performance: in the cavity it is not as effective as it should be.

Incredible numbers of complaints of illness caused by the foam led the Canadian government to give grants of $5,000 per home for UF foam removal. In 1982 the chairman of the House Commerce, Consumer and Monetary Affairs Subcommittee considered legislation that would provide low interest loans, tax grants, or outright grants to homeowners for removal of UF foam. On August 15, 1982 the US Consumer Product Safety Commission (CPSC) banned UFFI. Many UF foam insulation contractors were forced out of business because they were forced to pay for the removal of the foam they had installed. Some, but not all, of the formaldehyde problems were caused by unqualified installers using improperly formulated UFF, and improper installation techniques. In April 1983, the US Fifth Circuit Court of Appeals reversed the CPSC, claiming there was not enough evidence to support the ban.

In 1980, the National Institute of Occupational Safety and Health (NIOSH) recommended that UF be regulated as a potential human carcinogen. In 1985, OSHA set the permissible exposure limit (PEL) at 1 part per billion (ppb). As of June 1990, the agreed-on PEL is now 0.75 parts per million (ppm). Some states, such as Massachusetts, have banned UFFI, and prohibit the use of any insulation containing UF. There are a few UFFI contractors left, some of whom claim their product reduces the formaldehyde emissions. Formaldehyde emissions are a problem in mobile homes and are some of the materials involved in the sick building syndrome. In January 1990 a jury in Centralia, Missouri awarded the Pinkerton family 16 million dollars for a *chemically induced immune dysfunction* (CIIF) caused by high formaldehyde levels.

There are still four companies selling UFF. Three of these companies restrict their sales to the commercial sector except for PolyMaster. Tailored Chemical Products,Inc., of Hickory, North Carolina is the largest manufacturer and installer of UFF. It manufactures CORE-FILL 500.

Tripolymer foam is manufactured by C. P. Chemical of White Plains, New York. It is a cold-setting foam with an R-value of R-4.6 at 72 degrees Fahrenheit. Rated Class 1, it is fire resistant and does not contain or use chlorofluorocarbons. According to the manufacturer, it is chemically different from urea formaldehyde. It is not a phenolic foam but a distant *cousin* of the old urea formaldehyde foam. The manufacturer claims that "Laboratory and field tests have determined less than de-

Fig. 17–8. Tripolymer foam. *(Courtesy C.P. Chemical)*

tectable levels of any aldehydes (including formaldehyde)." But the level that was considered detectable was not stated (Fig. 17–8).

PolyMaster of Knoxville, Tennessee markets urea formaldehyde foam insulation from a resin manufactured by Mt. Edison, of Milan, Italy. According to PolyMaster, Mt. Edison has adequately locked in the formaldehyde to where the extractable amount is .33 lbs. per 1000 lbs. of foam. PolyMaster spokesmen add that cured PolyMaster in essence poses no off-gassing problem. Their type R-505 designed for use in residential construction has a claimed R-value of R-4.58 per inch at 75 degrees Fahrenheit.

Cellular Glass insulation is a lightweight, rigid material made up of millions of closed glass cells. It is available in blocks, boards, and various shapes. At one inch, it has an R-value of 2.85. Cellular glass is completely inorganic, will not burn, is immune to acid attack, and impervious to moisture because water wets only the surface. It has compressive strength as high as 175 psi, as we saw in the section in Chapter 3 on *Frost-Protected Shallow Foundations.* FOAMGLAS®, manufactured by Pittsburgh-Corning guarantees its product for 20 years (Fig. 17–9).

Fig. 17–9. Foamglas®. *(Courtesy Pittsburgh-Corning)*

Vermiculite is a mica-like layered mineral which contains water. When subjected to high temperatures, the water turns to steam. The steam is driven off, which causes the layers to separate. A range of densities is available and controlled by the amount of expansion. The lower density material is used as loose-fill insulation and concrete aggregate. The higher density material is used as a plaster aggregate and in high-temperature applications. Vermiculite is mixed with portland cement and sand to produce vermiculite concrete. It has an R-value from R-1.0 to 1.7 per inch.

Vermiculite is treated to make it water repellent. It does not burn, melts at 2400 degrees Fahrenheit, and gives off no odors. Because it is

Fig. 17–10. Zonolite® being installed in concrete blocks. *(Courtesy W. R. Grace)*

inorganic, it will not rot, corrode, or deteriorate from age, temperature, or humidity. It is immune to termites.

Zonolite Masonry Insulation is the brand name of vermiculite manufactured by W. R. Grace & Company. It is widely used as insulation in concrete masonry blocks (Fig. 17–10).

Perlite is a natural volcanic glass containing between two and five percent water by weight. When perlite is heated to 1800 degrees Fahrenheit, it expands to between 4 and 20 times its original volume, vaporizes the trapped water, and forms vapor cells in the heat-softened glass.

It is used in the commercial sector as roofing board material. A lightweight insulating concrete is made by mixing perlite with portland cement. Perlite insulating cement is used primarily for roof decks, floor slabs, and wall systems. Low density perlite is used as a loose-fill insulation.

Perlite is nontoxic, does not burn, and is immune to weathering, insect attack, rot, and corrosion. Its R-value depends on its density and ranges between R-2.5 and R-3.7.

Reflective Insulation has been in use for many years, but sales of these products represent but a fraction of the market. There are about seventeen firms marketing or producing reflective insulations.

Fiberglass batts and gypsum board can be purchased with aluminum foil facings. The PIR rigid boards have aluminum foil on both sides. Reflective insulations (RIs) are systems made up of two or more layers of aluminum foil, or aluminum foil laminated to one or both sides of Kraft paper, air spaces and air. There may be a single air gap or as many as ten air gaps. The long edges of these insulations are sealed together to form nailing flanges (Fig. 17–11). Shipped in flattened rolls, they are designed to be installed between studs in a wall, between floor joists, or rafters. The five- and seven-layer systems are designed for 2 × 6 or deeper studs.

As the name suggests, RIs work as insulation by reflecting radiant heat energy that strikes the foil surface. In a vertical stud wall cavity, about 60 percent of heat transfer is due to radiation. The series of re-

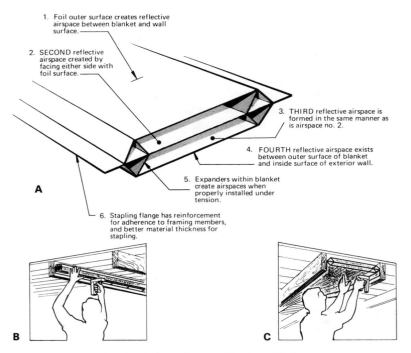

1. Foil outer surface creates reflective airspace between blanket and wall surface.

2. SECOND reflective airspace created by facing either side with foil surface.

3. THIRD reflective airspace is formed in the same manner as is airspace no. 2.

4. FOURTH reflective airspace exists between outer surface of blanket and inside surface of exterior wall.

5. Expanders within blanket create airspaces when properly installed under tension.

6. Stapling flange has reinforcement for adherence to framing members, and better material thickness for stapling.

A

B   C

Fig. 17–11.   Reflective insulation and its installation.

flective surfaces limits the transfer of radiant energy across adjoining air space. Although these systems greatly reduce radiant energy transfer, conduction and convection can take place across the air spaces, and conduction through the flanges and foils. Aluminum foil has little effect in reducing conductive or convective heat losses. Therefore, it works best where conductive/convective heat loss is small, for example, in a floor cavity.

The thermal resistance, R-value, of the system depends on (1) the number of cells or air spaces, (2) the position of the foil in the cavity, (3) the direction of the heat flow, (4) the quality of the installation. Tears, missing sections, uneven spacing, degrade the performance of the system. *The performance of any insulation is only as good as the care with which it was installed.* The published R-values for RIs tested alone, that is, not attached to framing, is considerably higher than when the RI system and the framing is tested together, even when an allowance for the framing has been made. Do not use the R-value for foil alone when doing heat loss calculations. The foil-only R-values will result in significant errors when doing heat load calculations or comparing different insulations.

Foil-Faced Insulations and Gypsum Board—Just as convective loops within the air space of RIs can degrade their performance, convective loops in an exterior wall can also degrade its performance. An improperly installed R-20 batt of insulation will have an *effective value* that is lower than the *nominal* value, that is, the value stamped on the bag.

Because of the aluminum foil facing on the insulation or on the back of the gypsum board, it is tempting to think that by leaving an air space between the gypsum board and the insulation, an increase in R-value will result from the *dead* air space. In the imperfect world of construction, there may be no dead air spaces.

Convection takes place within a wall when the insulation is porous, does not completely fill the cavity, and there is a large temperature difference across the wall. In a completely filled cavity, plumbing, wiring, and stapling the Kraft paper flaps to the insides of the studs (Fig. 17–12) compresses the insulation and leaves spaces. The compression reduces its material R-value and the system R-value for the wall in which it is installed. Even if the insulation were to block the air flow, the compressions leave openings and provide air paths around the insulation. This can result as a continuous loop of rising and falling air:

Fig. 17–12. Improperly installed insulation. *(Courtesy Owens-Corning)*

warm air at the top, cool air at the bottom. This is a common problem in hollow—unfilled—concrete masonry blocks (Fig. 17–13).

If fiberglass batts are not protected from air circulation inside the wall cavity by a convective retarder, there will be a substantial reduction in the R-value of the insulation. One study (in ASTM publication STP 789), "Effectiveness of Wall Insulation," reported a 34 percent reduction with *unfaced* fiberglass batts, caused by air spaces created by electrical workboxes, wiring, or compression of the edges of the insulation.

Do not compress the insulation with wiring. Use the Arkansas House technique and run the wire along the bottom of the studs. Keep duct work and plumbing out of outside walls. Consider using surface-mounted wiring, such as Wiremold, or ElectroStrip, both of which are permitted by the *National Electrical Code (NEC)*. Surface-mounted wiring eliminates wiring, workboxes, and hole drilling at the bottom of outside wall stud.

Radiant Barriers—Radiant barriers (as they are presently called) and reflective insulation look exactly alike because they are the same mate-

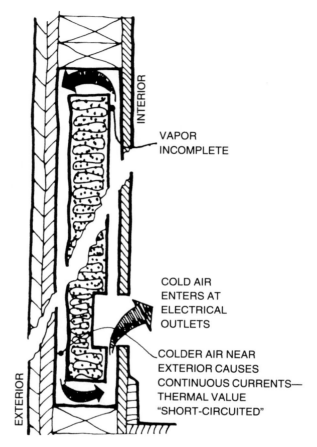

Fig. 17–13.   Convective loop in exterior wall.

rial. However, when the foil material is installed in a wall cavity with a closed air space, it is called reflective insulation. When the air space and the foil are open to air circulation, as in an attic, the reflective foil is called a *radiant barrier*.

Insulating Concrete—Insulating concrete can be made with aggregates of perlite, vermiculite, or expanded polystyrene beads. Sparfil International of Ontario, Canada has been manufacturing an insulated concrete masonry block since 1976. Called SPARFIL (Fig. 17–14), 60 percent of the volume of the block is EPS beads. A 10-inch block has an R-value of R-10. This can be increased to R-18.5 by adding polystyrene inserts. Insulating concrete can also be made by foaming a portland cement mixture. Adding a foaming agent such as aluminum dust causes a chemical reaction that generates a gas which produces a closed-cell porous material. The density of the concrete depends on the amount of gas produced.

Another cementitious-based foam insulation, one widely used and known, is AIR KRETE®. Air Krete is a non-structural, magnesium oxychloride-based insulating foam that contains no CFCs, formaldehyde, or other known toxic chemicals. It is cold-setting, and there is no further expansion after it leaves the hose (Fig. 17–15). It does not burn, and does not smolder after exposure to flames. Even when exposed to intense heat (2500 degrees Fahrenheit), it remains stable and does not give off hazardous fumes. Because it does not stick to surfaces, it cannot be sprayed in: it must be trowelled or foamed in place. Trowelling with Air Krete is easy because it does not slump. The foam is not temperature sensitive: it is an all-weather material. As long as the materials

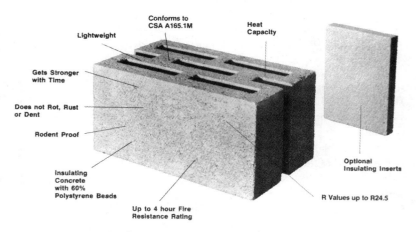

Fig. 17–14. Sparfil® insulated concrete block.

Fig. 17–15. Air Krete® cement-based insulating foam being installed in concrete blocks. *(Courtesy Palmer Industries, Inc.)*

are between 40 and 70 degrees Fahrenheit when they are delivered to the application gun, the outside temperature does not matter. When installed at its rated density of 2.07 lbs/ft^3, the R-value is R-3.9 per inch.

Cotton Insulation

Cotton Unlimited Incorporated, of Post, Texas has announced a new insulation made from cotton. Called INSUL-COT, it has an R-value of R-3.2 per inch at a density of 0.42 lbs/ft^3. According to a spokesperson at the factory, the R-value could be as high as R-4 per inch with certain types of cotton fibers.

Southwest Research Institute rates Insul-Cot as Class A, with a flame spread of 20. Insul-Cot is treated with fire inhibitors that are absorbed into the fibers so that they should not leach out. It is also treated with insect and rodent repellents, and complies with OSHA Hazard Communications Standards, because it contains no hazardous materials. Unlike fiberglass insulation, it contains no *itchy* dust. The only negative at this point is that it could irritate the eyes if it contacts them directly. It is basically an organic material and does not come with a causes-cancer warning.

Vapor Diffusion Retarders

Vapor diffusion retarders, along with attic ventilation, roof felt underlayment, and wood siding over rigid foam board, are one of the least understood and most hotly debated topics in residential construction.

For many years the rule has been: to keep moisture out of the exterior walls, a vapor barrier must be placed over the studs on the interior (warm) side of the wall. Should moisture get into the wall, there should be nothing on the exterior of the studs that would prevent the moisture from escaping. The rule-of-thumb is that the exterior sheathing and cladding should be five times more permeable than the inner vapor barrier. The more permeable a material, the more easily water vapor moves through it. Therefore, use a highly *impermeable* material on the warm interior side and highly *permeable* materials on the exterior of the studs.

Moisture can affect the performance of insulation and lead to the rotting of the framing if the moisture percentage reaches 30 percent, the so-called *fiber saturation point*. This was discussed in Chapter 7, **Wood.**

From about the early 1920s on, it was believed that water vapor got into exterior walls by *diffusion*—the movement of water molecules through porous materials. The water vapor movement is not caused by air. The water vapor is driven by vapor pressure that results when the humidity is higher on one side of the material than on the other side. Polyethylene is placed on the warm side because the vapor pressure is highest on the warm side. Norwegian and Canadian building scientists, working in colder areas of the world with low sloped roofs, began to question the role of diffusion. Although air leakage and diffusion could be happening at the same time, these researchers concluded that *air leakage* was the major cause of moisture in exterior walls. It is tempting to say that at this point the *air barrier* was born. But the air barrier already existed. What did not exist was a clear understanding of the difference between an air barrier and a vapor barrier, as they were then called.

At about the time of the designing and building of the Saskatchewan Conservation House (SCH)—1976—the term *air/vapor barrier* began to be used. With reference to the SCH, it was said that the polyethylene was both a vapor barrier and an air barrier. The vapor barrier polyethylene kept moisture out of the walls, and the air barrier polyethylene kept any wind that got into the cavity out of the house. R. L. Quirouette in *Building Practice Note No. 54*, "The Difference Between a Vapour Barrier and an Air Barrier," comments,

> *A 4-mil polyethylene sheet will make a good quality vapour barrier and twelve inches of cast-in-place concrete will make a good quality air barrier. This would be too restrictive a definition, but the functions of these two barriers are as different as polyethylene and concrete.*

The term *barrier* says we are trying to use a material to stop wind or moisture penetration. Few building materials are *totally* impervious to wind or moisture penetration. The term diffusion *retarder,* or air *retarder* is used to show that we are trying to slow down, control, or retard wind, air, and moisture movement.

Just because air carrying moisture is the major cause of moisture in walls does not mean that vapor diffusion retarders are unnecessary. They have their place, but their function and how they function is quite a bit different than what we have been led to believe.

To retard the movement of moisture-laden air requires an *air retarder.* In residential construction the exterior wall sheetrock serves that purpose. Of course, most builders and building inspectors see only gypsum board, not an air retarder. Even if the exterior walls had no penetrations, they will still insist that a vapor barrier be installed. Those builders who still believe that VDRs cause moisture problems will not install them if they can get away with it.

Are Vapor Diffusion Retarders Necessary?

If the air retarder can be sealed against air movement, a VDR is not necessary. If a VDR is full of penetrations because of openings in the air retarder that allow warm moist air to move into the wall cavity, it is useless.

Water vapor diffusion is a very slow process and large amounts of moisture are not moved. How much diffusion takes place depends on the area. The greater the area the greater the amount of diffusion that takes place, all other things being equal. If 90 percent of the wall area is covered with a VDR, it is said to be 90 percent effective. Therefore, the VDR need not cover the entire wall. Holes, penetrations, minor cuts need not be patched, because they do not increase the overall moisture diffusion rate into the wall.

As an example, air leakage will move 100 times more water vapor through a 2-inch square hole than will be moved by vapor diffusion through the same hole. A 10' × 10' sheetrock wall will allow 1 cup of water per year to diffuse through it. A 1-inch hole in the same wall will allow 20 gallons of water per year to enter by air leakage (Fig. 17–16).

Many materials, such as polyethylene, are used as both air retarders and vapor diffusion retarders. But not all materials are long lasting or stable enough in all applications. An air retarder is not just a piece of material, but an assembly, a system. Any material, or system of materials, may be used as an air retarder if the following requirements are met:

1. All the various parts of the air retarder system must be joined together so that the system is continuous and unbroken throughout

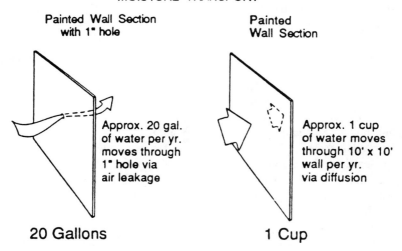

MOISTURE TRANSPORT

Painted Wall Section with 1" hole

Approx. 20 gal. of water per yr. moves through 1" hole via air leakage

20 Gallons

Painted Wall Section

Approx. 1 cup of water moves through 10' x 10' wall per yr. via diffusion

1 Cup

Fig. 17–16. Air leakage moves 100 more times water vapor through a 1-inch hole than will vapor diffusion.

the structure. Although the air retarder could be located within the wall, in residential construction the gypsum board serves this purpose.

2. The material must be secured to a supporting structure—sheetrock to studs, for example—to resist peak wind loads, pressure from mechanical ventilation equipment, and stack effect. The air retarder's prime function is to stop the wind that is trying to get into the building. Therefore, it must be rigid enough to resist movement or rupture from inward or outward air pressure loads.

3. The air retarder system must be very nearly zero impermeable.

4. The deflection of the air retarder between supports must be minimized to prevent it from moving other materials, such as insulation in a cavity.

5. It must be constructed of a material that is known to be durable and long lasting, or be so located that it may be serviced when necessary.

Any system or material that meets these criteria may be used as an air retarder. These criteria rule out many materials: concrete block, EPS (beadboard), acoustical insulation, fiberboard.

Polyethylene is incorrectly called an air/vapor retarder. Although it does meet the first and third criteria, it does not meet the remaining criteria when it is installed behind the air retarder, in this case, the sheetrock. Even if all but the fifth criterion were met, the polyethylene is not accessible for repair.

The Airtight Drywall Approach (ADA) of using gaskets and caulking sealing techniques, makes the use of a VDR unnecessary, although it does not do away with the polyethylene. A VDR and an air retarder are necessary, but the VDR does not have to be polyethylene. That it is predominantly polyethylene says only that it is easier to deal with. A vapor diffusion paint will work, but raises the question of whether or not the coverage and thickness of the paint are adequate.

A comprehensive study of paints used as vapor diffusion retarders was conducted in Canada. The study concluded that alkyd sealers and enamels are effective water vapor diffusion retarders but latex primers and paints cannot be considered adequate vapor diffusion retarders.

In summary then, a VDR is not necessary as long as there are no holes in the gypsum board air retarder. If someone puts a hole in the air retarder, the VDR will also be punctured; moisture-laden air can enter and moisture damage can result. If the VDR has holes, tears, or rips in it, but the air retarder is continuous and unbroken, the holes do not matter.

Polyethylene is usually the choice because it can be seen. Although vapor diffusion retarder alkyd sealers and enamels can be used instead of polyethylene, they must be carefully and thoroughly applied.

Although the air retarder can be located anywhere in the wall, in residential construction it is usually the exterior wall sheetrock. The use of the sheetrock as the air retarder complies with the five criteria. It has the added advantage that any damage to it can be seen and easily and quickly repaired.

Caution—Do not assume that the buyer of the house *knows* that the exterior wall gypsum board is the *air retarder*. Its function, and the importance of not putting holes into it, must be fully explained.

Heat Flow Basics

R-value is usually defined as resistance to heat flow. But in order to understand R-value, other basic terms, such as thermal conductivity or k, must first be understood.

Conduction, convection, and radiation were discussed. We know, from sometimes painful experience, that some materials are better conductors than others. A poor conductor is called an insulator. We use a pot holder—insulator—to pick up a hot metal pot—conductor.

The most important thing about an insulator is its thermal conductivity (k). All materials can be described in terms of their k-value. Insulators have low k: porcelain, glass, wood are good examples. Conductors, such as aluminum, silver, copper, cast iron, have high k. Table 4 of the 1989 ASHRAE *Handbook of Fundamentals* lists the conductivity of many common materials. Notice that k is given as Btu-in/hr-ft^2-F.

There are four factors that affect the conduction of heat from one area to another:

(1) The difference in temperature (ΔT) between the warmer side and the colder side,

(2) The area (A) of the warmer and colder sides,

(3) The resistance (R) to heat flow and conductance (U) between the warmer and colder areas,

(4) The length of time (t) over which the transfer takes place.

Laboratory testing of the k factor is done by measuring the amount of heat (BTU) that passes through a 1-inch-thick piece of material 1 square foot in area, in 1 hour, with a ΔT of 1 degree Fahrenheit (Fig. 17–17). The piece must be made of one material (homogeneous). If the sample is not homogeneous, or if other than 1 inch thick, the value is reported as C, the thermal conductance. For example, aluminum-foil-faced fiberglass blankets would be measured on a conductance basis.

Fig. 17–17 has the symbols T_{in} on one side and T_{out} on the other side of the material being measured. Delta (Δ) is a Greek letter and means *the difference in*. $T_{in} - T_{out}$ is the difference between the temperature on one side—or on the inside—and the temperature on the other side—or outside. Rather than always saying the difference in temperature, it is shortened to Delta T or ΔT.

The rate at which heat flows through a material depends on what it is made of. Some materials transmit heat more readily than others. Resistance is a material's resistance to the flow of heat; conductivity is a measure of how easily heat flows through a material.

When the measurement of a building wall section is made, the conductivity is expressed as the U-factor of that wall. It is the overall

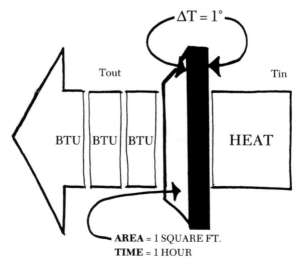

Fig. 17–17. Heat flow through a material.

(combines heat transfer by conduction, convection, and radiation) co-efficient of transmission. The U-factor is the number of BTUs that will pass through one square foot of the wall in one hour with a one degree temperature difference between the two surfaces. The U-factor = Btus per square foot, per hour, per degree Fahrenheit, or U= Btu/ft²/h°F.

British thermal unit (Btu) is the amount of heat required to raise the temperature of one pound of water 1 degree Fahrenheit. Lighting a kitchen match and allowing it to burn completely generates 1 Btu. Sitting down while reading this book generates 300 Btus.

Resistance (R) and conductivity (U) are the opposites of each other. They are inversely proportional; that is, if a material's resistance is high, its conductivity is low. If its conductivity is high, its resistance is low.

$$\text{Resistance (R)} = \frac{1}{\text{Conductivity (U)}}$$

$$\text{Conductivity (U)} = \frac{1}{\text{Resistance (R)}}$$

The U-factor is the thermal transmittance, a measure of how readily heat moves through a material. In looking at Fig. 17–17, we can see that heat flow is a function of the area, the ΔT, and the U factor. We can now write an equation to express this:

$$Q = UA(T_i - T_o)$$

or $\quad Q = UA\Delta T$

- Q   is the rate of heat loss in Btu/hr
- U   is the thermal transmittance in Btu/ft²/h– °F
- A   is the surface area (ft²)
- T_i and T_o are the inside and outside temperatures (°F)

If we want to calculate the heat loss through a wall using the formula $Q=UA\Delta T$, we will need the total U of the wall elements and the wall area. The total U-factor of the wall elements can be found if we know their R-values because U = 1/R. However, if the R values are unknown, they can be found in Table 4 in the 1989 *ASHRAE Handbook of Fundamentals* If the local library does not have a copy, try the larger HVAC contractors.

Problem—Assume a wall section as follows:

Item	R-value
Outside air film	0.17
$\frac{1}{2}" \times 8"$ bevel siding	0.81
$\frac{1}{2}"$ plywood	0.62
$3\frac{1}{2}"$ fiberglass	11.00
$\frac{1}{2}"$ gypsum board	0.45
Inside air film	0.68
Total R-value	13.73

Assume a wall of 1080 ft² and a ΔT of 70 degrees Fahrenheit, that is, T_{in} is 70 degrees Fahrenheit, and T_{out} is 0 degrees Fahrenheit outside. Calculate the heat loss of the wall.

Given R_t, $U = 1/13.73 = 0.073$ Btu/ft²/hr °F.

$Q = UA (t_i - t_0) = 0.073 \times 1080 \times (70 - 0) = 5519$ Btu/hr.

There are 5519 Btus flowing through the wall per hour.

Attic Radiant Barriers*

This fact sheet was developed with the assistance of a Radiant Barrier Systems Technical Panel, which included representatives from the

*LEGAL NOTICE   This document was prepared by the Oak Ridge National Laboratory (ORNL) as part of work sponsored by the United States Department of Energy (DOE) in carrying out their statutory authorities. Neither ORNL, DOE, the United States Government, nor any of their agents or employees, makes any warranty, express or implied, or assumes any legal liability or responsibility for the accuracy, completeness, or usefulness of any information, apparatus, product, or process disclosed, or represents that its use would not infringe privately owned rights. Reference herein to any specific commercial product, process, or service by trade name, trademark, manufacturer, or otherwise, does not necessarily constitute or imply its endorsement, recommendation, or favoring by ORNL, DOE, the United States Government, or any of their agents or employees. The views and opinions of the authors expressed herein do not necessarily state or reflect those of ORNL, DOE, or the United States Government or any agency thereof. [DOE, June 1991]

U.S. Department of Energy, the Oak Ridge National Laboratory, the Reflective Insulation Manufacturers Association, the Mineral Insulation Manufacturers Association, the Tennessee Valley Authority, the Electric Power Research Institute, and the Florida Solar Energy Center.

Introduction

What is a radiant barrier?

Radiant barriers are materials that are installed in buildings to reduce summer heat gain and winter heat loss, and hence to reduce building heating and cooling energy usage. The potential benefit of attic radiant barriers is primarily in reducing air-conditioning cooling loads in warm or hot climates. Radiant barriers usually consist of a thin sheet or coating of a highly reflective material, usually aluminum, applied to one or both sides of a number of substrate materials. These substrates include kraft paper, plastic films, cardboard, plywood sheathing, and air infiltration barrier material. Some products are fiber-reinforced to increase the durability and ease of handling.

Radiant barriers can be used in residential, commercial, and industrial buildings. However, this fact sheet was developed only for applications of radiant barriers in ventilated attics of residential buildings. For information on other applications, see the references at the end of the Fact Sheet.

How are radiant barriers installed in a residential attic?

Radiant barriers may be installed in attics in several configurations. The simplest is to lay the radiant barrier directly on top of existing attic insulation, with the reflective side up. This is often called the attic floor application.

Another way to install a radiant barrier is to attach it near the roof. The roof application has several variations. One variation is to attach the radiant barrier to the bottom surfaces of the attic truss chords or rafter framing. Another is to drape the radiant barrier over the tops of the rafters before the roof deck is applied. Still another variation is to attach the radiant barrier directly to the underside of the roof deck.

How do radiant barriers work?

Radiant barriers work by reducing heat transfer by thermal radiation across the air space between the roof deck and the attic floor, where conventional insulation is usually placed. All materials give off, or emit, energy by thermal radiation as a result of their temperature. The amount of energy emitted depends on the surface temperature and a property called the *emissivity* (also called the *emittance*). The emissivity is a number between zero (0) and one (1). The higher the emissivity, the greater the emitted radiation.

A closely related material property is the *reflectivity* (also called the *reflectance*). This is a measure of how much radiant heat is reflected by a material. The reflectivity is also a number between 0 and 1 (sometimes, it is given as a percentage, and then it is between 0 and 100%). For a material that is opaque (that is, it does not allow radiation to pass directly through it), when the emissivity and reflectivity are added together, the sum is one (1). Hence, a material with a high reflectivity has a low emissivity, and vice versa. **Radiant barrier materials must have high reflectivity (usually 0.9, or 90%, or more) and low emissivity (usually 0.1 or less), and must face an open air space to perform properly.**

On a sunny summer day, solar energy is absorbed by the roof, heating the roof sheathing and causing the underside of the sheathing and the roof framing to radiate heat downward toward the attic floor. When a radiant barrier is placed on the attic floor, much of the heat radiated from the hot roof is reflected back toward the roof. This makes the top surface of the insulation cooler than it would have been without a radiant barrier and thus reduces the amount of heat that moves through the insulation into the rooms below the ceiling.

Under the same conditions, a roof-mounted radiant barrier works by reducing the amount of radiation incident on the insulation. Since the amount of radiation striking the top of the insulation is less than it would have been without a radiant barrier, the insulation surface temperature is lower and the heat flow through the insulation is reduced.

Radiant barriers can also reduce indoor heat losses through the ceiling in the winter. Radiant barriers reduce the amount of energy radiated from the top surface of the insulation, but can also reduce beneficial heat gains due to solar heating of the roof. The net benefits of radiant barriers for reducing winter heat losses are still being studied.

How does a radiant barrier differ from conventional attic insulation?

Radiant barriers perform a function that is similar to that of conventional insulation, in that they reduce the amount of heat that is transferred from the attic into the house. They differ in the way they reduce the heat flow. A radiant barrier reduces the amount of heat radiated across an air space that is adjacent to the radiant barrier. The primary function of conventional insulation is to trap still air within the insulation, and hence reduce heat transfer by air movement (convection). The insulation fibers or particles also partially block radiation heat transfer through the space occupied by the insulation.

Conventional insulations are usually rated by their R-value. Since the performance of radiant barriers depends on many variables, simple R-value ratings have not been developed for them.

What are the characteristics of a radiant barrier?

All radiant barriers have at least one reflective (or low emissivity) surface, usually a sheet or coating of aluminum. Some radiant barriers have a reflective surface on both sides. Both types work about equally well, but if a one-sided radiant barrier is used, the reflective surface must face the open air space. For example, if a one-sided radiant barrier is laid on top of the insulation with the reflective side facing down and touching the insulation, the radiant barrier will lose most of its effectiveness in reducing heating and cooling loads.

Emissivity is the property that determines how well a radiant barrier will perform. This property is a number between 0 and 1, with lower numbers indicating better potential for performance. The emissivity of typical, clean, unperforated radiant barriers is about 0.03 to 0.05. Hence they will have a reflectivity of 95 to 97 percent. Some materials may have higher emissivities. It is not always possible to judge the emissivity just by visual appearance. Measured emissivity values should be part of the information provided by the manufacturer.

A radiant barrier used in the attic floor application must allow water vapor to pass through it. This is necessary because, during the winter, if there is no effective vapor retarder at the ceiling, water vapor from the living space may condense and even freeze on the underside of a radiant barrier lying on the attic floor. In extremely cold climates or during prolonged periods of cold weather, a layer of condensed water could build up. In more moderate climates, the condensed water could evaporate and pass through the radiant barrier into the attic space. While most uniform aluminum coatings do not allow water vapor to pass through them, many radiant barrier materials do allow passage of water vapor. Some allow water vapor passage through holes or perforations, while others have substrates that naturally allow water vapor passage without requiring holes. However, excessively large holes will increase the emissivity and cause a reduction in the radiant barrier performance. The ability to allow water vapor to pass through radiant barrier materials is not needed for the roof applications.

What should a radiant barrier installation cost?

Costs for an attic radiant barrier will depend on several factors, including the following

- Whether the radiant barrier is installed by the homeowner or by a contractor.
- Whether the radiant barrier will be installed in a new home (low cost) or in an existing home (possibly higher cost if done by a contractor).
- What extra features are desired; e.g., a radiant barrier with perforations and reinforcements may be more expensive than a basic radiant barrier.
- Any necessary retrofit measures such as adding venting (soffit, ridge, etc.)
- Whether the radiant barrier is installed on the attic floor or on the rafters.

Radiant barrier costs vary widely. As with most purchases, some comparison shopping can save you money. A survey of nine radiant barrier manufacturers and contractors representing 14 products, taken by the Reflective Insulation Manufacturers Association (RIMA) in 1989, shows the installed costs of radiant barriers to range as shown in Table 17–1.

Table 17–1. Costs for Radiant Barriers Installed by Contractors

Costs per Square Foot of Material*		
Type of Application	**New Construction**	**Existing Home**
Attic Floor	$0.15–0.30	$0.15–0.30
Roof: stapled to bottom of faces of rafters	$0.15–0.30	$0.20–0.45
Roof: draped over rafters	$0.12–0.35	—
Roof: underside of roof deck	$0.12–0.30	—

*The cost figures in this table are the costs per square foot of radiant barrier. Since the total area of the roof and gables is larger than the area of the ceiling, roof applications will require about 7 to 50 percent more material than an attic floor application, depending upon the shape of the roof.
(Source: Reflective Insulation Manufacturers Association.)

In some cases, radiant barriers are included in a package of energy-saving features sold to homeowners. When considering a "package deal", you may want to ask for an itemized list that includes material and installation costs for all measures included. Then shop around to see what each item would cost if purchased individually before you make a decision.

What should conventional insulation cost?

Heating and cooling bills can also be reduced by adding conventional attic insulation. So that you can have some basis for comparison shopping, typical installed costs for adding various levels of insulation are given in Table 17–2. These costs are typical for insulation installed by contractors. Actual insulation costs will vary from region to region of the country, will vary with the type of insulation selected (blown, or loose-fill, insulation is usually lower in price than batt insulation), and may vary from one local contractor to another. You can expect to deduct 20% to 50% for a do-it-yourself application.

You should always check with your local or state energy office or building code department for current insulation recommendations or see the DOE INSULATION FACT SHEET.

Effect of Radiant Barriers on Heating and Cooling Bills

Have heating and cooling effects been tested?

At present, there is no standardized method for testing the effectiveness of radiant barriers in reducing heating and cooling bills. But numerous field tests have been performed that show, depending on the amount of existing conventional insulation and other factors, radiant

Table 17–2. Costs for Conventional Attic Insulation Installed by Contractors

R-Value	Cost per Square Foot
R-11	$0.27–0.30
R-19	$0.38–0.47
R-22	$0.48–0.51
R-30	$0.54–0.68
R-38	$0.68–0.95

Note: The higher the R-value, the greater the insulating power.
(Source: Residential Construction and Utility Cost data base, developed by NAHB National Research Center, 1986.)

barriers are effective in reducing cooling bills, and also possibly heating bills.

Most of these field tests have been performed in warm climates where a large amount of air-conditioning is used. The Florida Solar Energy Center (FSEC) at Cape Canaveral has performed tests for a number of years using attic test sections, and has also performed tests with full-size houses. A test using a duplex house in Ocala, Florida has been performed by the Mineral Insulation Manufacturers Association. The Tennessee Valley Authority has performed a number of winter and summer tests using small test cells in Chattanooga, Tennessee. The Oak Ridge National Laboratory (ORNL) has performed a series of tests using three full-size houses near Knoxville, Tennessee. The ORNL tests included summer and winter observations. So far, very little testing has been done in climates colder than that of Knoxville. Also, little testing has been done in hot, arid climates such as the southwestern United States.

The tests to date have shown that in attics with R-19 insulation, radiant barriers can reduce summer *ceiling heat gains* by about 16 to 42 percent compared to an attic with the same insulation level and no radiant barrier. These figures are for the average reduction in heat flow through the insulation path. They do not include effects of heat flow through the framing members. See Tables 17–3 and 17–4 for a comparison of measured performance.

Table 17–3. Average Reductions in Ceiling Heat Flow Due to Addition of Radiant Barrier to R-19 Attic Floor Insulation: Summer Cooling Conditions

Radiant Barrier Location	Whole House Tests		Test Cell Tests	
	MIMA	ORNL	FSEC*	TVA
Roof: attached to roof deck	—	—	36–42%	16%
Roof: draped over rafters	20%**	—	—	—
Roof: stapled between rafters	—	—	38–43%	—
Roof: stapled under rafters	24%	25–30%	—	23–30%
Attic Floor***	35%	32–35%	38–44%	40–42%

*Tested at attic air space ventilation rate of five air changes per hour. Typical average ventilation rates are somewhat lower.

**Test was a simulation of draped configuration. The radiant barrier did not extend over the rafters, but was stapled near the joints between the rafters and the roof deck.

***Values are for new and undusted radiant barrier installations; percentages will be lower for aged radiant barriers.

Table 17–4. Average Reductions in Ceiling Heat Flow Due to Addition of Radiant Barrier to R-19 Attic Floor Insulation: Winter Heating Conditions

Radiant Barrier Location	Whole House Tests ORNL	Test Cell Tests TVA
Roof: attached to roof deck	—	4%
Roof: draped over rafters	—	—
Roof: stapled between rafters	—	—
Roof: stapled under rafters	5% to 8%	8%
Attic Floor	1% to 19%	15%

Notes for Tables 17–3 and 17–4:
Caution: These % values do not represent utility bill savings and cannot be represented as such.
NOTE: All measurements represent average heat flows through the insulation path, and do not include effects of heat flow through framing.
Key to Abbreviations:
FSEC: Florida Solar Energy Center
MIMA: Mineral Insulation Manufacturers Association
ORNL: Oak Ridge National Laboratory
TVA: Tennessee Valley Authority

THIS DOES NOT MEAN THAT A 16 TO 42 PERCENT SAVINGS IN UTILITY BILLS CAN BE EXPECTED. Since the ceiling heat gains represent about 15 to 25 percent of the total cooling load on the house, a radiant barrier would be expected to reduce the space cooling portion of summer utility bills by less than 15 to 25 percent. Multiplying this percentage (15 to 25 percent) by the percentage reduction in ceiling heat flow (16 to 42 percent) would result in a 2 to 10 percent reduction in the cooling portion of summer utility bills. However, under some conditions, the percentage reduction of the cooling portion of summer utility bills may be larger, perhaps as large as 17 percent. The percentage reduction in total summer utility bills, which also include costs for operating appliances, water heaters, etc., would be smaller. Tests have shown that the percentage reductions for winter heat losses are lower than those for summer heat gains.

Experiments with various levels of conventional insulation show that the percentage reduction in ceiling heat flow due to the addition of a radiant barrier is larger with lower amounts of insulation. Since the fraction of the whole-house heating and cooling load that comes from the ceiling is larger when the amount of insulation is small, radiant barriers produce the most energy savings when used in combination with lower levels of insulation. Similarly, radiant barriers produce significantly less energy savings when used in combination with high levels of insulation.

Most of the field tests have been done with clean radiant barriers. Laboratory measurements have shown that dust on the surface of aluminum foil increases the emissivity and decreases the reflectivity. This means that dust or other particles on the exposed surface of a radiant barrier will reduce its effectiveness. Radiant barriers installed in locations that collect dust or other surface contaminants will have a decreasing benefit to the homeowner over time.

The attic floor application is most susceptible to accumulation of dust, while downward-facing reflective surfaces used with many roof applications are not likely to become dusty. When radiant barriers are newly installed, some testing shows that the attic floor application will work better than the roof applications. As dust accumulates on the attic floor application, its effectiveness will gradually decrease. After a long enough period of time, a dusty attic floor application will lose much of its effectiveness. Predictive modeling results, based on testing, suggest that a dusty attic floor application will lose about half of its effectiveness after about one to ten years.

Testing of radiant barriers has been primarily concerned with the effect of radiant barriers on the heat gains or losses through the ceiling. Another aspect of radiant barriers may be important when air-conditioning ducts are installed in the attic space. The roof applications of radiant barriers can result in lowered air temperatures within the attic space, which in turn can reduce heat gains by the air flowing through the ducts, thus increasing the efficiency of the air-conditioning system. These changes in heat gains to attic ducts have not been tested; however, computer models have been used to make estimates of the impact on cooling bills.

Not all field tests have been able to demonstrate that radiant barriers or even attic insulation are effective in reducing cooling bills. In a field test performed by ORNL in Tulsa, Oklahoma, using 19 full-sized, occupied houses, neither radiant barriers nor attic insulation produced air-conditioning electricity savings that could be measured. As in all field tests, these results are applicable only to houses with similar characteristics as those tested. Unique characteristics of the houses used in this field test included the facts that the houses were cooled by only one or two window air-conditioning units, that the units were able to cool only a portion of the house, and that the occupants chose to limit their use of the units (initial air-conditioning electricity consumption averaged 1664 kilowatt-hours per year or about $119 per year).

How much will I save on my heating and cooling bills?

Your savings on heating and cooling bills will vary, depending on many factors. Savings will depend on the type of radiant barrier application, the size of your house, whether it is a ranch-style or a two-story house, the amount of insulation in the attic, effectiveness of attic ventilation, the color of the roof, the thermostat settings, the tightness of the building envelope, the actual weather conditions, the efficiency of the heating and cooling equipment, and fuel prices.

Research on radiant barriers is not complete. Estimates of expected savings, however, have been made using a computer program that has been checked against some of the field test data that have been collected. These calculations used weather data from a number of locations to estimate the reductions in heating and cooling loads for a typical house. These load reductions were then converted to savings on fuel bills using average gas furnace and central air-conditioner efficiencies and national average prices for natural gas and electricity.

ASSUMPTIONS. For these calculations, the house thermostat settings were taken to be 78°F in the summer and 70°F in the winter. In the summer, it was assumed that windows would be opened when the outdoor temperature and humidity were low enough to take advantage of free cooling. Also, it was assumed that the roof shingles were dark, and that the roof was not shaded. The furnace efficiency used was 65 percent, and the air-conditioner coefficient of performance (COP) was 2.34. Fuel prices used were 52.7 cents per therm (hundred cubic feet) for natural gas and 7.86 cents per kilowatt-hour for electricity.

Factors that could make your savings less than the ones calculated would be: a summer thermostat setting lower than 78°F, a winter thermostat setting higher than 70°F, keeping the windows closed at all times, lower efficiency furnace or air-conditioner, or higher fuel prices. Factors that could make your savings larger than the ones calculated would be: a summer thermostat setting higher than 78°F, a winter thermostat setting lower than 70°F, light colored roof shingles, shading of the roof by trees or nearby structures, higher efficiency furnace or air-conditioner, and lower fuel prices.

A standard economic calculation was then performed that converts the dollar savings from periods in the future to a "present value". The dollar savings were also adjusted to account for estimates of how prices for natural gas and electricity are predicted to rise in future years. This calculation gives a present value savings in terms of dollars per square foot of ceiling area. When this value is multiplied by the total ceiling area, the result is a number that can be compared with the cost of installing a radiant barrier. If the present value savings for the whole ceiling is greater than the cost of a radiant barrier, then the radiant barrier will be cost effective. A real discount rate of 7 percent, above and beyond inflation, and a life of 25 years were used in the calculations.

Tables 17–5 through 17–7 give present value savings for radiant barriers based on average prices and equipment efficiencies. Table 17–5 applies to the attic floor application, where the effects of dust accumulation have been taken into account. Since dust will accumulate at different rates in different houses, and since the effect of dust on performance is not well known, ranges of values are given for this application. Table 17–6 applies to radiant barriers attached to the bottoms of the rafters, while Table 17–7 applies to radiant barriers either draped over the tops of the rafters or attached directly to the underside of the roof deck. For comparison purposes, the same computer program has also been used to estimate present value savings for putting additional insulation in the attic; these values are listed in Table 17–8. By examining several options, the consumer can compare the relative savings that may be obtained versus the cost of installing the option. Generally, the option with the largest net savings (that is, the present value savings minus the cost) would be the most desirable. However, personal preferences will also enter into a final decision.

If you want a better estimate based on your local fuel prices or other equipment efficiencies, you may use the worksheet in the Appendix. Local fuel prices may be obtained from your local utilities.

Examples of Use of Present Value Tables
Example 1

I live in Orlando, Florida in an 1800-square-foot ranch-style house. I have R-11 insulation in my attic, and the air-conditioning ducts are in the attic. A contractor has quoted a price for a radiant barrier installed on the bottoms of my rafters and on the gable ends for $400. Would this be a good investment?

For this type of radiant barrier, the appropriate table is Table

Table 17–5. Present Value Savings for Dusty Radiant Barrier on Attic Floor

| City | Present Value Savings, Dollars per Square Foot of Attic Floor | | | |
	R-11	R-19	R-30	R-38
Albany, NY	0.04–0.13	0.02–0.06	0.01–0.03	0.01–0.03
Albuquerque, NM	0.05–0.18	0.03–0.10	0.02–0.06	0.01–0.05
Atlanta, GA	0.05–0.17	0.02–0.08	0.01–0.05	0.01–0.04
Bismarck, ND	0.05–0.14	0.02–0.06	0.01–0.04	0.01–0.03
Chicago, IL	0.04–0.13	0.02–0.06	0.01–0.04	0.01–0.03
Denver, CO	0.05–0.15	0.02–0.07	0.01–0.05	0.01–0.04
El Toro, CA	0.04–0.15	0.02–0.07	0.01–0.05	0.01–0.04
Houston, TX	0.05–0.19	0.03–0.10	0.02–0.06	0.01–0.04
Knoxville, TN	0.05–0.17	0.02–0.08	0.02–0.05	0.01–0.04
Las Vegas, NV	0.07–0.24	0.03–0.12	0.02–0.07	0.02–0.06
Los Angeles, CA	0.03–0.08	0.02–0.05	0.01–0.03	0.01–0.02
Memphis, TN	0.05–0.18	0.02–0.09	0.01–0.05	0.01–0.04
Miami, FL	0.06–0.23	0.03–0.12	0.02–0.07	0.01–0.06
Minneapolis, MN	0.04–0.13	0.02–0.06	0.01–0.03	0.01–0.03
Orlando, FL	0.05–0.21	0.03–0.10	0.02–0.07	0.01–0.05
Phoenix, AZ	0.08–0.29	0.04–0.14	0.02–0.08	0.02–0.07
Portland, ME	0.04–0.10	0.02–0.04	0.01–0.02	0.01–0.02
Portland, OR	0.04–0.11	0.02–0.05	0.01–0.03	0.01–0.02
Raleigh, NC	0.05–0.16	0.02–0.08	0.01–0.05	0.01–0.04
Riverside, CA	0.06–0.21	0.03–0.10	0.02–0.06	0.01–0.05
Sacramento, CA	0.05–0.18	0.03–0.09	0.02–0.06	0.01–0.05
Salt Lake City, UT	0.05–0.16	0.02–0.08	0.01–0.05	0.01–0.04
St. Louis, MO	0.05–0.16	0.02–0.08	0.01–0.05	0.01–0.04
Seattle, WA	0.03–0.08	0.01–0.03	0.01–0.02	0.00–0.01
Topeka, KS	0.05–0.17	0.02–0.09	0.02–0.05	0.01–0.04
Waco, TX	0.06–0.21	0.03–0.10	0.02–0.06	0.01–0.05
Washington, DC	0.05–0.15	0.02–0.07	0.01–0.04	0.01–0.04

Note: R-11, R-19, R-30, and R-38 refer to the existing level of conventional insulation.

Note: Values represent range of savings due to variations in rate of dusting and to uncertainties in effect of dust on heat flows. This level of degradation would be typical over 25 years of exposure.

Figures in table are based on a radiant barrier that had an emissivity of 0.05 or less when clean. Savings are for a 25-year period.

Table 17–6. Present Value Savings for Radiant Barrier Attached to Bottoms of Rafters

City	Present Value Savings, Dollars per Square Foot of Attic Floor			
	R-11	**R-19**	**R-30**	**R-38**
Albany, NY	0.17–0.19	0.08–0.09	0.04–0.05	0.03–0.04
Albuquerque, NM	0.24–0.27	0.12–0.15	0.08–0.10	0.06–0.08
Atlanta, GA	0.21–0.25	0.10–0.13	0.06–0.08	0.05–0.07
Bismarck, ND	0.18–0.20	0.09–0.10	0.05–0.06	0.04–0.05
Chicago, IL	0.17–0.19	0.08–0.10	0.05–0.06	0.04–0.05
Denver, CO	0.19–0.22	0.10–0.12	0.06–0.08	0.05–0.07
El Toro, CA	0.19–0.22	0.10–0.12	0.06–0.08	0.05–0.07
Houston, TX	0.23–0.28	0.12–0.15	0.07–0.10	0.05–0.08
Knoxville, TN	0.22–0.25	0.11–0.13	0.07–0.09	0.05–0.07
Las Vegas, NV	0.30–0.36	0.15–0.19	0.09–0.12	0.07–0.10
Los Angeles, CA	0.11–0.12	0.06–0.07	0.04–0.05	0.03–0.04
Memphis, TN	0.23–0.27	0.11–0.14	0.07–0.09	0.06–0.08
Miami, FL	0.28–0.36	0.15–0.20	0.09–0.13	0.07–0.10
Minneapolis, MN	0.18–0.19	0.08–0.10	0.05–0.06	0.03–0.04
Orlando, FL	0.26–0.32	0.13–0.17	0.08–0.12	0.07–0.10
Phoenix, AZ	0.36–0.43	0.17–0.23	0.10–0.14	0.08–0.12
Portland, ME	0.14–0.15	0.06–0.06	0.03–0.04	0.03–0.03
Portland, OR	0.14–0.16	0.07–0.08	0.04–0.05	0.03–0.04
Raleigh, NC	0.20–0.24	0.10–0.12	0.06–0.08	0.05–0.07
Riverside, CA	0.27–0.37	0.13–0.17	0.07–0.10	0.06–0.08
Sacramento, CA	0.23–0.26	0.12–0.14	0.07–0.10	0.06–0.08
Salt Lake City, UT	0.21–0.24	0.10–0.12	0.06–0.08	0.05–0.07
St. Louis, MO	0.21–0.24	0.10–0.13	0.06–0.08	0.05–0.07
Seattle, WA	0.11–0.12	0.05–0.05	0.03–0.03	0.02–0.02
Topeka, KS	0.22–0.26	0.11–0.13	0.07–0.09	0.05–0.07
Waco, TX	0.26–0.31	0.13–0.17	0.08–0.11	0.06–0.09
Washington, DC	0.20–0.23	0.09–0.12	0.06–0.07	0.05–0.06

Note: R-11, R-19, R-30, and R-38 refer to the existing level of conventional insulation.

Note: First value applies to houses with no air-conditioning ducts in attics. Second value applies to houses with air-conditioning ducts in attics.

Figures in table are based on a radiant barrier with an emissivity of 0.05 or less, with the radiant barrier covering the insides of the gables. Savings are for a 25-year period.

Table 17–7. Present Value Savings for Radiant Barrier Draped over Tops of Rafters or Attached to Roof Deck

| City | Present Value Savings, Dollars per Square Foot of Attic Floor | | | |
	R-11	R-19	R-30	R-38
Albany, NY	0.16–0.17	0.07–0.08	0.04–0.05	0.03–0.04
Albuquerque, NM	0.21–0.24	0.11–0.14	0.07–0.09	0.06–0.07
Atlanta, GA	0.19–0.22	0.09–0.12	0.06–0.07	0.04–0.06
Bismarck, ND	0.17–0.18	0.08–0.09	0.05–0.06	0.03–0.04
Chicago, IL	0.15–0.17	0.07–0.09	0.04–0.05	0.03–0.04
Denver, CO	0.17–0.19	0.09–0.10	0.05–0.07	0.05–0.06
El Toro, CA	0.17–0.20	0.09–0.10	0.05–0.07	0.05–0.06
Houston, TX	0.20–0.25	0.10–0.14	0.06–0.09	0.05–0.07
Knoxville, TN	0.19–0.22	0.10–0.12	0.06–0.08	0.05–0.07
Las Vegas, NV	0.27–0.32	0.14–0.17	0.08–0.11	0.06–0.09
Los Angeles, CA	0.10–0.11	0.06–0.06	0.03–0.04	0.03–0.04
Memphis, TN	0.20–0.24	0.10–0.13	0.06–0.08	0.05–0.07
Miami, FL	0.25–0.31	0.13–0.18	0.08–0.11	0.06–0.09
Minneapolis, MN	0.16–0.18	0.07–0.09	0.04–0.05	0.03–0.04
Orlando, FL	0.23–0.28	0.11–0.15	0.07–0.10	0.06–0.09
Phoenix, AZ	0.31–0.38	0.15–0.20	0.09–0.13	0.07–0.11
Portland, ME	0.13–0.13	0.06–0.06	0.03–0.03	0.02–0.03
Portland, OR	0.13–0.14	0.06–0.07	0.04–0.04	0.03–0.04
Raleigh, NC	0.18–0.21	0.09–0.11	0.06–0.07	0.04–0.06
Riverside, CA	0.24–0.33	0.11–0.15	0.07–0.09	0.05–0.07
Sacramento, CA	0.20–0.23	0.10–0.13	0.06–0.08	0.06–0.07
Salt Lake City, UT	0.19–0.21	0.09–0.11	0.05–0.07	0.04–0.06
St. Louis, MO	0.18–0.21	0.09–0.11	0.05–0.07	0.04–0.06
Seattle, WA	0.10–0.11	0.04–0.05	0.02–0.03	0.02–0.02
Topeka, KS	0.20–0.23	0.10–0.12	0.06–0.08	0.05–0.06
Waco, TX	0.23–0.28	0.11–0.15	0.07–0.09	0.05–0.08
Washington, DC	0.18–0.21	0.08–0.10	0.05–0.06	0.04–0.05

Note: R-11, R-19, R-30, and R-38 refer to the existing level of conventional insulation.

Note: First value applies to houses with no air-conditioning ducts in attics. Second value applies to houses with air-conditioning ducts in attics.

Figures in table are based on a radiant barrier with an emissivity of 0.05 or less, with the radiant barrier covering the insides of the gables. Savings are for a 25-year period.

Table 17–8. Present Value Savings for Additional Insulation

| City | Present Value Savings, Dollars per Square Foot of Attic Floor | | | |
	R-11 + R-8*	R-11 + R-19	R-19 + R-11	R-19 + R-19
Albany, NY	0.76	1.10	0.35	0.48
Albuquerque, NM	0.53	0.80	0.28	0.37
Atlanta, GA	0.50	0.71	0.21	0.29
Bismarck, ND	0.90	1.35	0.45	0.61
Chicago, IL	0.69	1.02	0.33	0.45
Denver, CO	0.64	0.96	0.32	0.44
El Toro, CA	0.33	0.48	0.15	0.20
Houston, TX	0.31	0.49	0.18	0.24
Knoxville, TN	0.53	0.78	0.24	0.34
Las Vegas, NV	0.47	0.70	0.23	0.32
Los Angeles, CA	0.22	0.33	0.11	0.15
Memphis, TN	0.52	0.74	0.22	0.31
Miami, FL	0.22	0.34	0.11	0.15
Minneapolis, MN	0.80	1.21	0.42	0.57
Orlando, FL	0.25	0.37	0.12	0.17
Phoenix, AZ	0.53	0.77	0.24	0.33
Portland, ME	0.73	1.09	0.37	0.50
Portland, OR	0.50	0.77	0.27	0.36
Raleigh, NC	0.50	0.72	0.22	0.31
Riverside, CA	0.49	0.70	0.21	0.29
Sacramento, CA	0.44	0.65	0.22	0.29
Salt Lake City, UT	0.65	0.97	0.32	0.44
St. Louis, MO	0.63	0.92	0.29	0.40
Seattle, WA	0.52	0.80	0.28	0.37
Topeka, KS	0.61	0.92	0.31	0.42
Waco, TX	0.41	0.62	0.21	0.28
Washington, DC	0.60	0.88	0.28	0.38

*Denotes existing level of conventional attic insulation (for example, R-11), and additional amount (for example, R-8). Savings are for a 25-year period.

17–6. For Orlando with R-11 insulation, the present value savings is listed as $0.32 when the air-conditioning ducts are in the attic. Multiplying this value by 1800 square feet gives a total of $576. This value exceeds the quoted cost of the radiant barrier of $400, and thus this would be a good investment.

Example 2

I live in Minneapolis, Minnesota in a 2400-square-foot two-story house. I have R-19 insulation in my attic, and have no air-conditioning ducts in the attic. A contractor has quoted a price for a radiant barrier installed on the bottoms of my rafters and on the gable ends for $250. Would this be a good investment? Would investment in another layer of R-19 insulation be a better investment? A contractor has quoted a price of $564 for adding this insulation.

For this type of radiant barrier, the appropriate table is Table 17–6. For Minneapolis with R-19 insulation, the present value savings is listed as $0.08 when there are no air-conditioning ducts in the attic. Since the house is two-story, the ceiling area is 1200 square feet. Multiplying $0.08 by 1200 gives a total of $96. This value is less than the quoted cost of the radiant barrier of $250 and thus this would not be a good investment.

For adding another layer of insulation, the appropriate table is Table 17–8. For Minneapolis, this table gives a present value savings of $0.57 for adding a layer of R-19 insulation to an existing layer of R-19 insulation. Multiplying this value by 1200 square feet gives a total of $684. This value exceeds the quoted cost of the insulation, and thus this would be a good investment.

Important Non-Energy Considerations

Potential for Moisture Condensation—Condensation of moisture can be a concern when a radiant barrier is installed on the attic floor directly on top of conventional insulation. During cold weather, water vapor from the interior of a house may move into the attic. In most cases, this water vapor will not cause any problem because attic ventilation will carry excess vapor away. During cold weather, a radiant barrier on top of the insulation could cause water vapor to condense on the barrier's underside.

Condensation of large amounts of water could lead to the following problems: 1) the existing insulation could become wet and lose some of its insulating value, 2) water spots could appear on the ceiling, and 3) under severe conditions, the ceiling framing could rot.

Some testing has been performed to determine the potential for moisture condensation with perforated radiant barriers laid on top of the insulation. A test was conducted during the winter near Knoxville, Tennessee, using houses that were operated at much higher-than-normal indoor relative humidities. Since this testing did not reveal any significant moisture condensation problems, it is expected that moisture condensation will not be a problem in climates warmer than that of Knoxville. Further testing of radiant barriers is needed to determine if moisture condensation is a problem in climates colder than that of Knoxville.

One precaution for preventing potential moisture problems is the use of perforated or naturally permeable radiant barriers. The higher the perm rating, the less potential for problems. Avoiding high indoor relative humidities, sealing cracks and air leaks in the ceiling, using a vapor retarder below the attic insulation, and providing for adequate attic ventilation are additional precautions.

Attic Ventilation—Attic ventilation is an important consideration. With adequate ventilation, radiant barriers will perform better in summer and excess water vapor will be removed in winter. Unfortunately, specific recommendations for the best type and amount of attic ventilation for use with radiant barriers are not available. Model building codes have established guidelines for the amount of attic ventilation area per square foot of attic floor area to minimize the occurrence of condensation. These guidelines specify one square foot of net free ventilation area for each 150 square feet of attic floor area. This ratio may be reduced to 1 to 300 if a ceiling vapor retarder is present or if high (for example, ridge or gable vents) and low (soffit vents) attic ventilation is used. Since part of the vent area is blocked by meshes or louvers, the net free area of a vent is smaller than its overall dimensions.

Effect of Radiant Barriers on Roof Temperatures—Field tests have shown that radiant barriers can cause a small increase in roof temperatures. Roof-mounted radiant barriers may increase shingle temperatures by 2 to 10°F, while radiant barriers on the attic floor may cause smaller increases of 2°F or less. The effects of these increased temperatures on roof life, if any, are not known.

Fire Ratings—The fire ratings of radiant barriers are important because flame and smoke characteristics of materials exposed to ambient air are critical.

TO MEET CODE, A RADIANT BARRIER MUST BE RATED EITHER CLASS A BY THE NATIONAL FIRE PROTECTION ASSOCIATION (NFPA) OR CLASS I BY THE UNIFORM BUILDING CODE (UBC).

To obtain these ratings, a material must have an ASTM E-84 Flame Spread Index of 25 or less and a Smoke Developed Index of 450 or less. Look for these ratings either printed on the product, or listed on material data sheets provided by the manufacturer.

Installation Procedures

Most residential roofs provide some type of attic or air space that can accommodate an effective radiant barrier system. In new residential construction, it is fairly easy to install a radiant barrier system. Fig. 17–18 shows five possible locations for the installation of an attic radiant barrier system.

Location 1 is a relatively new application, where the radiant barrier material is attached directly to the underside of the roof deck.

Location 2 may offer advantages to the builder during construction of a new house. Before the roof sheathing is applied, the radiant barrier is draped over the rafters or trusses in a way that allows the product to droop 1½ to 3 inches between each rafter.

In Locations 3 and 4, the radiant barrier is attached to either the faces or bottoms of the rafters or top chords of the roof trusses. Locations 3 and 4 may be used with either new construction, or with retrofit of an existing house. With either Location 2, 3 or 4, the space between the roof sheathing and the radiant barrier provides a channel through which warm air can move freely, as shown in Figure 17–19.

In Location 5, the radiant barrier is laid out on the attic floor over the top of existing attic insulation. As discussed previously, this location is susceptible to the effects of dust accumulation. This location is not appropriate when a large part of the attic is used for storage, since the radiant barrier surface must be exposed to the attic space. Also, kitchen and bathroom vents and recessed lights should not be covered with the radiant barrier.

To obtain the best performance with radiant barriers installed in Locations 1 through 4, radiant-barrier material should also be installed over the gable ends. For attics that are open to the space over garages or carports, the radiant barrier should extend eight feet or more into the garage or carport to achieve the same effect as installing a radiant barrier on the gable ends. It is not necessary to cover the gable ends with Location 5.

Radiant barriers that are reflective on one or both sides may be used with any of these locations. However, if the radiant barrier is reflective on only one side, the reflective side *must* face toward the main

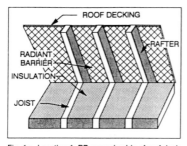

Fig. 1a. Location 1: RB on underside of roof deck. Reflective side must face downwards.

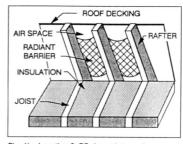

Fig. 1b. Location 2: RB draped over rafters. It is recommended that the reflective side face downwards.

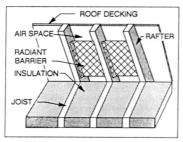

Fig. 1c. Location 3: RB attached between the rafters. It is recommended that the reflective side face downwards.

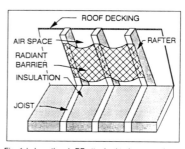

Fig. 1d. Location 4: RB attached to bottoms of rafters. It is recommended that the reflective side face downwards.

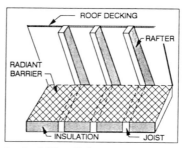

Fig. 1e. Location 5: RB on attic floor over conventional insulation. Reflective side must face upwards.

Fig. 17–18.   Five possible locations for an attic radiant barrier (RB).

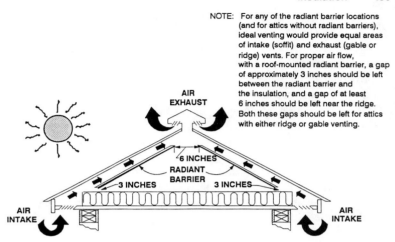

NOTE: For any of the radiant barrier locations
       (and for attics without radiant barriers),
       ideal venting would provide equal areas
       of intake (soffit) and exhaust (gable or
       ridge) vents. For proper air flow,
       with a roof-mounted radiant barrier, a gap
       of approximately 3 inches should be left
       between the radiant barrier and
       the insulation, and a gap of at least
       6 inches should be left near the ridge.
       Both these gaps should be left for attics
       with either ridge or gable venting.

Fig. 17–19.   Cross section of attic showing attic ventilation paths.

attic space for Locations 1 and 5. Since a surface facing downwards is
less likely to have dust settle on it, it is also recommended that the reflective side face downwards toward the main attic space for Locations
2, 3, and 4.

Since proper attic venting is important to obtain the best performance of the radiant barrier, some modification in the attic vents may
be required to achieve expected performance. Where no ridge or gable
vents exist, it is recommended that one or the other be installed. Always
check existing ridge vent systems to ensure that roofing paper is not
blocking the vent opening, and check the soffit vents to ensure that
they have not been covered with insulation.

When installing a radiant barrier, care should be taken not to compress existing insulation present in the attic. The effectiveness of the
existing insulation is dependent upon its thickness, so if it is compressed, its R-value is decreased. For instance, an R-19 batt compressed to 3½ inches (to top of 2 × 4 attic floor joists) would now be
approximately an R-13 batt.

Safety Considerations

• The installer should wear proper clothing and equipment as recommended by the radiant barrier manufacturer. Handling conven-

tional insulation may cause skin, eye, and respiratory system irritation. If in doubt about the effects of the insulation, protective clothing, gloves, eye protection, and breathing protection should be worn.

- Be especially careful with electrical wiring, particularly around junction boxes and old wiring. Never staple through, near, or over electrical wiring. Repair any obvious frayed or defective wiring in advance of radiant barrier installation.
- Work in the attic only when temperatures are reasonable.
- Work with a partner. Not only does it make the job go faster, it also means there is assistance should a problem occur.
- If the attic is unfinished, watch where you walk. If you step in the wrong place, you could fall through the ceiling. Step and stand only on the attic joists or trusses or the center of a strong movable working surface.
- Watch your head. In most attics, roofing nails penetrate through the underside of the roof. A hard hat may be of some use.
- Make sure that the attic space is well ventilated and lighted.
- Do not cover any recessed lights or vents with radiant barrier material (attic floor application).

Interior Walls and Ceilings

Plasterboard

Plasterboard (sometimes referred to as drywall or wallboard) is available in solid sheet material. Standard width is usually 4 feet, with a few types in 2-foot widths. Lengths are from 6 to 16 feet and are available with thicknesses from ¼ to ¾ inch. Edges are tapered to permit smooth edge finishing, although some boards can be purchased with square edges.

The most common length is 8 feet to permit wasteless mountings to the wall studs. This particular size will fit standard 8-foot ceiling heights when vertically mounted. However, recommended mounting for easier handling is horizontal. Two lengths, one above the other, fit the 8-foot ceiling heights when mounted horizontally. In addition to regular wallboard, many other types are available for special purposes.

Regular wallboard comes in thicknesses of ¼, ⅜, ½, and ⅝ inch. The ¼-inch board is used in two- or three-layer wallboard construction and generally has square edges. The most commonly used thickness is ⅜ inch, which can be obtained in lengths up to 16 feet. All have a smooth cream-colored paper covering that takes any kind of decoration.

Insulating (foil-backed) wallboards have aluminum foil which is laminated to the back surface. The aluminum foil creates a vapor barrier and provides reflected insulation value. It can be used in a single-layer construction or as a base layer in two-layer construction. Thicknesses are ⅜, ½, and ⅝ inch, with lengths from 6 to 14 feet. All three thicknesses are available with either square or tapered edges.

Wallboard with a specially formulated core, which provides increased fire-resistance ratings when used in recommended wall and ceiling systems, is made by many companies. National Gypsum Co. calls its product FIRE-SHIELD, and it achieves a one-hour fire rating in single-layer construction over wood studs. It is manufactured in ½- and ⅝-inch thicknesses with lengths from 6 to 14 feet, with tapered edges.

Some suppliers have a lower cost board called backer board. These are used for the base layer in two-layer construction. Thicknesses are ⅜, ½, and ⅝ inch, with a width of 2 feet as well as the usual 4 feet. Length is 8 feet only. The edges are square but the 2-foot wide, in ½-, and ⅝-inch thickness can be purchased with tongue-and-groove edges.

Another type of backer board has a vinyl surface. It is used as a waterproof base for bath and shower areas. The size is 4′ × 11′, which is the right size for enclosing around a standard bathtub. It is available in ½- and ⅝-inch thicknesses with square edges.

A moisture-resistant board is specially processed for use as a ceramic tile base. Both core and facing paper are treated to resist moisture and high humidity. Edges are tapered and the regular tile adhesive seals the edges. The boards are 4 feet wide with lengths of 8 feet and 12 feet.

There is also vinyl-covered decorator plasterboard, which eliminates the need for painting or wallpapering. The vinyl covering is available in a large number of colors and patterns to fit many decorator needs. Many colors in a fabric-like finish, plus a number of wood grain appearances, can be obtained. Only mild soap and water are needed to keep the finish clean and bright. Usual installation is by means of an adhesive, which eliminates nails or screws except at the top and bottom where decorator nails are generally used. The square edges are butted together (Fig. 18–1). They are available in ½- and ⅜-inch thicknesses, 4 feet wide, and 8 feet, 9 feet, and 10 feet long.

A similar vinyl-covered plasterboard, called monolithic, has an extra width of vinyl along the edges. The boards are installed with adhesive or nails. The edges are brought over the fasteners and pasted

CHAPTER 18

Fig. 18–1. Home interior design utilizes vinyl-finish plasterboard. *(Courtesy National Gypsum Co.)*

down then cut flush at the edges. This method hides the fasteners. Extruded aluminum bead and trim accessories, covered with matching vinyl, can be purchased to finish off vinyl-covered plasterboard inside, outside, and ceiling corners.

Plasterboard Ratings

The principal manufacturers of plasterboard maintain large laboratories for the testing of their products for both sound absorption ability and fire retardation. They are based on established industry-accepted national standards and all follow the same procedure. Tests for sound absorption result in figures of merit and those for fire retardation on the length of time a wall will retard a fire. Both vary depending on the material of the wallboard, the layers and thickness, and the method of wall construction.

Construction With Plasterboard

Several alternate methods of plasterboard construction may be used, depending on the application and desires of the customer, or yourself if you are the homeowner. Most residential homes use 2″ × 4″ wall studs so single-layer plasterboard walls are easily installed. Custom built homes, commercial buildings, and party walls between apartments should use double-layer plasterboard construction to provide better sound insulation and fire retardation. For the greatest protection against fire hazards, steel frame partitions, plus plasterboard, provide an all-noncombustible system of wall construction. All-wood or all-steel construction applies to both walls and ceilings.

Plasterboard may also be applied directly to either insulated or uninsulated masonry walls. Plasterboard may be fastened in place by any of several methods such as nails, screws, or adhesives. In addition, there are special resilient furrings and spring clips, both providing added sound deadening.

Single Layer on Wood Studs

The simplest method of plasterboard construction is a single layer of plasterboard nailed directly to the wood studs and ceiling joists (Fig. 18–2). For single-layer construction, ⅜-, ½-, or ⅝-inch plasterboard is recommended.

The ceiling panels should be installed first, then the wall material. Install plasterboard perpendicular to the studs for minimum joint treatment and greater strength. This applies to the ceiling as well. Either nails, screws, or adhesive may be used to fasten the plasterboard. Blue wallboard nails have annular rings for better grip than standard nails and the screws have Phillips heads for easy installation with a power driver. In single-nail installations, as shown in the illustration of Fig. 18–3, the nails should be spaced not to exceed 7 inches on the ceilings and 8 inches on the walls. An alternate method, one which reduces nail popping, is the double-nail method. Install the boards in the single-nail method first, but with nails 12 inches apart. Then, drive a second set of nails about 1½ to 2 inches from the first (Fig. 18–3), from the center out, but not at the perimeter. The first series of nails are then struck again to assure the board being drawn up tight.

Fig. 18–2. The most common interior finishing uses plasterboard, known as drywall construction. *(Courtesy National Gypsum Co.)*

Use 4d cooler-type nails for ⅜-inch regular foil backed or backer board if used, 5d for ½-inch board, and 6d for ⅝-inch board.

Screws are even better than nails for fastening plasterboard, since they push the board up tight against the studding and will not loosen. Where studs and joists are 16 inches on center, the screws can be 12 inches apart on ceilings and 16 inches apart on walls. If the studs are 24 inches oc, the screws should not be over 12 inches apart on the walls. Fig. 18–4 shows a power screwdriver in use. Strike the heads of all nails, and screws, to just below the surface of the board. The dimples will be filled in when joints and corners are finished.

A number of adhesives are available for installing plasterboard. Some are quick-drying, others are slower. Adhesives can be purchased in cartridge or bulk form, as shown in Fig. 18–5. The adhesive is applied in a serpentine bead, as shown in Fig. 18–6, to the facing edges of

Fig. 18–3. In the double-nail method of framing wallboard, the first set of nails is put in place and a second set is driven in place about 1½ to 2 inches from the first. *(Courtesy National Gypsum Co.)*

the studs and joists. Place the wallboard in position and nail it temporarily in place. Use double-headed nails or nails through a piece of scrap plasterboard. When the adhesive has dried, the nails can be removed and the holes filled when sealing the joints.

Plasterboard is butted together but not forced into a tight fit. The treatment of joints and corners is covered later. The adhesive method is ideal for prefinished wallboard, although special nails with colored heads can be purchased for the decorator boards. Do not apply more adhesive than can permit installation within 30 minutes.

Two-Layer Construction

As mentioned, two layers of drywall improves sound deadening and fire retardation. The first layer of drywall or plasterboard is applied as described before. Plasterboard may be less expensive used as a back-

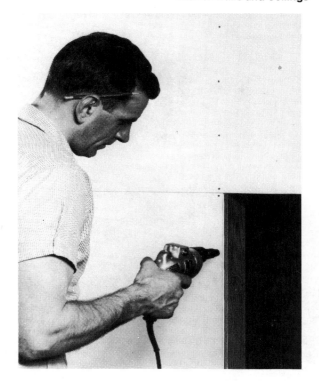

Fig. 18–4. Wallboard fastened to studs with a power screwdriver.
(Courtesy National Gypsum Co.)

ing board. Foil-backed or special sound reducing board may also be used. Nails need only be struck and screws driven with their heads just flush with the surface of the board. The joints will not be given special treatment since they will be covered by the second layer.

The second layer is cemented to the base layer. The facing layer is placed over the base layer and temporarily nailed at the top and bottom. Temporary nailing can be done with either double-headed nails or blocks of scrap wood or plasterboard under the heads. The face layer is left in place until the adhesive is dry.

To make sure of good adhesion, lay bracing boards diagonally from the center of the facing layer to the floor. Prebowing of the boards is

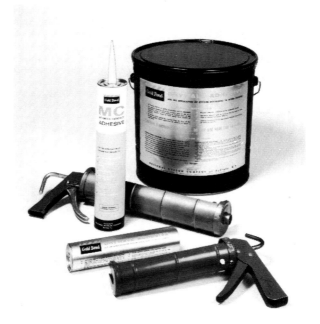

Fig. 18–5. Adhesives are available in bulk or cartridge form. *(Courtesy National Gypsum Co.)*

another method. Lay the panels finish face down across a 2″ × 4″ for a day or two. Let the ends hang free.

Joints of boards on each layer should not coincide but should be separated by about 10 inches. One of the best ways of doing this is to install the base layer horizontally and finish layer vertically. When using adhesive, a uniform temperature must be maintained. If construction is in the winter, the rooms should be heated to somewhere between 55 degrees Fahrenheit and 70 degrees Fahrenheit, and kept well ventilated.

Double-layer construction, with adhesive, is the method used with decorator vinyl finished panelling, except that special matching nails are used at the base and ceiling lines and left permanently in place.

Fig. 18–7 shows the basic double-layer construction, with details for handling corners, and an alternate method of wall construction for improved sound deadening.

Fig. 18–6. Using a cartridge to apply a bead of adhesive to wall studs. *(Courtesy National Gypsum Co.)*

Other Methods of Installation

Sound transmission through a wall can be further reduced if the vibrations of the plasterboard can be isolated from the studs or joists. Two methods are available to accomplish this. One uses metal furring strips whose edges are formed in such a way as to give them flexibility. The other is by means of metal push-on clips with bowed edges also for flexibility. These are used only on ceilings.

Fig. 18–8 shows the appearance and method of installing resilient furring channels. They are 12-foot long strips of galvanized steel and include predrilled holes spaced every inch. This permits nailing to studs 16 or 24 inches oc. Phillips head self-tapping screws are power driven through the plasterboard and into the surface of the furring strips.

As shown in Fig. 18–8, strips of $3'' \times \frac{1}{2}''$ plasterboard are fastened to the sole and plate at top and bottom to give the plasterboard a solid base. For best sound isolation, the point of intersection between the wall and floor should be caulked prior to application of the baseboard.

On ceiling joists, use two screws at each joist point to fasten the furring strips. Do not overlap ends of furring strips, leave about a $\frac{3}{8}$-

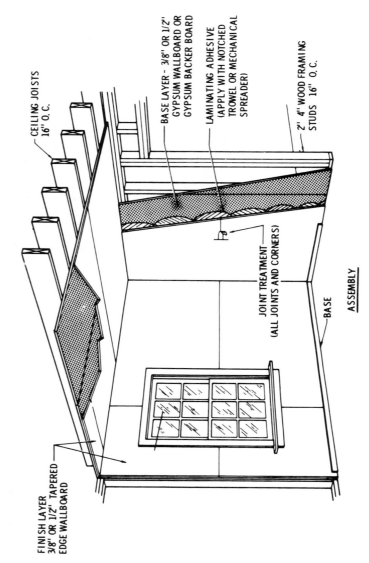

CEILING JOISTS
16" O. C.

BASE LAYER - 3/8" OR 1/2"
GYPSUM WALLBOARD OR
GYPSUM BACKER BOARD

LAMINATING ADHESIVE
(APPLY WITH NOTCHED
TROWEL OR MECHANICAL
SPREADER)

2" 4" WOOD FRAMING
STUDS 16" O. C.

JOINT TREATMENT
(ALL JOINTS AND CORNERS)

ASSEMBLY

BASE

FINISH LAYER
3/8" OR 1/2" TAPERED
EDGE WALLBOARD

Fig. 18–7. A cross-section view of double-layer wall construction, details for handling corners, and special wall construction for sound deadening. *(Courtesy National Gypsum Co.)*

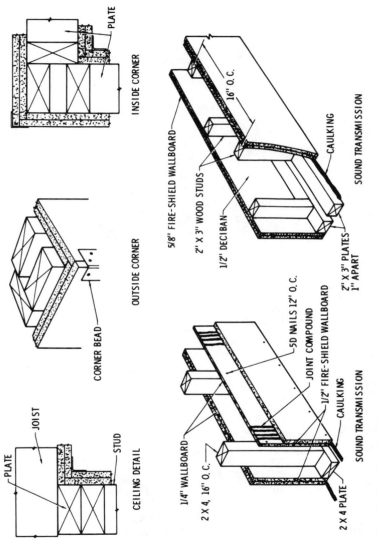

PLATE

INSIDE CORNER

CORNER BEAD

OUTSIDE CORNER

PLATE

JOIST

STUD

CEILING DETAIL

5/8" FIRE-SHIELD WALLBOARD

2" X 3" WOOD STUDS

1/2" DECIBAN

16" O. C.

CAULKING

SOUND TRANSMISSION

2" X 3" PLATES
1" APART

5D NAILS 12" O. C.

JOINT COMPOUND

1/2" FIRE-SHIELD WALLBOARD

CAULKING

SOUND TRANSMISSION

1/4" WALLBOARD

2 X 4, 16" O. C.

2 X 4 PLATE

Fig. 18–7. Continued.

471

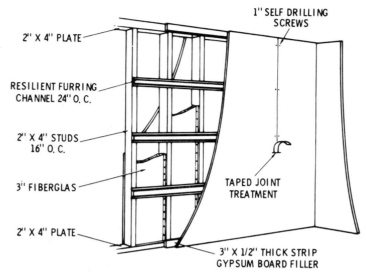

PERSPECTIVE

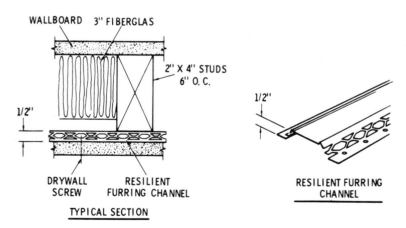

Fig. 18–8. Resilient furring strips, with expanded edges, may be installed to isolate the plasterboard from the stud. *(Courtesy National Gypsum Co.)*

inch space between ends. Use only ½- or ⅝-inch wallboard with this system.

Suspending plasterboard ceilings from spring clips is a method of isolating the vibrations from the ceiling joists and floor above. Details are shown in Fig. 18–9.

The push-on clips are placed on $1'' \times 2''$ wood furring strips. The clips are then nailed to the sides of the ceiling joists with short annular-ringed or cooler-type nails. With furring strips installed, they will be against the nailing edge of the joists. This provides a solid foundation for nailing the plasterboard to the clips. The weight of the plasterboard after installation will stretch the clips to a spring position.

The furring strips must be of high quality material so they will not buckle, twist, or warp. Nails for attaching the wallboard should be short enough so they will not go through the furring strips and into the joist edges. The wallboard must be held firmly against the strips while it is being nailed. Expansion joints must be provided every 60 feet or for every 2,400 ft^2 of surface.

It is essential that the right number of clips be used on a ceiling, depending on the thickness of the wallboard and whether single or double layered. The spring or efficiency of the clips is affected by the weight.

Wall Tile Backing

At least two types of plasterboard are available as backing material for use in tiled areas where moisture protection is important, such as tub enclosures, shower stalls, powder rooms, kitchen-sink splash boards, and locker rooms. One is a vinyl-covered plasterboard and the other is a specially designed board material to prevent moisture penetration.

The vinyl-covered board is $4' \times 11'$, the right length for a tub enclosure. By scoring it and snapping it to length, the vinyl covering can be a continuous covering all around the tub and no corner sealing is required. This is shown in one of the sketches in Fig. 18–10. This board has square edges and is nailed or screwed to the studs without the need to countersink the nail or screws. The waterproof tile adhesive is applied and the tile is installed over the board. This board is intended for full tile treatment and is not to extend beyond the edges of the tile.

Unlike vinyl-surfaced board, the moisture-resistant board may be

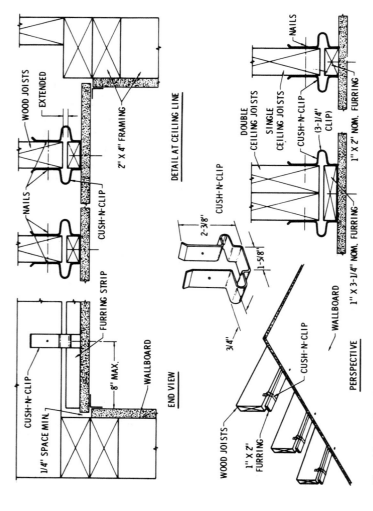

Fig. 18–9. Details for installing spring clips to ceiling joist. Clips suspend plasterboard from the stud. (*Courtesy National Gypsum Co.*)

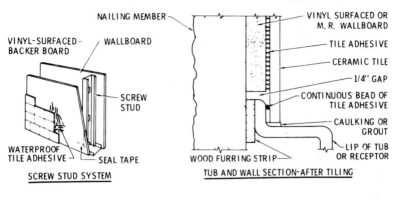

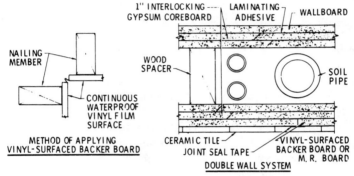

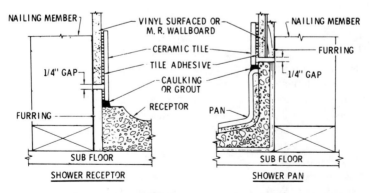

Fig. 18–10. Details for installing vinyl-locked and moisture-resistant wallboard for bath and shower use.

extended beyond the area of the tile. The part extending beyond the tile may be painted with latex, oil-based paint, or papered. It has tapered edges and is installed and treated in the same manner as regular plasterboard. Corners are made waterproof by the tile adhesive. The sketches in Fig. 18–10 show details for installation of either board.

Plasterboard over Masonry

Plasterboard may be applied to inside masonry walls in any one of several methods: directly over the masonry using adhesive cement; over wood or furring strips fastened to the masonry; polystyrene insulated walls by furring strips over the insulation; or by lamination directly to the insulation.

Regular or prefinished drywall may be laminated to unpainted masonry walls such as concrete or block interior partitions and to the interior of exterior walls above or below grade. While nearly all of the adhesive available may be used either above or below grade, the regular joint compounds are recommended for use only above grade. Exterior masonry must be waterproofed below grade and made impervious to water above grade.

The masonry must be clean and free of dust, dirt, grease, oil, loose particles, or water soluble particles. It must be plumb, straight, and in one plane. Fig. 18–11 shows the method of applying the adhesive to masonry walls. Boards may be installed either horizontally or vertically. Ceiling wallboard should normally be applied last to allow nailing temporary bracing to the wood joists. The wallboard should be installed with a clearance of ⅛ inch or more from the floor to prevent wicking. If there are expansion joints in the masonry, cut the wallboards to include expansion joints to match that of the masonry.

The sketches of Fig. 18–12 show the installation of wallboard to masonry with furring strips. Furring strips may be wood or U-shaped metal as used for foam insulation. The furring strips are fastened to the masonry with concrete nails. They may be mounted vertically or horizontally but there must be one horizontal strip along the base line. The long dimension of the wallboards should be perpendicular to the furring strips. Furring strips are fastened 24 inches oc. The wallboard is attached to the furring strips in the usual way. Use self-drilling screws on the metal furring strips and nails or screws on the wood strips.

Insulation may be applied between the plasterboard and the ma-

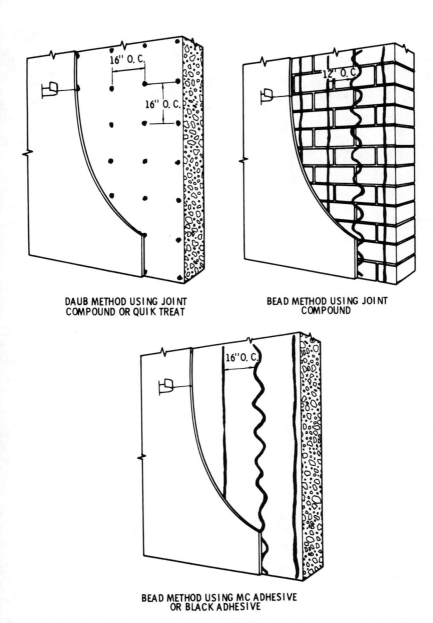

DAUB METHOD USING JOINT
COMPOUND OR QUIK TREAT

BEAD METHOD USING JOINT
COMPOUND

BEAD METHOD USING MC ADHESIVE
OR BLACK ADHESIVE

Fig. 18–11.   Method of applying adhesive directly to masonry walls.
(Courtesy National Gypsum Co.)

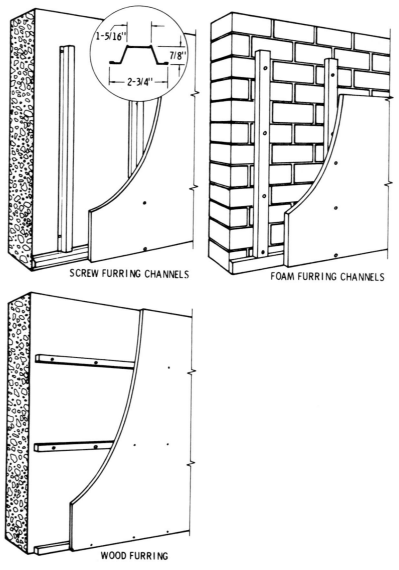

Fig. 18–12. Three types of furring strips used between masonry walls and plasterboard. *(Courtesy National Gypsum Co.)*

sonry, using urethane foam sheets or extruded polystyrene. Plasterboard is then secured to the wall either with the use of furring strips over the insulation or by laminating directly to the insulation.

Special U-shaped metal furring strips are installed over the insulation. This is shown in Fig. 18–13. The insulation pads may be held against the wall while the strips are installed, or they can to be held in place with dabs of plasterboard adhesive. Use a fast-drying adhesive. If some time is to elapse before the furring strips and plasterboard are to be applied, dot the insulation with adhesive every 24 inches in both directions on the back on the insulation pads.

The metal furring strips are fastened in place with concrete nails. Place them a maximum of 24 inches apart, starting about 1 inch to 1½ inches from the ends. The strips may be mounted vertically or horizontally, as shown in Fig. 18–13. Fasten the wallboard to the furring strips

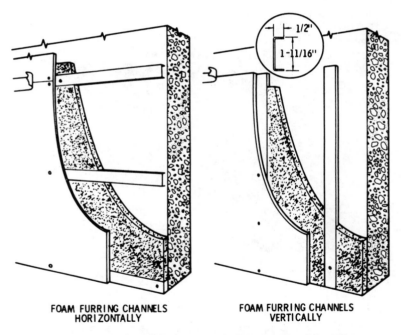

FOAM FURRING CHANNELS
HORIZONTALLY

FOAM FURRING CHANNELS
VERTICALLY

Fig. 18–13.   U-shaped furring channels installed horizontally or vertically over foam insulating material. *(Courtesy National Gypsum Co.)*

with a power driver (Fig. 18–14), using self-drilling screws. Screws must not be any longer than the thickness of the wallboard plus the ½-inch furring strip. Space screws 24 inches oc for ½-inch and ⅝-inch board, 16 inches oc for ⅜-inch board.

Plasterboard may be laminated directly to the foam insulation. The sketches in Fig. 18–15 show how this is done for both horizontally and vertically mounted boards. Install wood furring strips onto the masonry wall, for the perimeter of the wallboard. The strips should be 2 inches wide and ⅟₃₂ inch thicker than the foam insulation. Include furring where the boards join, as shown in the lefthand sketch of Fig. 18–15.

Apply a ⅜-inch diameter bead over the back of the foam insulation and in a continuous strip around the perimeter. Put the bead dots about 16 inches apart. Apply the foam panels with a sliding motion and hand press the entire panel to ensure full contact with the wall surface. For some adhesives, it is necessary to pull the panel off the wall to allow flash-off of the solvent. Then reposition the panel. Read the instructions with the adhesive purchased.

Fig. 18–14. Special self-drilling screws are power-driven through the plasterboard and into the metal furring strip. *(Courtesy National Gypsum Co.)*

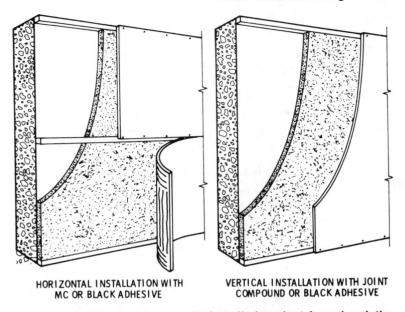

HORIZONTAL INSTALLATION WITH
MC OR BLACK ADHESIVE

VERTICAL INSTALLATION WITH JOINT
COMPOUND OR BLACK ADHESIVE

Fig. 18–15. Plasterboard may be installed against foam insulation with adhesive. *(Courtesy National Gypsum Co.)*

Coat the back of the wallboard and press it against the foam insulation. Nail or screw the board to the furring strips. The nails or screws must not be too long or they will penetrate the furring strips and press against the masonry wall. The panels must clear the floor by ⅛ inch. About any type of adhesive may be used with urethane foam but some adhesives cannot be used with polystyrene. Also, in applying prefinished panels, be careful on the choice of adhesive. Some of the quick-drying types are harmful to the finish. Read the instructions on the adhesive before you buy.

Fireproof Walls and Ceilings

For industrial applications, walls or ceilings may be made entirely of noncombustible materials, using galvanized steel wall and ceiling construction faced with plasterboard. Assembly is by self-drilling screws, driven by a power screwdriver with a Phillips bit.

Wall Partitions

Construction consists of U-shaped track which may be fastened to existing ceilings, or the steel member ceilings, and to the floors. U-shaped steel studs are screwed to the tracks. Also available are one-piece or three-piece metal door frames. The three-piece frames are usually preferred by contractors since they permit finishing the entire wall before the door frames are installed.

The sketches in Fig. 18–16 show details of wall construction as well as the cross section of the track and stud channels. For greater sound deadening, double-layer board construction with resilient furring strips between the two layers is recommended. This is shown in a cutaway view in the lower righthand corner of Fig. 18–16. Heavier material and wider studs are required for high wall construction.

Fig. 18–17 and 18–18 show details for intersecting walls, jambs, and other finishing needs. Chase walls are used (Fig. 18–19), where greater interior wall space is needed (between the walls) for equipment.

Fig. 18–20 shows how brackets are attached for heavy loads. Table 18–1 lists the allowable load for bolts installed directly to the plasterboard.

Steel Frame Ceilings

Three types of furring members are available for attaching plasterboard to ceilings. They are screw-furring channels, resilient screw-furring channels, and screws studs. Self-drilling screws attach the wallboard to the channels. These are illustrated in Fig. 18–21. Any of the three may be attached to the lower chord of steel joists or carrying channels in suspended ceiling construction. Either special clips or wire ties are used to fasten the channels. Fig. 18–22 shows the details in a complete ceiling assembly.

Semisolid Partitions

Complete partition walls may be made of all plasterboard, at a considerable savings in material and time over the regular wood and plas-

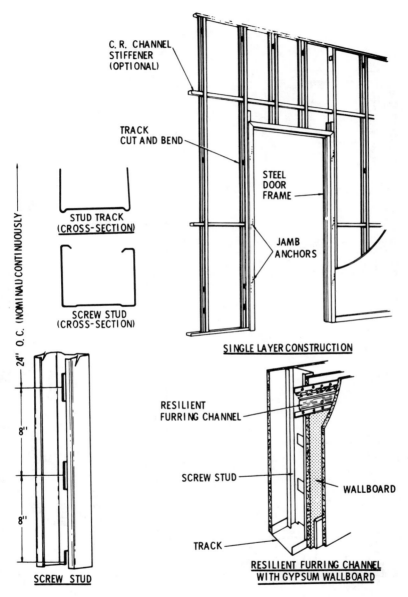

C. R. CHANNEL
STIFFENER
(OPTIONAL)

TRACK
CUT AND BEND

STEEL
DOOR
FRAME

JAMB
ANCHORS

STUD TRACK
(CROSS-SECTION)

SCREW STUD
(CROSS-SECTION)

24" O. C. (NOMINAL) CONTINUOUSLY

8"

8"

SCREW STUD

SINGLE LAYER CONSTRUCTION

RESILIENT
FURRING CHANNEL

SCREW STUD

WALLBOARD

TRACK

RESILIENT FURRING CHANNEL
WITH GYPSUM WALLBOARD

Fig. 18–16. Details of metal wall construction using metal studs and tracks. *(Courtesy National Gypsum Co.)*

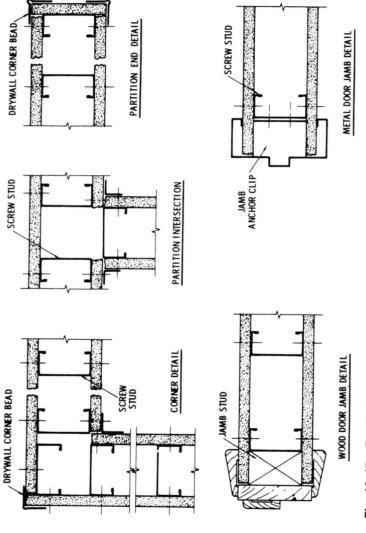

DRYWALL CORNER BEAD

SCREW STUD

PARTITION END DETAIL

SCREW STUD

PARTITION INTERSECTION

DRYWALL CORNER BEAD

SCREW STUD

CORNER DETAIL

JAMB ANCHOR CLIP

SCREW STUD

METAL DOOR JAMB DETAIL

JAMB STUD

WOOD DOOR JAMB DETAIL

Fig. 18–17. Details for intersections and jambs using metal wall construction. *(Courtesy National Gypsum Co.)*

terboard types. These walls are thinner than the usual $2'' \times 4''$ stud wall, but they are not load-bearing walls.

Fig. 18–23 shows cross-sectional views of the all-plasterboard walls, with details for connecting them to the ceiling and floor. Also shown is a door frame and a section of the wall using the baseboard. Partitions may be 2¼, 2⅝, or 2⅞ inches thick, depending on the thickness of the wallboard used and the number of layers. The center piece is not solid, but a piece of 1- or 1⅝-inch plasterboard about 6 inches wide acting as a stud element.

Fig. 18–24 shows how the wall is constructed. The boards are vertically mounted, so their length must be the same distance as from the floor to the ceiling, but not exceeding 10 feet. The layers are prelaminated on the job, then raised into position.

Place two pieces of wallboard on a flat surface with face surfaces together. These must be the correct length for the height of the wall. Cut two studs 6 inches wide and a little shorter than the length of the large wallboard. Spread adhesive along the entire length of the two plasterboard studs and put one of them, adhesive face down, in the middle of the top large board. It should be 21 inches from the edges and equidistant from the ends. Set the other stud, adhesive face up, flush along one edge of the large plasterboard.

Place two more large panels on top of the studs, with edge of the bundle in line with the outer edge of the uncoated stud. Temporarily place a piece of plasterboard under the opposite edge for support. In this system each pair of boards will not be directly over each other, but alternate bundles will protrude. Continue the procedure until the required number of assemblies are obtained. Let dry, with temporary support under the overhanging edges. Fig. 18–24 shows the lamination process.

To raise the wall, first install a 24-inch wide starter section of wallboard with one edge plumbed against the intersecting wall. Spread adhesive along the full length of the plasterboard stud of one wall section and erect it opposite to the 24-inch starter section. The free edge of the starter section should center on the stud piece. Apply adhesive to the plasterboard stud of another section and erect it adjacent to the starter panel. Continue to alternate sides as you put each section in place.

Some plasterboard manufacturers make thick solid plasterboards for solid partition walls. Usually they will laminate two pieces of 1-inch-thick plasterboard to make a 2-inch-thick board. Chase walls, elevator

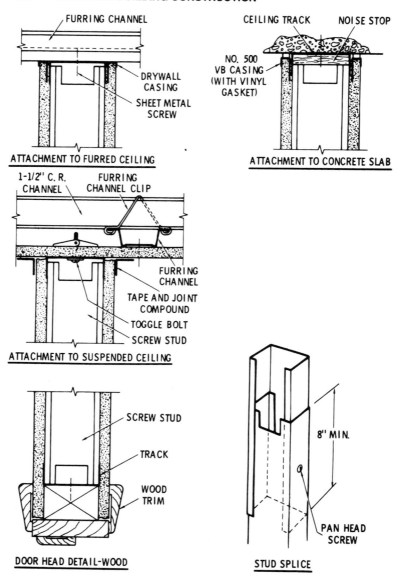

Fig. 18–18. **How to handle ceiling and base finishing.** *(Courtesy National Gypsum Co.)*

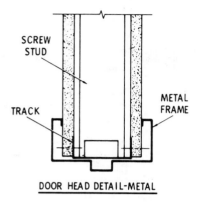

DOOR HEAD DETAIL-METAL

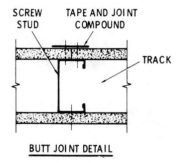

BUTT JOINT DETAIL

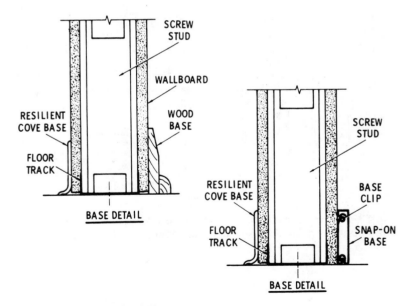

BASE DETAIL

BASE DETAIL

Fig. 18–18.   (Continued)

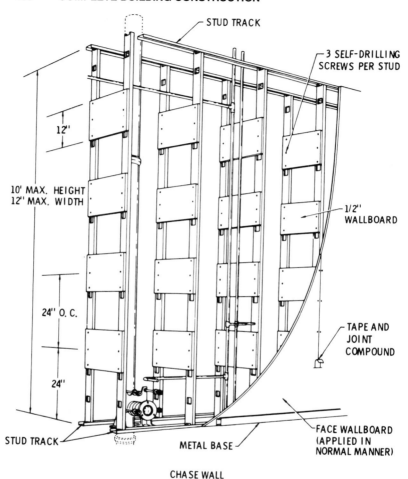

STUD TRACK

3 SELF-DRILLING SCREWS PER STUD

12"

10' MAX. HEIGHT
12" MAX. WIDTH

1/2" WALLBOARD

24" O. C.

TAPE AND JOINT COMPOUND

24"

STUD TRACK

METAL BASE

FACE WALLBOARD (APPLIED IN NORMAL MANNER)

CHASE WALL

Fig. 18–19. Deep-wall construction where space is needed for reasons such as plumbing and heating pipes. *(Courtesy National Gypsum Co.)*

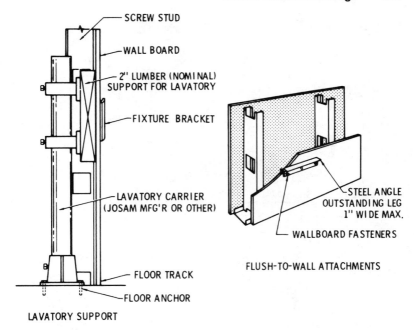

SCREW STUD

WALL BOARD

2" LUMBER (NOMINAL) SUPPORT FOR LAVATORY

FIXTURE BRACKET

LAVATORY CARRIER (JOSAM MFG'R OR OTHER)

FLOOR TRACK

FLOOR ANCHOR

LAVATORY SUPPORT

STEEL ANGLE OUTSTANDING LEG 1" WIDE MAX.

WALLBOARD FASTENERS

FLUSH-TO-WALL ATTACHMENTS

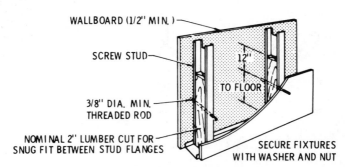

WALLBOARD (1/2" MIN.)

SCREW STUD

12" TO FLOOR

3/8" DIA. MIN. THREADED ROD

NOMINAL 2" LUMBER CUT FOR SNUG FIT BETWEEN STUD FLANGES

SECURE FIXTURES WITH WASHER AND NUT

FOR WALL HUNG FURNITURE (BOTH SIDES OF PARTITION) ALLOWABLE 60 FT. LBS. PER FASTENER- (2'-0" O.C. STUD SPACING)

Fig. 18–19. (Continued)

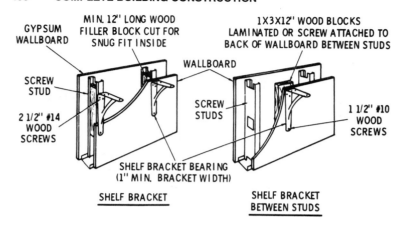

SHELF BRACKET

SHELF BRACKET
BETWEEN STUDS

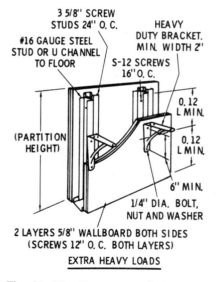

EXTRA HEAVY LOADS

Fig. 18–20. Recommended method for fastening shelf brackets to plasterboard. *(Courtesy National Gypsum Co.)*

Table 18–1. Allowable Carrying Loads for Anchor Bolts

TYPE FASTENER	SIZE	ALLOWABLE LOAD	
		½″ WALLBOARD	⅝″ WALLBOARD
HOLLOW WALL SCREW ANCHORS	⅛″ dia. SHORT	50 LBS.	--
	³⁄₁₆″ dia. SHORT	65 LBS.	--
	¼″, ⁵⁄₁₆″, ⅜″ dia. SHORT	65 LBS.	--
	³⁄₁₆″ dia. LONG	--	90 LBS.
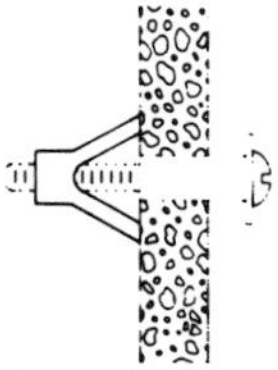	¼″, ⁵⁄₁₆″, ⅜″ dia. LONG	--	95 LBS.
COMMON TOGGLE BOLTS	⅛″ dia.	30 LBS.	90 LBS.
	³⁄₁₆″ dia.	60 LBS.	120 LBS.
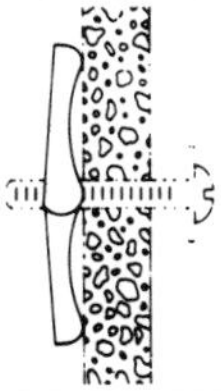	¼″, ⁵⁄₁₆″, ⅜″ dia.	80 LBS.	120 LBS.

(Courtesy National Gypsum Co.)

shafts, and stairwells usually require solid and thick walls which can be of all plasterboard at reduced cost of construction. There is no limit to the thickness that can be obtained—it depends on the requirements.

Joint Finishing

Flat and inside corner joints are sealed with perforated tape embedded in joint compound and with finishing coats of the same compound. Outside corners are protected with a metal bead, nailed into place and finished with joint compound, or a special metal-backed tape, which is also used on inside corners.

Two types of joint compound are generally available—regular, which takes about 24 hours to dry, and quick-setting, which takes about 2½ hours to dry. The installer must make the decision on which type to

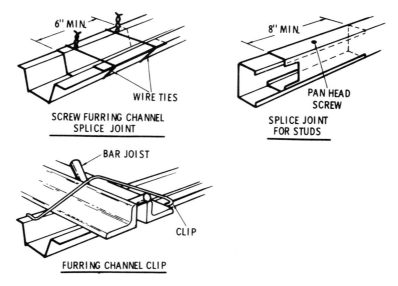

Fig. 18–21. Metal ceiling channels may be fastened to metal joists by wire ties or special clips. *(Courtesy National Gypsum Co.)*

use, depending on the amount of jointing work he has to do and when he will apply subsequent coats.

Begin by spotting nail heads with joint compound. Use a broad knife to smooth out the compound. Apply compound over the joint (Fig. 18–25). Follow this immediately by embedding the perforated tape over the joint (Fig. 18–26). Fig. 18–27 is a closeup view showing the compound squeezed through the perforations in the tape for good keying. Before the compound dries, run the broad knife over the tape to smooth down the compound and to level the surface (Fig. 18–28).

After the first coat has dried, apply a second coat (Fig. 18–29) thinly and feather it out 3 to 4 inches on each side of the joint. Also apply a second coat to the nail spots. When the second coat has dried, apply a third coat also thinly. Feather it out to about 6 or 7 inches from the joint (Fig. 18–30). Final nail spotting is also done at this time.

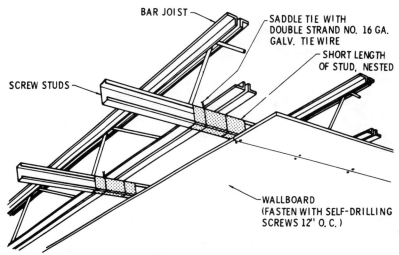

BAR JOIST

SADDLE TIE WITH
DOUBLE STRAND NO. 16 GA.
GALV. TIE WIRE

SHORT LENGTH
OF STUD, NESTED

SCREW STUDS

WALLBOARD
(FASTEN WITH SELF-DRILLING
SCREWS 12" O. C.)

SCREW STUDS IN CEILING SYSTEMS

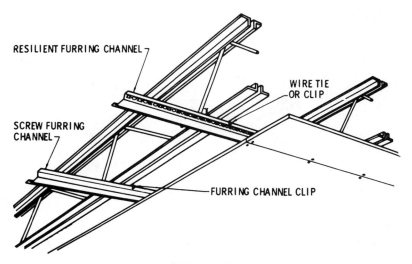

RESILIENT FURRING CHANNEL

WIRE TIE
OR CLIP

SCREW FURRING
CHANNEL

FURRING CHANNEL CLIP

FURRING CHANNELS

Fig. 18–22. Complete details showing two methods of mounting wallboard to suspended ceiling structures. *(Courtesy National Gypsum Co.)*

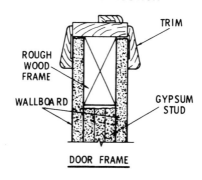

DOOR FRAME

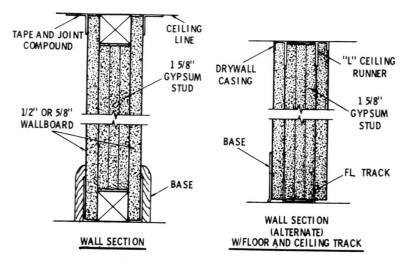

Fig. 18–23. Cross-sectional view of a semisolid all-plasterboard wall. *(Courtesy National Gypsum Co.)*

Inside Corners

Inside corners are treated in the same way as flat joints, with one exception. The tape must be cut to the proper size and creased down the middle. Apply it to the coated joint (Fig. 18–31) and follow with the treatment mentioned above, but to one side at a time. Let the joint dry before applying the second coat of compound and the same for the third coat.

LAMINATING PANEL ASSEMBLIES

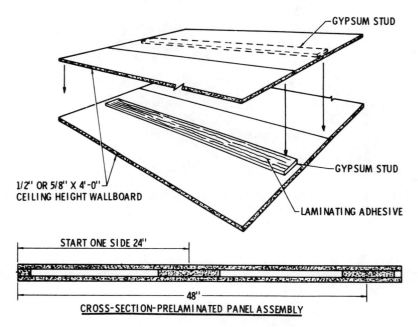

Fig. 18–24. **Plasterboards are laminated together with wide pieces of plasterboard acting as wall studs.** *(Courtesy National Gypsum Co.)*

Outside Corners

Outside corners need extra reinforcement because of the harder knocks they may take. Metal corner beads (Fig. 18–32) are used. Nail through the bead into the plasterboard and framing. Apply joint compound over the beading, using a broad knife, as shown in Fig. 18–29. The final treatment is the same as for other joints. The first coat should be about 6 inches wide and the second coat about 9 inches. Feather out the edges and work the surface smooth with the broad knife and a wet sponge.

Fig. 18–25. The first step in finishing joints between plasterboard. *(Courtesy National Gypsum Co.)*

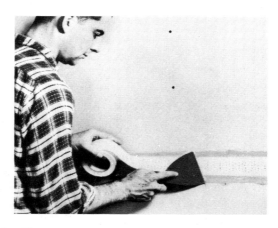

Fig. 18–26. Place perforated tape over the joint and embed it in the joint compound. *(Courtesy National Gypsum Co.)*

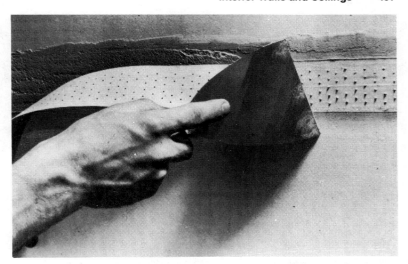

Fig. 18–27. Close-up view showing how a properly embedded tape will show beads of the compound through the perforations. *(Courtesy National Gypsum Co.)*

Fig. 18–28. Using a broad knife to smooth out the compound and feather edges. *(Courtesy National Gypsum Co.)*

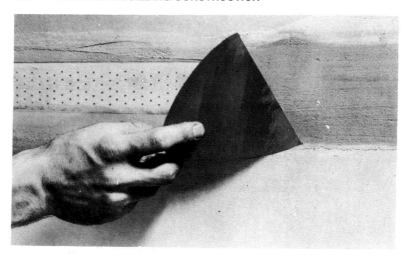

Fig. 18–29. Applying second coat of compound over perforated tape after first coat has dried. This coat must be applied smooth with a good feathered edge. *(Courtesy National Gypsum Co.)*

Fig. 18–30. The final coat is applied with a wide knife, carefully feathering edges. A wet sponge will eliminate need for further smoothing when dry. *(Courtesy National Gypsum Co.)*

Fig. 18–31.   Inside corners are handled in the same manner as flat joints. A specially shaped corner knife is also good. *(Courtesy National Gypsum Co.)*

Prefinished Wallboard

Prefinished wallboard is surfaced with a decorative vinyl material. Since it is not painted or papered after installation, in order to maintain a smooth and unmarred finish, its treatment is slightly different than that of standard wallboard. In most installations, nails and screws are avoided (except at top and bottom) and no tape and joint compound are used at the joints.

Three basic methods are used to install prefinished wallboard:

Fig. 18–32. Outside corners are generally reinforced with a metal bead. It is nailed in place through the plasterboard into the stud. *(Courtesy National Gypsum Co.)*

1. Nailing to 16 inches oc studs or to furring strips, but using colored and matching nails available for the purpose.
2. Cementing to the studs or furring strips with adhesive, nailing only at the top and bottom. These nails may be matching colored nails or plain nails, which are covered with matching cove molding and base trim.
3. Laminating to a base layer on regular wallboard, or to old wallboard in the case of existing wall installations.

Before installing prefinished wallboard, a careful study of the wall arrangement should be made. Joints should be centered on architectural features such as fireplaces and windows. End panels should be of equal width. Avoid narrow strips as much as possible.

Fig. 18–33. Applying compound over the corner bead. *(Courtesy National Gypsum Co.)*

Decorator wallboard is available in lengths to match most wall heights without further cutting. It should be installed vertically, and should be about ⅛ inch a shorter than the actual height, so it will not be necessary to force it into place. Prefinished wallboards have square edges and are butted together at the joints, with or without the vinyl surface lapped over the edges.

As with standard wallboard, prefinished wallboard is easily cut into narrow pieces by scoring and snapping. Place the board on a flat surface with the vinyl side up. Score the vinyl side with a dimension about 1 inch wider than the width of the panel (Fig. 18–34). Turn vinyl surface face down and score the back edge to the actual dimension desired. In both cases use a good straight board as a straightedge. Place

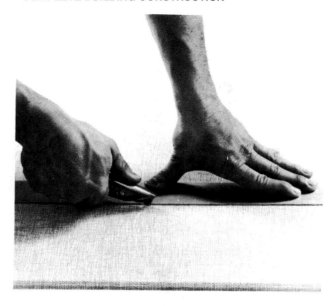

Fig. 18–34. When cutting vinyl-finished wallboard, the cut should be about 1 inch wider than the desired width. This is to allow an inch of extra vinyl for edge treatment. *(Courtesy National Gypsum Co.)*

the board over the edge of a long table and snap the piece off (Fig. 18–35). This will leave a piece of vinyl material hanging over the edge. Fold the material back and tack it into place onto the back of the wallboard (Fig. 18–36).

Cutouts are easily made on wallboard with a fine-toothed saw. Where a piece is to be cut out, as for a window, saw along the narrower cuts, then score the longer dimension and snap off the piece (Fig. 18–37). Circles are cut out by first drilling a hole large enough to insert the end of the keyhole saw (Fig. 18–38). Square cutouts for electrical outlet boxes need not be sawed but can be punched out. Score through the vinyl surface, as shown in Fig. 18–39. Give the area a sharp blow and it will break through.

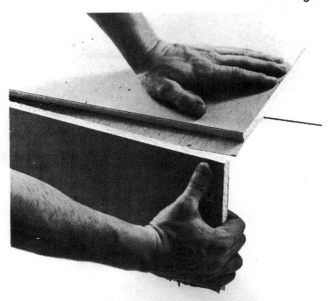

Fig. 18–35. Score the back of the wallboard to the actual width desired, leaving a 1-inch width of vinyl. *(Courtesy National Gypsum Co.)*

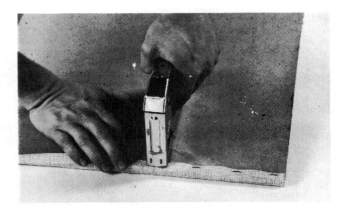

Fig. 18–36. Fold the 1-inch vinyl material over the edge of the wallboard and tack in place. This will provide a finished edge without further joint treatment. *(Courtesy National Gypsum Co.)*

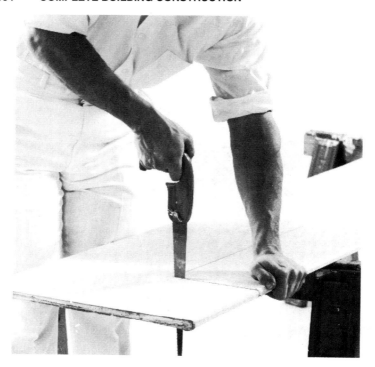

Fig. 18–37. Cutting vinyl-covered wallboard for other openings.
(Courtesy National Gypsum Co.)

Installing Prefinished Wallboard

Prefinished wallboard may be nailed to studs or furring strips with decorator or colored nails. In doing so, however, the job must be done carefully so the nails make a decorative pattern. Space them every 12 inches and not less than ⅜ inch from the panel edges (Fig. 18–40). When nailing directly to the studs, to be sure the studs are straight and flush. If warped, they may require shaving down at high spots or shimming up at low spots. To avoid extra work, carpenters should be instructed, on new construction, to select the best 2″ × 4″ lumber and do a careful job of placing it. Where studs are already in place, it may be easier to install furring strips horizontally and shim them during instal-

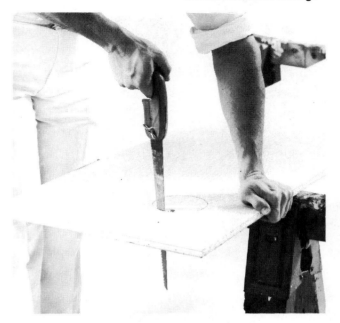

Fig. 18–38.   Circle cutouts are made by first drilling a hole large enough to insert a keyhole saw. *(Courtesy National Gypsum Co.)*

lation (Fig.18–41). Use 1″ × 3″ wood spaced 16 inches apart. Over an existing broken plaster wall, or solid masonry, install furring strips as shown in Fig. 18–42. Use concrete nails to fasten furring strips on solid masonry.

A good way of installing prefinished wallboard is with adhesive, either to vertical studs or horizontal furring strips, or to existing plaster-board or solid walls. In order to get effective pressure over the entire area of the boards for good adhesion, the panels must be slightly bowed. This is done in the manner shown in Fig. 18–43. A stock of wallboard is placed either face down over a center support, or face up across two end supports. The supports can be a 2″ × 4″ lumber, but if they are in contact with the vinyl finish, they must be padded to prevent marring. It may take one day or several days to get a moderate bow, depending on the weather and humidity.

The adhesive may be applied directly to the studs (Fig. 18–44) or

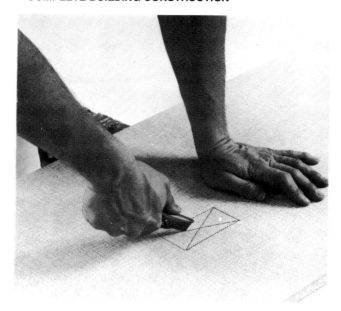

Fig. 18–39. For rectangular cutouts, score deeply on the vinyl side for the outline of the cutout. Tap the section to be removed. *(Courtesy National Gypsum Co.)*

to furring strips (Fig.18–45). Where panel edges join, run two adhesive lines on the stud, one for each edge of a panel. These lines should be as close to the edge of the 2″ × 4″ stud as possible to prevent the adhesive from oozing out between the panel joints. For the same reason, leave a space of 1 inch with no adhesive along the furring strips where panel edges join.

Place the panels in position with a slight sliding motion and nail the top and bottom edges. Be sure long edges of each panel are butted together evenly. Nail to sill and plate at top and bottom only. The bowed panels will apply the right pressure for the rest of the surface.

The top and bottom nails may be matching colored nails or nails finished with a cove and wall trim. Special push-on trims are available which match the prefinished wall. Inside corners may be left with panels butted but one panel must overlap the other. Outside corners must include solid protection. For inside corners, if desired, and for outside

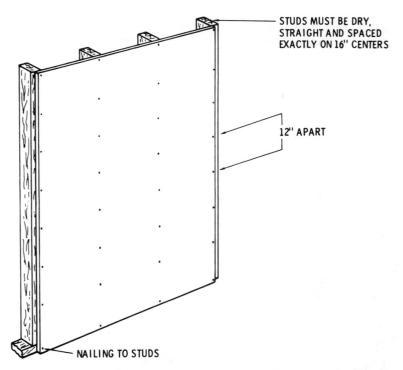

STUDS MUST BE DRY,
STRAIGHT AND SPACED
EXACTLY ON 16" CENTERS

12" APART

NAILING TO STUDS

Fig. 18–40. Prefinished walls may be nailed directly to studs as shown. *(Courtesy National Gypsum Co.)*

corners, snap-on trim and bead matching the panels can be obtained (Fig. 18–46). They are applied by installing retainer strips first (vinyl for inside corners and steel for outside corners), which holds the finished trim material in place.

Prefinished panels may be placed over old plaster walls or solid walls by using adhesive. If on old plaster, the surface must be clean and free of dust or loose paint. If wallpapered, the paper must be removed and walls completely washed. The new panels are bowed, as described before, then lines of adhesive are run down the length of the panels about 16 inches apart (Fig. 18–47). Keep edge lines ¾ to 1 inch from the edges. Apply the boards to the old surface with a slight sliding motion and fasten at top and bottom with 6d nails, or matching colored nails. If boards will not stay in proper alignment, add more bracing

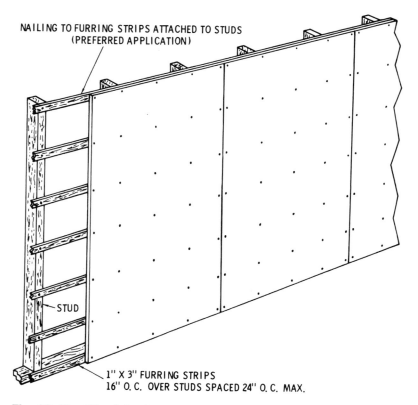

NAILING TO FURRING STRIPS ATTACHED TO STUDS
(PREFERRED APPLICATION)

STUD

1" X 3" FURRING STRIPS
16" O. C. OVER STUDS SPACED 24" O. C. MAX.

Fig. 18–41. Wood furring strips may be installed over wall studs that are not straight. *(Courtesy National Gypsum Co.)*

against the surface and leave for 24 hours. The top and bottom may be finished with matching trim, as explained before.

Gypsonite and FiberBond

GYPSONITE™ and FIBERBOND™ are two products new to the USA that originated in Europe in the early 1980s. They account for about 25 percent of Germany's wallboard market. Gypsonite is manufactured by Highland American Corporation of East Providence,

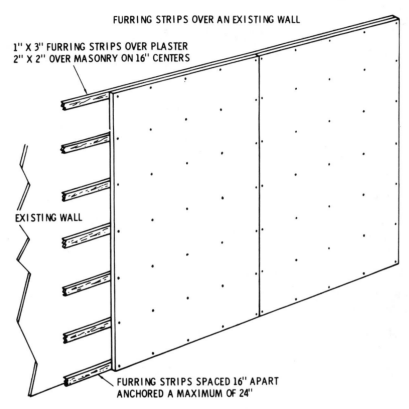

FURRING STRIPS OVER AN EXISTING WALL

1" X 3" FURRING STRIPS OVER PLASTER
2" X 2" OVER MASONRY ON 16" CENTERS

EXISTING WALL

FURRING STRIPS SPACED 16" APART
ANCHORED A MAXIMUM OF 24"

**Fig. 18–42.   Wood furring strips installed over solid masonry or old
existing walls.** *(Courtesy National Gypsum Co.)*

Rhode Island. FiberBond is a product of Louisiana-Pacific, manufactured in Nova Scotia, Canada.

Gypsonite is made from both natural and recovered gypsum, cellulose fiber—recycled newspaper—and perlite. The materials are
bonded together without glue by heat and pressure. Gypsonite is a
solid material, and not a layered material covered with paper. It is available in thicknesses of ⅜, ⁷⁄₁₆, ½, and ⅝ inch, and lengths of 8, 10, and
12 feet, and up to 8′ × 12′ and odd lengths by special order.

FiberBond is layered with a center core of perlite between two

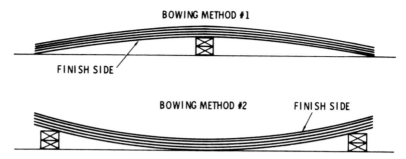

Fig. 18–43. Bowing wallboard for the adhesive method of installation. *(Courtesy National Gypsum Co.)*

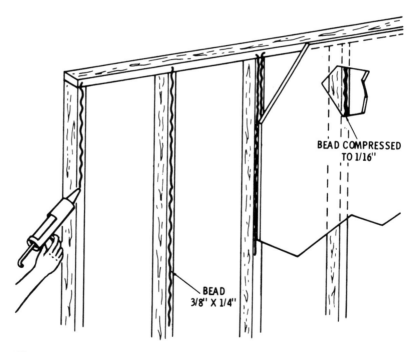

Fig. 18–44. Adhesive applied directly to wall studs. The adhesive is applied in a wavy line the full length of the stud. *(Courtesy National Gypsum Co.)*

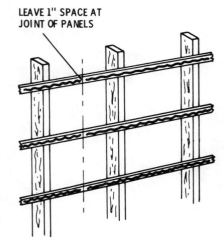

LEAVE 1" SPACE AT
JOINT OF PANELS

Fig. 18–45. Adhesive applied directly to wood furring strips. *(Courtesy National Gypsum Co.)*

Fig. 18–46. **Snap-on matching trim is available for prefinished wallboard to be installed over old plaster walls.** *(Courtesy National Gypsum Co.)*

Fig. 18–47. Applying adhesive to prefinished wallboard to be installed over old plaster walls. *(Courtesy National Gypsum Co.)*

layers of fiber-reinforced gypsum. Panels are available in 4' × 8', 4' × 10', 4' × 12', and in ½-inch and ⅝-inch thicknesses. Widths of 8 feet and lengths to 24 feet are also available. It has a 1-hour type X fire rating, for the ⅝-inch and a 45-minute rating for the ½-inch-thick board.

Gypsonite weighs about 2¼ pounds per ft², a 4 × 8½-inch sheet weighs 72 pounds compared to 58 pounds for the same size gypsum wall board. The same size FiberBond weighs 72 pounds. Gypsonite bevels all four edges, which tends to make them a bit fragile. Fiber-Bond bevels only the long edges, and the taper is about ¾ inch wide.

Gypsonite is difficult to cut, and conventional drywall tools will not do the job. It is important not to treat Gypsonite as just another drywall. Blades dull quickly, and several passes with the knife are necessary. Because there is no paper, the back of the board does not have to be scored. The break is rough, and smoothing the edges with a Surform does not work too well. A drywall grater or coarse wood-rasp may work better. Or the edges can be left rough and filled with Step 1 caulk. The greater density of the material allows the use of pneumatic nailers—use 6d annular ring nails. Staples do not pull the board tightly against the studs. Conventional screw guns can be used but more force will have to be used. Because the board does not dimple, the gun may have to be adjusted to full depth to force the screw below the board surface. FiberBond is a little less difficult to cut and work.

Mudding

One of the advantages of these products is the lack of taping. The stability of the panels makes taping unnecessary. The joints are filled with Step 1 caulk (Figs. 18–48 and 18–49) and mudded over with two coats of a special joint compound. Because the caulk is a glue, the sheets must be gapped in order to glue the sheets together. The Gypsonite compound contains a coarse sand. The FiberBond compound

Fig. 18–48. Applying caulk-glue to Gypsonite panels. *(Courtesy Highland American Corporation)*

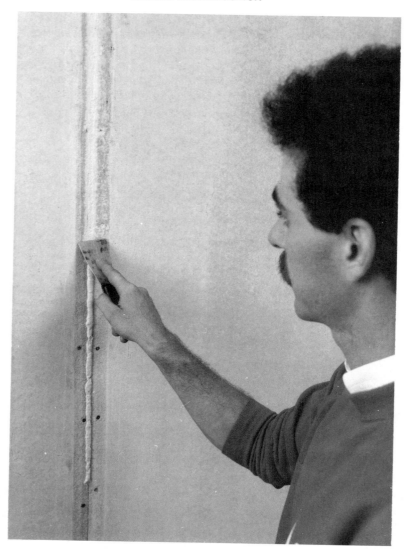

Fig. 18–49. Spreading the compound with a putty knife. *(Courtesy of Highland American Corporation)*

looks more like conventional mud and is much smoother. Step 1 caulk is very hard; care must be used not to go beyond the bevel edge as it may be very difficult to sand down.

Because the compound tends to shrink, a thick bead should be laid in the joint, and left to cure for two days or so. It can be spread with a putty knife or an angular trowel.

The finished wall using either product is much harder, more like a plaster than drywall, and not susceptible to the settlement cracks typical of plaster. These new materials are allowing drywall contractors to compete with plasterers. The manufacturers are improving the compounds and reducing the board weight. Better sound deadening, mildew resistance, better screw and nail holding—no screw anchors necessary—and a plaster-like finish at a drywall cost, are just some of the advantages of these new products.

Paneling

A major development in wallcovering in recent years is paneling. It can take walls in any condition—from ones that are slightly damaged to ones that are virtually decimated—and give them a sparkling new face. Paneling is a key improvement material for the carpenter and do-it-yourselfer.

Paneling is available in an almost limitless variety of styles, colors, and textures (Fig. 18–50). Prefinished paneling may be plywood or processed wood fiber products. Paneling faces may be real hardwood or softwood veneers finished to enhance their natural texture, grain, and color. Other faces may be printed, paper overlaid, or treated to simulate wood grain. Finishing techniques on both real wood and simulated wood-grain surfaces provide a durable and easily maintained wall. Cleaning usually consists of wiping with a damp cloth.

Hardwood and Softwood Plywood

Plywood panelings are manufactured with a face, core, and back veneer of softwood, hardwood, or both (Fig. 18–51). The face and back veneer wood grains run vertically, the core horizontally. This lends strength and stability to the plywood. Hardwood and softwood plywood

Fig. 18–50. This paneling looks like it came off a barn. Paneling is available in an array of colors and styles. *(Courtesy Georgia Pacific)*

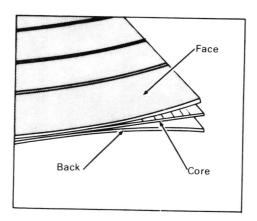

Fig. 18–51. Typical panel with core, back, and face.

wall paneling is normally ¼ inch thick with some paneling ranging up to 7⁄16 inch thick.

The most elegant and expensive plywood paneling has real hardwood or softwood face veneers—walnut, birch, elm, oak, cherry, cedar, pine, fir. Other plywood paneling may have a veneer of tropical hardwood. Finishing techniques include embossing, antiquing, or color toning to achieve a wood grain or decorative look. Panels may also have a paper overlay with wood grain or patterned paper laminated to the face. The panel is then grooved and finished. Most of these panelings are 5⁄32 inch thick.

Processed wood fiber—particleboard or hardboard—wall paneling is also available with grain-printed paper overlays or printed surfaces. These prefinished panels are economical yet attractive. Thicknesses available are 5⁄32, 3⁄16, and ¼ inch. Wood fiber paneling requires special installation techniques. Look for the manufacturer's installation instructions printed on the back of each panel.

Other paneling includes hardboard and hardboard with a tough plastic coating that makes the material suitable for use in high-moisture areas, such as the bathroom.

Groove Treatment

Most vertical wall paneling is *random grooved* with grooves falling on 16-inch centers so that nailing over studs will be consistent. A typical random-groove pattern may look like the one illustrated in Fig. 18–52.

Other groove treatments include uniform spacing (4, 8, 12, or 16 inches) and cross-scored grooves randomly spaced to give a *planked* effect. Grooves are generally striped darker than the panel surface. They are cut or embossed into the panel in V-grooves or channel grooves. Less expensive paneling sometimes has a groove striped on the surface.

Buying Tips

Like most products for the home, paneling is available in a wide range of prices, from as little as $5 per panel to over $30 per panel retail. Contractors and carpenters get a discount (usually 10 percent). Generally, paneling with a simulated wood-grain finish is less expensive than real wood-surfaced panels. Printed or plywood overlaid paneling

Fig. 18–52. Random groove paneling. *(Courtesy Georgia Pacific)*

is available for about $9 to $15 per panel. Paneling with wood fiber substrate costs $5 to $8 per sheet.

How to Figure the Amount of Material Needed—There are several ways to estimate the amount of paneling you will need. One way is to make a plan of the room. Start your plan on graph paper that has ¼-inch squares. Measure the room width, making note of window and door openings. Translate the measurements to the graph paper.

Example—Let us assume that you have a room 14′ × 16′ in size. If you let each ¼-inch square represent 6 inches, the scale is 1 inch equals 2 feet, so you end up with a 7″ × 8″ rectangle. Indicate window positions, doors (and the direction in which they open), a fireplace, and other structural elements. Now begin figuring your paneling requirements. Total all four wall measurements: 14′ + 16′ + 14′ + 16′ = 60′. Divide the total perimeter footage by the 4-foot panel width: 60′ divided by 4′ equals 15 panels. To allow for door, window, and fireplace cutouts, use the deductions as follows (approximate): door, ½ panel (A); window, ¼ panel (B), fireplace, ½ panel (C) (Fig. 18–53).

To estimate the paneling needed, deduct the cutout panels from your original figure: 15 panels – 2 panels = 13 panels. If the perimeter of the room falls between the figures in Table 18–2, use the next higher

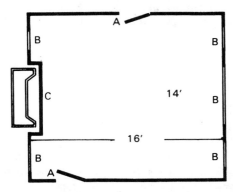

Fig. 18–53. See the text for details on calculating panel requirements as depicted in this drawing.

Table 18–2. Paneling Requirements

Perimeter	Number of 4' × 8' Panels Needed (Without Deductions)
36'	9
40'	10
44'	11
48'	12
52'	13
56'	14
60'	15
64'	16
68'	17
72'	18
92'	23

number to determine panels required. These figures are for rooms with 8-foot wall heights or less. For higher walls, add in the additional materials needed above the 8-foot level.

New paneling should be stored in a dry location. In new construction, freshly plastered walls must be allowed to dry thoroughly before panel installation. Prefinished paneling is moisture-resistant, but like all wood products, it is not waterproof and should not be stored or installed in areas subject to excessive moisture. Ideally the paneling should be stacked flat on the floor with sticks between sheets (Fig. 18–54) to allow air circulation, or it should be propped on the 8-foot edge. Panels should remain in the room 48 hours prior to installation to permit them to acclimatize to temperature and humidity.

Installation

Installing paneling over existing straight walls above grade requires no preliminary preparation. First, locate studs. Studs are usually spaced 16 inches oc, but variations of 24 inches oc centers and other spacing may be found. If you plan to replace the present molding, remove it carefully—you may reveal nailheads used to secure plaster lath or drywall to the studs. These nails mark stud locations. If you cannot locate the studs, start probing with a nail or small drill into the wall surface to be paneled, until you hit solid wood. Start 16 inches from one

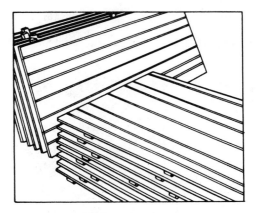

Fig. 18–54. If possible, store panels for a day or so with sticks between panels so that they can become accustomed to room conditions.

corner of the room with your first try, making test holes at ¾-inch intervals on each side of the initial hole until you locate the stud (Figs. 18–55 and 18–56).

Studs are not always straight, and so it is a good idea to probe at several heights. Once a stud is located, make a light pencil mark at floor

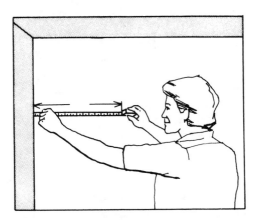

Fig. 18–55. Measure 16 inches from corner of room.

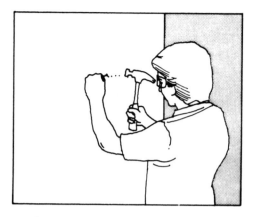

Fig. 18–56. **. . . then make test holes ¾ inch apart to locate stud.**

and ceiling to position all studs, then snap a chalk line at 4-foot intervals (the standard panel width).

Measuring and Cutting—Start in one corner of the room and measure the floor-to-ceiling height for the first panel. Subtract ½ inch to allow for clearance top and bottom. Transfer the measurements to the first panel and mark the dimensions in pencil, using a straightedge for a clean line. Use a sharp crosscut handsaw with ten or more teeth to the inch, or a plywood blade in a table saw, for all cutting. Cut with the panel face-up. If you are using a portable circular saw or a saber saw, mark and cut panels from the back.

Cutouts for door and window sections, electrical switches, and outlets or heat registers require careful measurements. Take your dimensions from the edge of the last applied panel for width, and from the floor and ceiling line for height. Transfer the measurements to the panel, checking as you go. Unless you plan to add molding, door and window cutouts should fit against the surrounding casing. If possible, cutout panels should meet close to the middle over doors and windows.

An even easier way to make cutouts is with a router. You can tack a panel over an opening—a window, say—after removing the trim. Then use the router to trim out the waste paneling in the opening, using the edge of the opening as a guide for the router bit to ride on. Then you can tack the molding or trim back in place.

For electrical boxes, shut off the power and unscrew the protective plate to expose the box. Then paint or run chalk around the box edges and carefully position the panel. Press the paneling firmly over the box area, transferring the outline to the back of the panel (Fig. 18–57). To replace switchplate, a ¼-inch spacer or washer may be needed between the box screw hole and the switch or receptacle.

Drive small nails in each corner through the panel until they protrude through the face. Turn the panel over, drill two ¾-inch holes just inside the corners, and use a keyhole or saber saw to make the cutout (Fig. 18–58). The hole can be up to ¼ inch oversize and still be covered when the protective switchplate is replaced.

Securing Panels—Put the first panel in place and butt to an adjacent wall in the corner. Make sure it is plumb and that both left and right panel edges fall on solid stud backing. Most corners are not perfectly true, however, so you will probably need to trim the panel to fit into the corner properly. Fit the panel into the corner, checking with a level to be sure the panel is plumb vertically. Draw a mark along the panel edge, parallel to the corner. On rough walls like masonry, or on walls adjoining a fireplace, scribe or mark the panel with a compass on the inner panel edge, then cut on the scribe line to fit (Fig. 18–59). Scribing and cutting the inner panel edge may also be necessary if the outer edge of the panel does not fit directly on a stud. The outer edge must

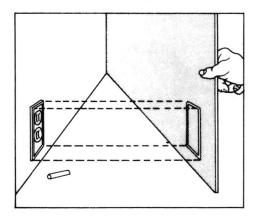

Fig. 18–57. A good way to get the outline is to transfer the chalk mark from plate to back of panel.

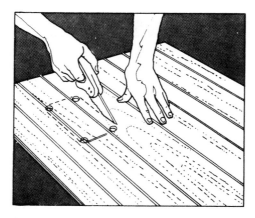

Fig. 18–58. Drill starter holes, then cut out plate section with a keyhole or saber saw.

fall on the center of a stud to allow room for nailing your next panel. Before installing the paneling, paint a stripe of color to match the paneling groove color on the wall location where panels meet. The gap between panels will not show.

Most grooved panels are random grooved to create a solid lumber

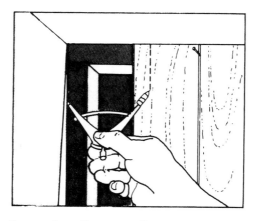

Fig. 18–59. On rough walls, use scriber as shown to get wall outline on paneling. Trim as required.

effect, but there is usually a groove located every 16 inches. This allows most nails to be placed in the grooves, falling directly on the 16-inch stud spacing. Regular small-headed finish nails or colored paneling nails can be used. For paneling directly onto studs 3d (1¼-inch) nails are recommended; but if you must penetrate backer board, plaster, or drywall, 6d (2-inch) nails are needed to get a solid bite in the stud. Space nails 6 inches apart on the panel edges and 12 inches apart in the panel field (Fig. 18–60). Nails should be countersunk slightly below the panel surface with a nailset, then hidden with a matching putty stick. Colored nails eliminate the need to do this. Use 1-inch colored nails to apply paneling to studs, 1⅝-inch nails to apply paneling through plasterboard or plaster.

Adhesive Installation—Using adhesive to install paneling eliminates countersinking and hiding nailheads. Adhesive may be used to apply paneling directly to studs or over existing walls as long as the surface is sound and clean.

Paneling must be cut and fitted prior to installation. Make sure the panels and walls are clean—free from dirt and particles—before you start. Once applied, the adhesive makes adjustments difficult.

A caulking gun with adhesive tube is the simplest method of application. Trim the tube end so that a ⅛-inch wide adhesive bead can be squeezed out. Once the paneling is fitted, apply beads of adhesive in a continuous strip along the top, bottom, and both ends of the panel. On

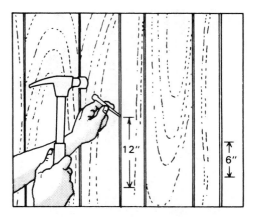

Fig. 18–60. Nails should be 12 inches apart vertically.

intermediate studs, apply beads of adhesive 3 inches long and 6 inches apart (Fig. 18–61).

With scrap plywood or shingles used as a spacer at the floor level, set the panel in place and press firmly along the stud lines spreading the adhesive on the wall. Using a hammer with a padded wooden block or rubber mallet, tap over the glue lines to assure a sound bond between panel and backing.

Some adhesives require panels to be placed against the adhesive, then gently pulled away and allowed to stand for a few minutes for the solvent to "flash off" and the adhesive to set up. The panel is then repositioned and tapped home.

Uneven Surfaces—Most paneling installations require no preliminary preparations. When you start with a sound level surface, paneling is quick and easy. Not every wall is a perfect wall, however. Walls can have chipped, broken, and crumbling plaster, peeling wallpaper, or gypsum board punctured by a swinging door knob or a runaway tricycle on a rainy afternoon. Or walls can be of rough-poured concrete or cinderblock. Walls must be fixed before you attempt to cover them with paneling.

Most problem walls fall into one of two categories. Either you are dealing with plaster or gypsum board applied to a conventional wood stud wall, or you are facing an uneven masonry wall—brick, stone, ce-

Fig. 18–61.   Apply adhesive as shown.

ment, or cinderblock, for example. The solution to both problems is the same, but getting there calls for slightly different approaches.

On conventional walls, clean off any obviously damaged areas. Remove torn wallpaper, scrape off flaking plaster and any broken gypsum board sections. On masonry walls, chip off any protruding mortar. Do not bother making repairs; there is an easier solution.

The paneling will be attached to furring strips. Furring strips are either 1″ × 2″ lumber or ⅜- or ½-inch plywood strips cut 1½ inches wide. The furring strips are applied horizontally 16 inches apart on center on the wall (based on 8-foot ceiling) with vertical members at 48-inch centers where the panels butt together.

Begin by locating the high spot on the wall. To determine this, drop a plumb line (Fig. 18–62). Fasten your first furring strip, making sure that the thickness of the furring strips compensates for the protrusion of the wall surface. Check with a level to make sure that each furring strip is flush with the first strip (Fig. 18–63). Use wood shingles or wedges between the wall and strips to assure a uniformly flat surface (Fig. 18–64). The furred wall should have a ½-inch space at the floor and ceiling with the horizontal strips 16 inches apart on center and the vertical strips 48 inches apart on center (Fig. 18–65). Remember to fur around doors, windows, and other openings. (Fig. 18–66).

On stud walls, the furring strips can be nailed directly through the shims and gypsum board or plaster into the studs. Depending on the

Fig. 18–62.   Wall must be plumbed before you apply furring strips.

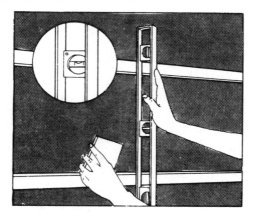

Fig. 18–63.   Use level to determine levelness of wall. Use shim as shown.

thickness of the furring and wallcovering, you will need 6d (2-inch) or 8d (2½-inch) common nails.

Masonry walls are a little tougher to handle. Specially hardened masonry nails can be used, or you can drill a hole with a carbide-tipped bit, insert wood plugs or expansion shields, and nail or bolt the shimmed frame into place (Fig. 18–67).If the masonry wall is badly

Fig. 18–64.   An overall view of shimmed furring strips.

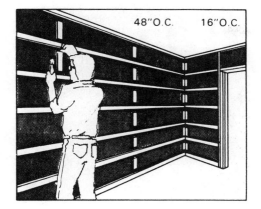

Fig. 18–65.   Secure vertical nailers as shown.

damaged, construct a 2″ × 3″ stud wall to install paneling. Panels may be directly installed on this wall (Fig. 18–68).

Damp Walls

Masonry walls, besides being uneven, often present a more difficult problem: dampness. Usually, damp walls are found partially or

Fig. 18–66.   Don't forget to install furring around windows.

Fig. 18-67. If wall is very rough, a 2″ × 3″ wall can be built, plumbed up, and paneling attached to it.

fully below ground level. Damp basement walls may result from two conditions: (1) seepage of water from outside walls; or (2) condensation, moist warm air within the home that condenses or beads up when it comes in contact with cooler outside walls. Whatever the cause, the moisture must be eliminated before you consider paneling.

Fig. 18-68. Before installing panels, line them up for best color matching.

Seepage may be caused by leaky gutters, improper grade, or cracks in the foundation. Any holes or cracks should be repaired with concrete patching compound or with special hydraulic cement designed to plug active leaks.

Weeping or porous walls can be corrected with an application of masonry waterproofing paints. Formulas for wet and dry walls are available, but the paint must be scrubbed onto the surface to penetrate the pores, hairline cracks, and crevices.

Condensation problems require a different attack. Here, the answer is to dry out the basement air. The basement should be provided with heat in the winter and cross-ventilation in the summer. Wrap cold-water pipes with insulation to reduce sweating. Vent basement washer, dryer, and shower directly to the outside. It may pay big dividends to consider installing a dehumidifying system to control condensation.

When you have the dampness problem under control, install plumb level furring strips on the wall. Apply insulation if desired. Then line the walls with a polyvinyl vapor moisture barrier film installed over the furring strips. Rolls of this inexpensive material can be obtained and installed, but be sure to provide at least a 6-inch lap where sections meet.

Apply paneling in the conventional method. If the dampness is so serious that it cannot be corrected effectively, then other steps must be taken. Prefinished hardwood paneling is manufactured with interior glue and must be installed in a dry setting for satisfactory performance.

Problem Construction

Sometimes the problem is out-of-square walls, uneven floors, or studs placed out of sequence. Or trimming around a stone or brick fireplace may be required, or making cutouts for wall pipes, or handling beams and columns in a basement.

Any wall to be paneled should be checked for trueness. If the wall is badly out-of-plumb, it must be corrected before you install paneling. Furring strips should be applied so that they run at right angles to the direction of the panel application. Furring strips of $3/8'' \times 1\frac{1}{2}''$ plywood strips or $1'' \times 2''$ lumber should be used. Solid backing is required along all four panel edges on each panel. Add strips wherever needed to ensure this support. It is a good idea to place the bottom furring strip $\frac{1}{2}$

inch from the floor. Leave a ¼-inch space between the horizontal strips and the vertical strips to allow for ventilation.

If the plaster, masonry, or other type of wall is so uneven that it cannot be trued by using furring strips and shimming them out where the wall protrudes, 2″ × 3″ studs may be necessary. You can use studs flat against the wall to conserve space. Use studs for top and bottom plates and space vertical studs 16 inches oc. Apply paneling directly to studs or over gypsum or plywood backboard.

Out-of-sequence studs (i.e., studs not on regular centers) require a little planning, but usually they are not a serious problem. Probe to find the exact stud locations as described earlier, snap chalk-lines, and examine the situation. Usually you will find a normal stud sequence starting at one corner; then perhaps where the carpenter framed for a mid-wall doorway, the spacing abruptly changes to a new pattern. Start paneling in the normal manner, using panel adhesive and nails. When you reach the changeover point, cut a filler strip of paneling to bridge the odd stud spacing then pick up the new pattern with full-size panels.

Occasionally you will find a stud or two applied slanted. The combination of panel adhesive plus the holding power of the ceiling and baseboard moldings generally solve this problem.

Uneven floors can usually be handled with shoe molding. Shoe, ½″ × ¾″, is flexible enough to conform to moderate deviations. If the gap is greater than ¾ inch, the base molding or panel should be scribed with a compass to conform to the floor line. Spread the points slightly greater than the gap, hold the compass vertically, and draw the point along the floor, scribing a pencil line on the base molding. Remove the molding trim to the new line with a coping or saber saw and replace. If you have a real washboard of a floor, then you have a floor problem, not a wall problem. Either renail the floor flat or cover it with a plywood or particleboard underlayer before paneling.

The compass trick is also used to scribe a line where paneling butts into a stone or brick fireplace. Tack the panel in place temporarily, scribe a line parallel to the fireplace edge, and trim. Check to be sure that the opposite panel edge falls on a stud before applying.

CHAPTER 19

Stairs

All craftsmen who have tried to build stairs have found it (like boat building) to be an art in itself. This chapter is not intended to discourage the carpenter, but to stress the fact that unless the principle of stair layout is mastered, there will be many difficulties in the construction. Although stair building is a branch of millwork, the craftsman should know the principles of simple stair layout and construction, because porch steps, basement and attic stairs, and sometimes the main stairs are often called for in jobs. In order to follow the instructions intelligently, the carpenter should be familiar with the terms and names of parts used in stair building.

Stair Construction

Stairways should be designed, arranged, and installed so as to afford safety, adequate headroom, and space for the passage of furniture. In general, there are two types of stairs in a house—those serving as principal stairs, and those used as service stairs. The principal stairs are designed to provide ease and comfort, and are often made a feature of design, while the service stairs leading to the basement or attic are usually somewhat steeper and constructed of less expensive materials.

Stairs may be built in place, or they may be built as units in the shop and set in place. Both methods have their advantages and disadvantages, and custom varies with locality. Stairways may have a straight, continuous run, with or without an intermediate platform, or they may consist of two or more runs at angles to each other. In the best and safest practice, a platform is introduced at the angle, but the turn may be made by radiating risers called *winders*. Nonwinder stairways are most frequently encountered in residential planning, because winder stairways represent a condition generally regarded as undesirable. However, use of winders is sometimes necessary because of cramped space. Winders are permitted in means of egress stairways in one- and two-family dwelling units, under the *BOCA* code, but there are certain requirements. The *line of travel* is the point on the stairs that stairwalkers are likely to follow, as they move up the stairs. Its distance, measured from the narrow ends of the winders, varies from 12 to 16 inches. However, the *BOCA* code requires that this distance not exceed 12 inches, and at that point the tread depth shall not be less than 9 inches. The width of the treads at the narrow end must be 6 inches minimum.

Ratio of Riser to Tread

There is a definite relation between the height of a riser and the width of a tread, and all stairs should be laid out to conform to the well-established rules governing these relations. If the combination of run and rise is too great, the steps are tiring, placing a strain on the leg muscles and on the heart. If the steps are too short, the foot may kick the leg riser at each step and an attempt to shorten the stride may be tiring.

There are three traditional rules-of-thumb that have been used to establish the ratio of riser to tread:

1. The sum of two risers and one tread should be between 24″ and 25″. Therefore, a riser 7″ to 7½″, and a tread of 10″ to 11″ would be acceptable. Two risers = 7½″ + 7½″ = 15″ plus one 10″ tread = 25″.

2. The sum of one riser and one tread should be between 17″ and 18″. A 7½″ riser and a 10″ tread = 17½″.

3. Multiplying the riser height by the tread width should give a number between 70″ and 75″. A 7″ riser times a 10″ tread = 70″.

Rule-of-thumb number 1 was devised in 1672 by Francois Blondel, director of the Royal Academy of Architecture in Paris, France. Unfortunately, with risers higher or lower than usual, the treads will be either very narrow or very wide.

Rule-of-thumb number 2 is easier to remember and has been generally adopted. But testing of this combination showed that it results in steps that are too small for gradual stairs, and too large for steeper stairs.

The issue of which ratio is best, has been the subject of considerable research in recent times as well as heated debate. The concern has been to find the combinations which are least likely to cause falls, and which are more comfortable—require less energy—than others. The combinations that require least amount of energy, and have the lowest rate of missteps are 4- to 7-inch risers with 11- to 14-inch treads.

The 1990 *BOCA National Building Code* requires a minimum riser height of 4 inches, and a maximum riser of 7 inches. The minimum tread depth is 11 inches (the so-called 7–11 rule). However, there are a number of exceptions to this rule, one of which is Use Group R3, single and multi-family residences. Here the maximum riser height is 8¼ inches, and the minimum tread depth is 9 inches. The 1988 *Uniform Building Code* sets maximum riser height of 8 inches and a minimum tread width of 9 inches.

Design of Stairs

The location and the width of a stairway (together with the platforms) having been determined, the next step is to fix the height of the riser and width of the tread. After a suitable height of riser is chosen, the exact distance between the finish floors of the two stories under consideration is divided by the riser height. If the answer is an *even* number, the number of risers is thereby determined. It very often happens, that the result is *uneven,* in which case the story height is divided by the whole number next above or below the quotient. The result of this division gives the height of the riser. The tread is then proportioned by one of the rules for ratio of riser to tread.

Assume that the total height from one floor to the top of the next floor is 9'6", or 114 inches, and that the riser is to be approximately 7½ inches. The 114 inches would be divided by 7½ inches, which would give 15⅕ risers. However, the number of risers must be an *equal* or *whole number*. Since the nearest whole number is 15, it may be assumed that there are to be 15 risers, in which case 114 divided by 15 equals 7.6 inches, or approximately 7⁹⁄₁₆ inches for the height of each riser. To determine the width of the tread, multiply the height of the riser by 2 (2 × 7⁹⁄₁₆ = 15⅛), and deduct from 25 (25 − 15⅛ = 9⅞ inches).

There are probably more *headroom* code violations than any other type of violation. Both the *BOCA* and the *UBC* require a minimum headroom clearance of 6'-8". All stairways and all parts of the stairway, under the *BOCA* code, must have a minimum headroom clearance of 6'-8". The Massachusetts *CABO One And Two Family Dwelling Code* allows basement stairs to have a 6'-6" headroom clearance.

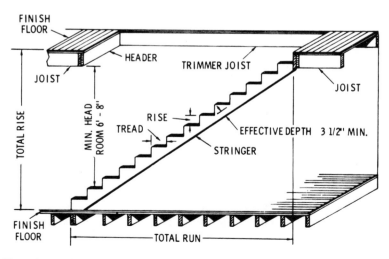

Fig. 19–1.   Stairway design.

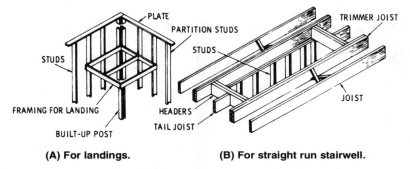

(A) For landings.　　　　　(B) For straight run stairwell.

Fig. 19–2.　Framing of stairways.

Framing of Stairwell

When large openings are made in the floor, such as for a stairwell, one or more joists must be cut. The location in the floor has a direct bearing on the method of framing the joists.

The framing members around openings for stairways are generally of the same depth as the joists. Fig. 19–2 shows the typical framing around a stairwell and landing.

The headers are the short beams at right angles to the regular joists at the end of the floor opening. They are doubled and support the ends of the joists that have been cut off. Trimmer joists are at the sides of the floor opening, and run parallel to the regular joists. They are also doubled and support the ends of the headers. Tail joists are joists that run from the headers to the bearing partition.

Stringers or Carriages

The treads and risers are supported upon stringers or carriages that are solidly fixed in place, and are level and true on the framework of the building. The stringers may be cut or ploughed to fit the outline of the tread and risers. The third stringer should be installed in the middle of the stairs when the treads are less than $1\frac{1}{8}$ inch thick and the stairs are more than 2′ 6″ wide. In some cases, rough stringers are used during the construction period. These have rough treads nailed across the stringers for the convenience of workmen until the wall finish is applied. There are several forms of stringers classed according to the method of attaching the risers and treads. These different types are *cleated, cut, built-up,* and *rabbeted.*

When the wall finish is complete, the finish stairs are erected or built in place. This work is generally done by a stair builder, who often operates as a member of separate specialized craft. The wall stringer may be ploughed out, or rabbeted, as shown in Fig. 19–3, to the exact profile of the tread, riser, and nosing, with sufficient space at the back to take the wedges. The top of the riser is tongued into the front of the tread and into the bottom of the next riser. The wall stringer is spiked to the inside of the wall, and the treads and risers are fitted together and forced into the wall stringer nosing, where they are set tight by driving and gluing the wood wedges behind them. The wall stringer shows above the profiles of the tread and riser as a finish against the wall and is often made continuous with the baseboard of the upper and lower landing. If the outside stringer is an open stringer, it is cut out to fit the risers and treads, and nailed against the outside carriage. The

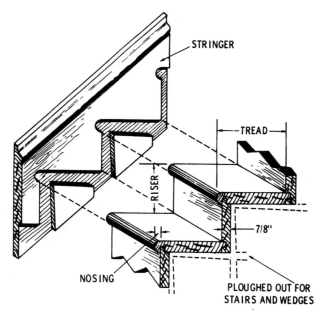

Fig. 19–3. The housing in the stringer board for the tread and riser.

edges of the riser are mitered with the corresponding edges of the stringer, and the nosing of the tread is returned upon its outside edge along the face of the stringer. Another method would be to butt the stringer to the riser and cover the joint with an inexpensive stair bracket.

Fig. 19–4 shows a finish stringer nailed in position on the wall, and the rough carriage nailed in place against the stringer. If there are walls on both sides of the staircase, the other stringer and carriage would be located in the same way. The risers are nailed to the riser cuts of the carriage on each side and butt against each side of the stringer. The treads are nailed to the tread cuts of the carriage and butt against the stringer. This is the least expensive of the types described and perhaps the best construction to use when the treads and risers are to be nailed to the carriages.

Another method of fitting the treads and risers to the wall stringers is shown in Fig. 19–5(A). The stringers are laid out with the same rise and run as the stair carriages, but they are cut out in reverse. The risers are butted and nailed to the riser cuts of the wall stringers, and the assembled stringers and risers are laid over the carriage. Sometimes the treads are allowed to run underneath the tread cut of the stringer. This makes it necessary to notch the tread at the nosing to fit around the stringer, as shown in Fig. 19–5(B).

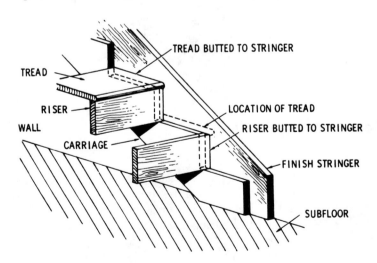

Fig. 19–4. Finished wall stringer and carriage.

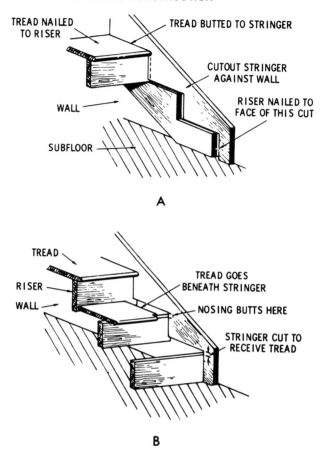

Fig. 19–5. Stringers and treads.

Another form of stringer is the cut and mitered type. This is a form of open stringer in which the ends of the risers are mitered against the vertical portion of the stringer. This construction is shown in Fig. 19–6, and is used when the outside stringer is to be finished and must blend with the rest of the casing or skirting board. A molding is installed on the edge of the tread and carried around to the side, making an overlap as shown in Fig. 19–7.

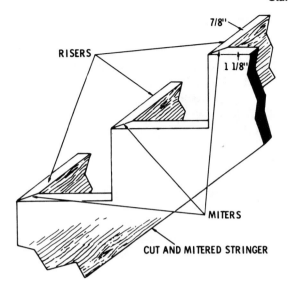

RISERS

7/8"

1 1/8"

MITERS

CUT AND MITERED STRINGER

Fig. 19–6.   Cut and mitered stringer.

Fig. 19–7.   Use of molding on the edge of treads.

Basement Stairs

Basement stairs may be built either with or without riser boards. Cutout stringers are probably the most widely used support for the treads, but the tread may be fastened to the stringers by cleats, as shown in Fig. 19–8. Fig. 19–9 shows two methods of terminating basement stairs at the floor line.

Newels and Handrails

All stairways should have a handrail from floor to floor. For closed stairways, the rail is attached to the wall with suitable metal brackets. The rails should be set 2'-10" (34 inches) minimum, and 3'-2" (38 inches) maximum above the tread at the riser line. Handrails and balusters are used for open stairs and for open spaces around stairs. The handrail ends against the newel post, as shown in Fig. 19–10.

Stairs should be laid out so that stock parts may be used for newels, rails, balusters, goosenecks, and turnouts. These parts are a matter of design and appearance, so they may be very plain or elaborate, but they should be in keeping with the style of the house. The balusters are doweled or dovetailed into the treads and, in some cases, are covered by a return nosing. Newel posts should be firmly anchored, and where half-

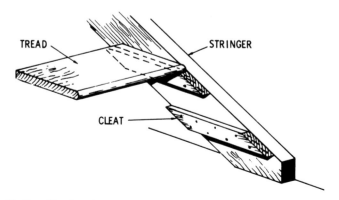

Fig. 19–8. Cleat stringer used in basement stairs.

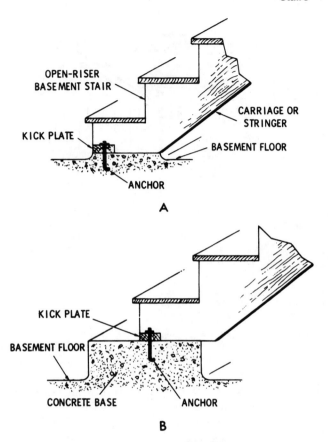

OPEN-RISER
BASEMENT STAIR

CARRIAGE OR
STRINGER

KICK PLATE

BASEMENT FLOOR

ANCHOR

A

KICK PLATE

BASEMENT FLOOR

CONCRETE BASE   ANCHOR

B

Fig. 19–9. Basement stair termination at floor line.

newels are attached to a wall, blocking should be provided at the time
the wall is framed.

Disappearing Stairs

Where attics are used primarily for storage, and where space for a
fixed stairway is not available, hinged or disappearing stairs are often
used. Such stairways may be purchased ready to install. They operate

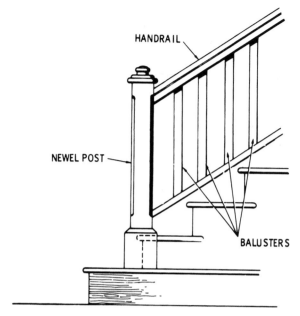

Fig. 19–10.   Newel post, balusters, and handrail.

through an opening in the ceiling of a hall and swing up into the attic space, out of the way when not in use. Where such stairs are to be provided, the attic floor should be designed for regular floor loading.

Exterior Stairs

Proportioning of risers and treads in laying out porch steps or approaches to terraces should be as carefully considered as the design of interior stairways. Similar riser-to-tread ratios can be used, however. The riser used in principal exterior steps should be between 6 and 7 inches. The need for a good support or foundation for outside steps is often overlooked. Where wood steps are used, the bottom step should be set in concrete. Where the steps are located over backfill or disturbed ground, the foundation should be carried down to undisturbed

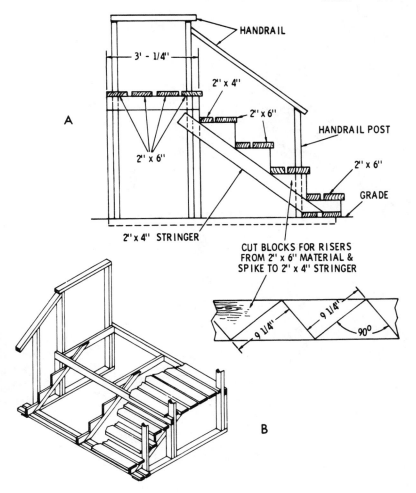

A

HANDRAIL

3' - 1/4"

2" x 4"

2" x 6"

HANDRAIL POST

2" x 6"

2" x 6"

GRADE

2" x 4" STRINGER

CUT BLOCKS FOR RISERS
FROM 2" x 6" MATERIAL &
SPIKE TO 2" x 4" STRINGER

9 1/4"

9 1/4"

90°

B

Fig. 19–11. Outside step construction.

ground. Fig. 19–11 shows the foundation and details of the step treads, handrail, and stringer, and method of installing them. This type of step is most common in field construction and outside porch steps. The materials generally used for this type of stair construction are 2 × 4s and 2 × 6s.

Fig. 19–12.   The baluster which supports the handrail.

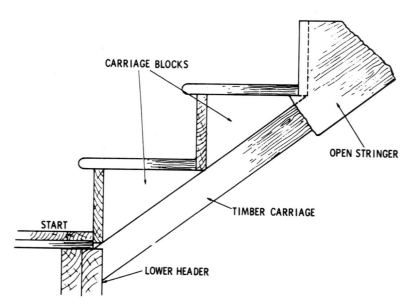

Fig. 19–13.   Carriage blocks connected to a stair stringer.

Glossary of Stair Terms

The terms generally used in stair design may be defined as follows.

Balusters—The vertical members supporting the handrail on open stairs (Fig. 19–12).

Carriage—The rough timber supporting the treads and risers of wood stairs, sometimes referred to as the string or stringer, as shown in Fig. 19–13.

Circular Stairs—A staircase with steps planned in a circle, all the steps being winders (Fig. 19–14).

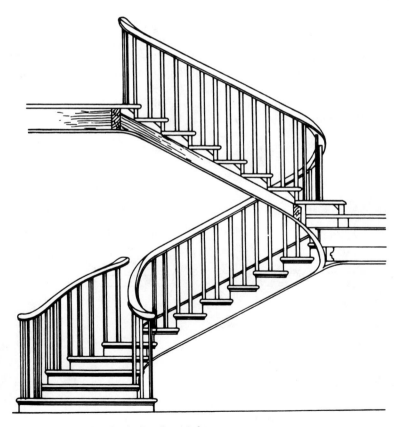

Fig. 19–14. A typical circular staircase.

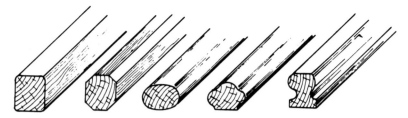

Fig. 19–15.   Various forms of handrails.

Flight of Stairs—The series of steps leading from one landing to another.

Front String—The string of that side of the stairs over which the hand rail is placed.

Fillet—A band nailed to the face of a front string below the curve and extending the width of a tread.

Flyers—Steps in a flight of stairs parallel to each other.

Half-Space—The interval between two flights of steps in a staircase.

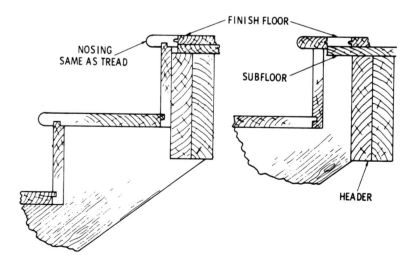

Fig. 19–16.   The nosing installed on the tread.

Handrail—The top finishing piece on the railing intended to be grasped by the hand in ascending and descending. For closed stairs where there is no railing, the handrail is attached to the wall with brackets. Various forms of hand rails are shown in Fig. 19–15.

Housing—The notches in the string board of a stair for the reception of steps.

Landing—The floor at the top or bottom of each story where the flight ends or begins.

Newel—The main post of the railing at the start of the stairs and the stiffening posts at the angles and platform.

Nosing—The projection of tread beyond the face of the riser (Fig. 19–16).

Rise—The vertical distance between the treads or for the entire stairs.

Riser—The board forming the vertical portion of the front of the step, as shown in Fig. 19–17.

Run—The total length of stairs including the platform.

Stairs—The steps used to ascend and descend from one story to another.

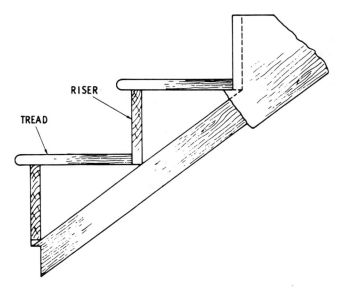

RISER

TREAD

Fig. 19–17. Tread and riser.

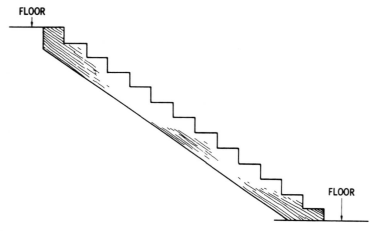

Fig. 19–18. The stair stringer.

Staircase—The whole set of stairs with the side members supporting the steps.

Straight Flight of Stairs—One having the steps parallel and at right angles to the strings.

String or Stringer—One of the inclined sides of a stair supporting the tread and riser. Also, a similar member, whether a support or not, such as finish stock placed exterior to the carriage on open stairs, and next to the walls on closed stairs, to give finish to the staircase. *Open stringers,* both rough and finish stock, are cut to follow the lines of the tread and risers. *Closed stringers* have parallel sides, with the risers and treads being housed into them (Fig. 19–18).

Tread—The horizontal face of a step, as shown in Fig. 19–17.

Winders—The radiating or wedge-shaped treads at the turn of a stairway.

CHAPTER 20

Flooring

Floor Coverings

There is a wide variety of finish flooring available, each having properties suited to a particular usage. Of these properties, durability and ease of cleaning are essential in all cases. Specific service requirements may call for special properties, such as resistance to hard wear in storehouses and on loading platforms; comfort to users in offices and shops; and attractive appearance, which is always desirable in residences.

Both hardwoods and softwoods are available as strip flooring in a variety of widths and thicknesses, as well as random-width planks, parquetry, and block flooring. Other materials include those mentioned above. A detailed round-up follows.

Wood Strip Flooring

Softwoods most commonly used for flooring are southern yellow pine, douglas fir, redwood, western larch, and western hemlock. It is customary to divide the softwoods into two classes:

1. Vertical or edge grain,
2. Flat grain.

Table 20–1. Plywood Thickness and Joist Spacing

Minimum thickness of five-ply subfloor	Medium thickness of finish flooring	Maximum joist spacing
†1/2 inch	25/32 inch wood laid at right angles to joists	24 inches
†1/2 inch	25/32 inch wood laid parallel to joists	20 inches
1/2 inch	25/32 inch wood laid at right angles to joists	20 inches
1/2 inch	Less than 25/32 inch wood or other finish	‡16 inches
†5/8 inch	Less than 25/32 inch wood or other finish	‡20 inches
†3/4 inch	Less than 25/32 inch wood or other finish	‡24 inches

†Installed with outer plies of subflooring at right angles to joists.
‡Wood strip flooring, 25/32 inch thick or less, may be applied in either direction.

Each class is separated into select and common grades. The select grades designated as "B and better" grades, and sometimes the "C" grade, are used when the purpose is to stain, varnish, or wax the floor. The "C" grade is well suited for floors to be stained dark or painted, and lower grades are for rough usage or when covered with carpeting. Softwood flooring is manufactured in several widths. In some places, the 2½-inch width is preferred, while in others, the 3½-inch width is more popular. Softwood flooring has tongue-and-groove edges, and may be hollow backed or grooved. Vertical-grain flooring stands up better than flatgrain under hard usage.

Hardwoods most commonly used for flooring are red and white oak, hard maple, beech, and birch. Maple, beech, and birch come in several grades, such as *first, second,* and *third.* Other hardwoods that are manufactured into flooring, although not commonly used, are walnut, cherry, ash, hickory, pecan, sweetgum, and sycamore. Hardwood flooring is manufactured in a variety of widths and thicknesses, some of which are referred to as standard patterns, others as special patterns. The widely used standard patterns consist of relatively narrow strips laid lengthwise in a room, as shown in Fig. 20–1. The most widely used standard pattern is 25/32 inch thick and has a face width of 2¼ inches. One edge has a tongue and the other edge has a groove, and the ends are similarly matched. The strips are random lengths, varying from 1 to

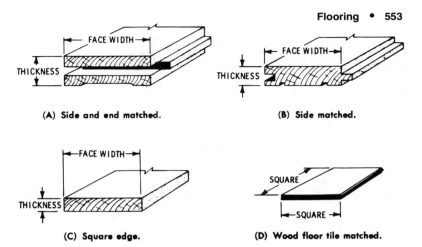

(A) Side and end matched.

(B) Side matched.

(C) Square edge.

(D) Wood floor tile matched.

Fig. 20–1.   Types of finished hardwood flooring.

16 feet in length. The number of short pieces will depend on the grade used. Similar patterns of flooring are available in thicknesses of $\frac{15}{32}$ and $\frac{11}{32}$ inch, width of $1\frac{1}{2}$ inch, and with square edges and a flat back.

The flooring is generally hollow backed. The top face is slightly wider than the bottom, so that the strips are driven tightly together at the top side but the bottom edges are slightly open. The tongue should fit snugly in the groove to eliminate squeaks in the floor.

Another pattern of flooring used to a limited degree is $\frac{3}{8}$ inch thick with a face width of $1\frac{1}{2}$ and 2 inches, with square edges and a flat back. Fig. 20–1(D) shows a type of wood floor tile commonly known as parquetry.

Installation of Wood Strip Flooring

Flooring should be laid after plastering and other interior wall and ceiling finish is completed, after windows and exterior doors are in place, and after most of the interior trim is installed. When wood floors are used, the subfloor should be clean and level, and should be covered with a deadening felt or heavy building paper, as shown in Fig. 20–2. This felt or building paper will stop a certain amount of dust, and will somewhat deaden the sound. Where a crawl space is used, it will in-

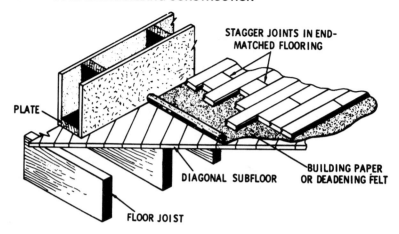

STAGGER JOINTS IN END-MATCHED FLOORING

PLATE

DIAGONAL SUBFLOOR

BUILDING PAPER OR DEADENING FELT

FLOOR JOIST

Fig. 20–2. Application of strip flooring showing the use of deadening felt or heavy building paper.

crease the warmth of the floor by preventing air infiltration. The location of the joists should be chalklined on the paper as a guide for nailing.

Strip flooring should be laid crosswise of the floor joist, and it looks best when the floor is laid lengthwise in a rectangular room. Since joists generally span the short way in a living room, that room establishes the direction for flooring in other rooms. Flooring should be delivered only during dry weather, and should be stored in the warmest and driest place available in the house. The recommended average moisture content for flooring at the time of the installation should be between 6 and 10 percent. Moisture absorbed after delivery to the house site is one of the most common causes of open joints between floor strips. It will show up several months after the floor has been laid.

Floor squeaks are caused by movement of one board against another. Such movement may occur because the floor joists are too light and not held down tightly, tongues fit too loosely in the grooves, or the floor was poorly nailed. Adequate nailing is one of the most important means of minimizing squeaks. When it is possible to nail the finish floor through the subfloor into the joist, a much better job is obtained than if the finish floor is nailed only to the subfloor. Various types of nails are used in nailing various thicknesses of flooring. For $^{25}/_{32}$-inch flooring, it is best to use 8d steel cut flooring nails; for ½-inch flooring, 6d nails should be used. Other types of nails have been developed in recent years for nailing of flooring, among these being the annularly-grooved and spirally-grooved nails. In using these nails, it is well to check with the floor manufacturer's recommendations as to size and diameter for a specific use.

Fig. 20–3 shows the method of nailing the first strip of flooring. The nail is driven straight down through the board at the groove edge. The nails should be driven into the joist and near enough to the edge so that they will be covered by the base or shoe molding. The first strip of flooring can be nailed through the tongue.

Fig. 20–4 shows the nail driven in at an angle of between 45 and 50 degrees where the tongue adjoins the shoulder. Do not try to drive the nail home with a hammer, as the wood may be easily struck and damaged. Instead use a nail set to finish off the driving. Fig. 20–5 shows a special nail set that is commonly used for the final driving. In order to avoid splitting the wood, it is sometimes necessary to pre-drill the holes through the tongue. This will also help to drive the nail easily into the joist. For the second course of flooring, select a piece so that the butt joints will be well separated from those in the first course. For floors to

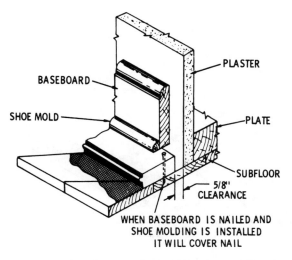

Fig. 20–3.   Method of laying the first strips of wood flooring.

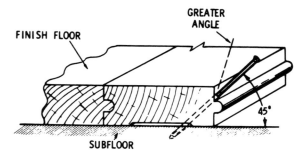

Fig. 20–4. Nailing method for setting nails in flooring.

be covered with rugs, the long lengths could be used at the sides of the room and the short lengths in the center where they will be covered.

Each board should be driven up tightly, but do not strike the tongue with the hammer, as this will crush the wood. Use a piece of scrap flooring for a driving block. Crooked pieces may require wedging to force them into alignment. This is necessary in order that the last piece of flooring will be parallel to the baseboard. If the room is not square, it may be necessary to start the alignment at an early stage.

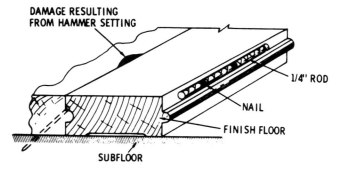

Fig. 20–5. Suggested method for setting nails in flooring.

Soundproof Floors

One of the most effective sound resistant floors is called a *floating floor*. The upper or finish floor is constructed on 2 × 2 joists actually floating on glass wool mats, as shown in Fig. 20–6. There should be absolutely no mechanical connection through the glass wool mat, not even a nail to either the subfloor or to the wall.

Parquet Flooring

Flooring manufacturers have developed a wide variety of special patterns of flooring, including parquet (Fig. 20–7), which is nailed in place or has an adhesive backing. One common type of floor tile is a block 9 inches square and 13⁄16-inch thick, which is made up of several individual strips of flooring held together with glue and splines. Two edges have a tongue and the opposite edges are grooved. Numerous

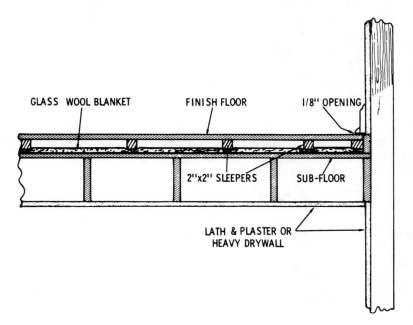

GLASS WOOL BLANKET   FINISH FLOOR   1/8" OPENING

2"x2" SLEEPERS   SUB-FLOOR

LATH & PLASTER OR HEAVY DRYWALL

Fig. 20–6.   A sound-resistant floor.

Fig. 20–7. Parquet flooring. *(Courtesy United Gilsonite)*

Fig. 20–8. Ceramic tile is used mainly in bathrooms. The tile comes in many colors and patterns.

Fig. 20–9. For the do-it-yourselfer, there are adhesive-backed tiles.

other sizes and thicknesses are available. In laying the floor, the direction of the blocks is alternated to create a checkerboard effect. Each manufacturer supplies instructions for laying its tile, and it is advisable to follow them carefully. When the tiles are used over a concrete slab, a vapor barrier should be used. The slab should be level and thoroughly aired and dried before the flooring is laid.

Fig. 20–10. Waxless flooring makes life easier for the homemaker.
(Courtesy Armstrong)

Ceramic Tile

Ceramic tile (Fig. 20–8) is made in different colors and with both glazed and unglazed surfaces. It is used as covering for floors in bathrooms, entryways, kitchens, and fireplace hearths. Ceramic tile presents a hard and impervious surface. In addition to standard sizes and plain colors, many tiles are especially made to carry out architectual effects. When ceramic-tile floors are used with wood-frame construction, a concrete bed of adequate thickness must be installed to receive the finishing layer.

Installation of tile is done with adhesive. See Chapter 22 for specific instructions on tile installations.

Fig. 20–11. Sheet flooring is not easy to install correctly. *(Courtesy Congoleum)*

Other Finished Floorings

The carpenter and do-it-yourselfer can select from a wealth of floor coverings. Perhaps the chief development has been resilient flooring, so called because it "gives" when you step on it. There are 12-inch tiles, available for installing with adhesive (Fig. 20–9) as well as adhesive-backed tiles. Today most tile is vinyl and comes in a tremendous variety of styles, colors, and patterns. The newest in flooring is the so-called waxless flooring (Fig. 20–10), which requires renewal with a waxlike material after a certain period of time.

Resilient flooring also comes in sheets or rolls 12 feet wide. It, too, is chiefly vinyl and comes in a great array of styles and colors. It is more difficult to install than tile (Fig. 20–11).

Resilient sheet flooring comes in several qualities, and you should check competing materials before you buy. Vinyl resilient flooring may be installed anywhere in the house, above or below grade, with no worry about moisture problems. An adequate subfloor, usually of particleboard or plywood, is required.

Also available is carpeting. Wall-to-wall carpet installation is a professional job, but carpet tiles are available for indoor and outdoor use. They are also popularly used in kitchens and bathrooms.

Other flooring materials include paint-on coatings and paints. Paint is normally used in areas where economy is most important.

CHAPTER 21

Doors

Doors can be obtained from the mills in stock sizes much more cheaply than they can be made by hand. Stock sizes of doors cover a wide range, but those most commonly used are 2′4″ × 6′8″, 2′8″ × 6′8″, 3′0″ × 6′8″, and 3′0″ × 7′0″. These sizes are either 1⅜-inch (interior) or 1¾-inch (exterior) thick.

Types of Doors

Paneled Doors

Paneled, or sash, doors are made in a variety of panel arrangements, horizontal, vertical, or combinations of both. A sash door has for its component parts of top rail, bottom rail, and two stiles, which form the sides of the door. Doors of the horizontal type have intermediate rails forming the panels; and panels of the vertical type have horizontal rails and vertical stiles forming the panels.

The rails and stiles of a door are generally mortised-and-tenoned, the mortise being cut in the side stiles as shown in Fig. 21–1. Top and bottom rails on paneled doors differ in width, the bottom rail being considerably wider. Intermediate rails are usually the same width as the top rail. Paneling material is usually plywood which is set in grooves

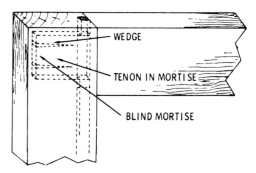

WEDGE

TENON IN MORTISE

BLIND MORTISE

Fig. 21–1. Door construction showing mortise joints.

or dadoes in the stiles and rails, with the molding attached on most doors as a finish.

Flush Doors

Flush doors are usually perfectly flat on both sides. Solid planks are rarely used for flush doors. Flush doors are made up with solid or hollow cores with two or more plies of veneer glued to the cores.

Solid-Core Doors

Solid-core doors are made of short pieces of wood glued together with the ends staggered very much like in brick laying. One or two plies of veneer are glued to the core. The first section, about ⅛-inch thick, is applied at right angles to the direction of the core, and the other section, ⅛ inch or less, is glued with the grain vertical. A ¾-inch strip, the thickness of the door is glued to the edges of the door on all four sides. This type of door construction is shown in Fig. 21–2.

Hollow-Core Doors

Hollow-core doors have wooden grids or other honeycomb material for the base, with solid wood edging strips on all four sides. The face of this type door is usually 3-ply veneer instead of two single plies. The hollow-core door has a solid block on both sides for installing door

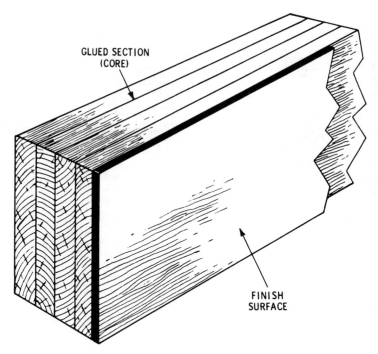

GLUED SECTION
(CORE)

FINISH
SURFACE

Fig. 21–2. Construction of a laminated or veneered door.

knobs and to permit the mortising of locks. The honeycomb-core door is for interior use only.

Louver Doors

This type of door has either stationary or adjustable louvers, and may be used as an interior door, room divider, or a closet door. The louver door comes in many styles, such as shown in Fig. 21–3. An exterior louver door may be used, which is called a *jalousie* door. This door has the adjustable louvers usually made of wood or glass. Although there is little protection against winter winds, a solid storm window is made to fit over the louvers to give added protection.

Fig. 21–3. Various styles of louver doors.

Installing Doors

Door Frames

Before a door can be installed, a frame must be built for it. There are numerous ways to do this. A frame for an exterior door consists of the following essential parts (Fig. 21–4):

1. Sill,
2. Threshold,
3. Side and top jamb,
4. Casing.

The preparation should be done before the exterior covering is placed on the outside walls. To prepare the openings, square off any uneven pieces of sheathing and wrap heavy building paper around the sides and top. Since the sill must be worked into a portion of the sub-

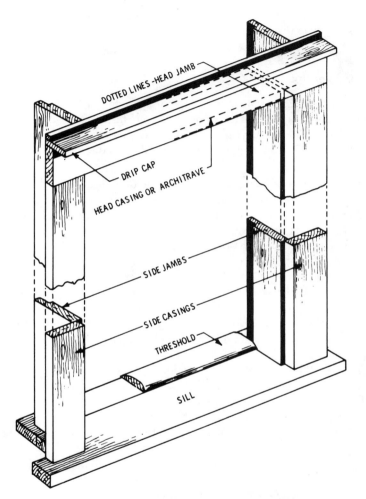

Fig. 21–4. View of door frame showing the general construction.

flooring, no paper is put on the floor. Position the paper from a point even with the inside portion of the stud to a point about 6 inches on the sheathed walls and tack it down with small nails.

In quick construction, there will be no door frame (the studs on each side of the opening act as the frame). The inside door frame is constructed in the same manner as the outside frame, except there is no sill and no threshold.

Door Jambs

Door jambs are the lining of a door opening. Casings and stops are nailed to the jamb, and the door is securely fastened by hinges at one side. The width of the jamb will vary in accordance with the thickness of the walls. The door jambs are made and set in the following manner.

1. Regardless of how carefully the rough openings are made, be sure to plumb the jambs and level the heads when the jambs are set.
2. Rough openings are usually made 2½ inches larger each way than the size of the door to be hung. For example, a 2′8″ × 6′8″ door would need a rough opening of 2′10½″ × 6′10½″. This extra space allows for the jamb, the wedging, and the clearance space for the door to swing.
3. Level the floor across the opening to determine any variation in floor heights at the point where the jamb rests on the floor.
4. Cut the head jamb with both ends square, allowing for the width of the door plus the depth of both dadoes and a full ³⁄₁₆ inch for door clearance.
5. From the lower edge of the dado, measure a distance equal to the height of the door plus the clearance wanted at the bottom.
6. Do the same thing on the opposite jamb, only make additions or subtractions for the variation in the floor.
7. Nail the jambs and jamb head together through the dado into the head jamb, as shown in Fig. 21–5.
8. Set the jambs into the opening and place small blocks under each jamb on the subfloor just as thick as the finish floor will be. This will allow the finish floor to go under the door.
9. Plumb the jambs and level the jamb head.
10. Wedge the sides to the plumb line with shingles between the jambs and the studs, and then nail securely in place.
11. Take care not to wedge the jambs unevenly.
12. Use a straightedge 5 to 6 feet long inside the jambs to help prevent uneven wedging.
13. Check each jamb and the head carefully. If a jamb is not plumb, it

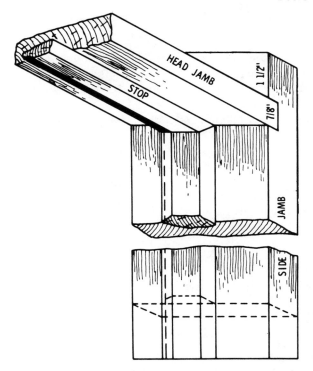

Fig. 21–5. Details showing upper head jambs dadoed into side jambs.

will have a tendency to swing the door open or shut, depending on the direction in which the jamb is out of plumb.

Door Trim

Door trim material is nailed onto the jambs to provide a finish between the jambs and the plastered wall. This is called *casing*. Sizes vary from ½ to ¾ inch in thickness, and from 2½ to 6 inches in width. Most casing material has a concave back, to fit over uneven wall material. In miter work, care must be taken to make all joints clean, square, neat, and well fitted. If the trim is to be mitered at the top corners, a miter box, miter square, hammer, nailset, and block plane will be needed. Door openings are cased up in the following manner.

1. Leave a ¼-inch margin between the edge of the jamb and the casing on all sides.
2. Cut one of the side casings square and even with the bottom of the jamb.
3. Cut the top or mitered end next, allowing ¼ inch extra length for the margin at the top.
4. Nail the casing onto the jamb and set it even with the ¼-inch margin line, starting at the top and working toward the bottom.
5. The nails along the outer edge will need to be long enough to penetrate the casing, plaster, and wall stud.
6. Set all nail heads about ⅛ inch below the surface of the wood.
7. Apply the casing for the other side of the door opening in the same manner, followed by the head (or top) casing.

Hanging Mill-Built Doors

If flush or sash doors are used, install them in the finished door opening as described below.

1. Cut off the stile extension, if any, and place the door in the frame. Plane the edges of the stiles until the door fits tightly against the hinge side and clears the lock side of the jamb about ¹⁄₁₆ inch. Be sure that the top of the door fits squarely into the rabbeted recess and that the bottom swings free of the finished floor by about ½ inch. The lock stile of the door must be beveled slightly so that the edge of the door will not strike the edge of the door jamb.
2. After the proper clearance of the door has been made, set the door in position and place wedges as shown in Fig. 21–6. Mark the position of the hinges on the stile and on the jamb with a sharp pointed knife. The lower hinge must be placed slightly above the lower rail of the door. The upper hinge of the door must be placed slightly below the top rail in order to avoid cutting out a portion of the tenons of the door rails. There are three measurements to mark—the location of the butt hinge on the jamb, the location of the hinge on the door, and the thickness of the hinge on both the jamb and the door.

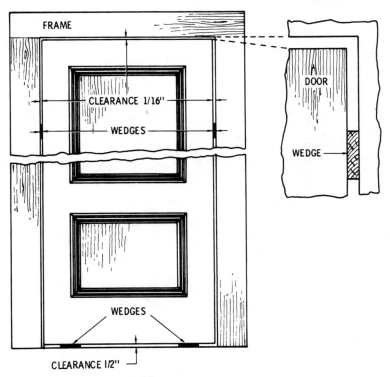

Fig. 21–6. Sizing a door for an opening.

3. Door butt hinges are designed to be mortised into the door and frame, as shown in Fig. 21–7. Fig. 21–8 shows recommended dimensions and clearances for installation. Three hinges are usually used on full-length doors to prevent warping and sagging.
4. Using the butt as a pattern, mark the dimension of the butts on the door edge and the fact of the jamb. The butts must fit snugly and exactly flush with the edge of the door and the face of the jamb. A device called a butt marker can be helpful here.

After placing the hinges and hanging the door, mark off the position for the lock and handle. The lock is generally placed about 36 inches from the floor. Hold the lock in position on the stile and mark

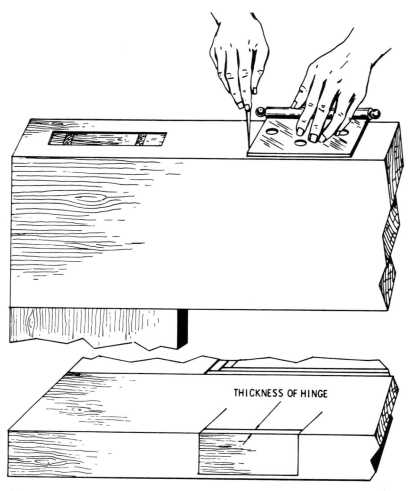

THICKNESS OF HINGE

Fig. 21–7. A method of installing hinges.

off with a sharp knife the area to be removed from the edge of the stile. Mark off the position of the doorknob hub. Bore out the wood to house the lock and chisel the mortises clean. After the lock assembly has been installed, close the door and mark the jamb for the striker plate.

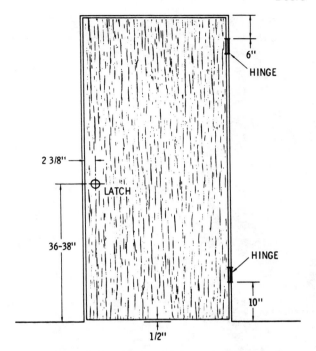

Fig. 21–8. Recommended clearances and dimensions after door is hung.

Installing a door is difficult. A recent trend is for manufacturers to provide not only the door but all surrounding framework (Fig. 21–9). This reduces the possibility of error.

Swinging Doors

Frequently it is desirable to hang a door so that it opens as you pass through from either direction, yet remains closed at all other times. For this purpose, you can use swivel-type spring hinges. This type of hinge attaches to the rail of the door and to the jamb like an ordinary butt hinge. Another type is mortised into the bottom rail of the door and is fastened to the floor with a floor plate. In most cases, the floor-plate

Fig. 21–9. Some companies provide the door and surrounding framework to ensure trouble-free installation.

hinge, as shown in Fig. 21–10, is best, because it will not weaken and let the door sag. It is also designed with a stop to hold the door open at right angles, if so desired.

Sliding Doors

Sliding doors are usually used for walk-in closets. They take up very little space, and they also allow a wide variation in floor plans. This type of door usually limits the access to a room or closet unless the doors are pushed back into a wall. Very few sliding doors are pushed

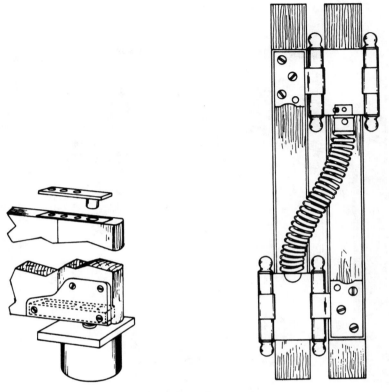

Fig. 21–10.   Two kinds of swivel spring hinges.

back into the wall because of the space and expense involved. Fig.
21–11 shows a double and a single sliding door track.

Garage Doors

Garage doors are made in a variety of sizes and designs. The principal advantage in using the garage door, of course, is that it can be
rolled up out of the way. In addition, the door cannot be blown shut
due to wind, and is not obstructed by snow and ice.

Although designed primarily for use in residential and commercial

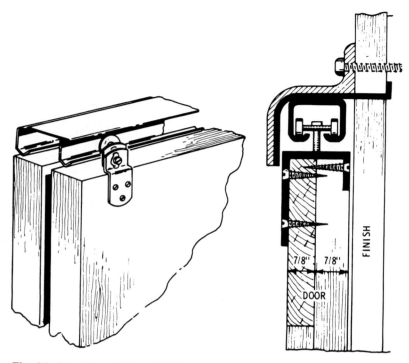

Fig. 21–11. Two kinds of sliding-door tracks.

garages, doors of this type are also employed in service stations, factory receiving docks, boathouses, and many other buildings. In order to permit overhead-door operation, garage doors of this type are built in suitable hinged sections. Usually 4 to 7 sections are used, depending upon the door height requirements. Standard residential garage doors are usually 9′ × 7′ for singles and 16′ × 7′ for a double. Residential garage doors are usually manufactured 1¾ inches thick unless otherwise requested.

When ordering doors for the garage, the following information should be forwarded to the manufacturer:

1. Width of opening between the finished jambs,
2. Height of the ceiling from the finished floor under the door to the underside of the finished header,

3. Thickness of the door,
4. Design of the door (number of glass windows and sections),
5. Material of jambs (they must be flush),
6. Headroom from the underside of the header to the ceiling, or to any pipes, lights, etc.,
7. Distance between the sill and the floor level,
8. Proposed method of anchoring the horizontal track,
9. Depth to the rear from inside of the upper jamb,
10. Inside face width of the jamb buck, angle, or channel.

This information applies for overhead doors only, and does not apply to garage doors of the slide, folding, or hinged type. Doors can be furnished to match any style of architecture and may be provided with suitable size glass windows if desired (Fig. 21–12).

If your garage is attached to your house, your garage door often represents from one-third to one-fourth of the face of your house. Style and material should be considered to accomplish a pleasant effect with masonry or wood architecture. Fig. 21–13 shows three types of overhead garage doors that can be used with virtually any kind of architectural design. Many variations can be created from combinations of

Fig. 21–12. Typical 16-foot overhead garage door. *(Courtesy Overhead Door Corporation)*

(A) Fiberglass.　　　　(B) Steel.

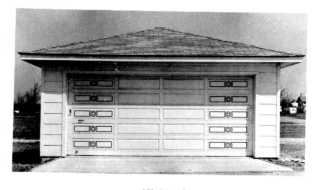

(C) Wood.

Fig. 21–13. Three kinds of garage doors. *(Courtesy Overhead Door Corporation)*

raised panels with routed or carved designs as shown in Fig. 21–14. These panels may also be combined with plain raised panels to provide other dramatic patterns and color combinations.

Automatic garage-door openers were once a luxury item, but in the past few years the price has been reduced and failure minimized the the extent that most new installations include this feature. Automatic garage-door openers save time, eliminate the need to stop the car and get out in all kinds of weather, and you also save the energy and effort required to open and close the door by hand.

The automatic door opener is a radio-activated motor-driven

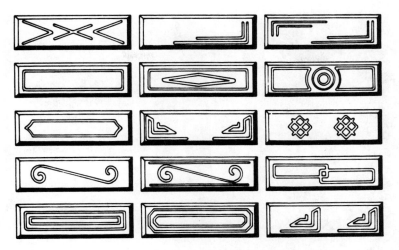

Fig. 21–14. Variations in carved or routed panel designs.

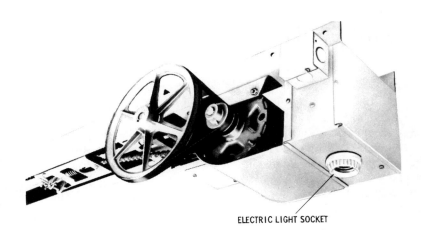

ELECTRIC LIGHT SOCKET

Fig. 21–15. Typical automatic garage-door opener. *(Courtesy Overhead Door Corporation)*

Fig. 21–16. Patio door.

power unit that mounts on the ceiling of the garage, and attaches to the inside top of the garage door. Electric impulses from a wall-mounted push button, or radio waves from a portable radio transmitter in your car, starts the door mechanism. When the door reaches its limit of travel (up or down), the unit turns itself off and awaits your next command. Most openers on the market have a safety factor built in. If the door encounters an obstruction in its travel, it will instantly stop, or stop and reverse its travel. The door will not close until the obstruction has been removed. When the door is completely closed, it is automatically locked and cannot be opened from the outside, making it burglar-resistant. Fig. 21–15 shows an automatic garage-door opener which can be quickly disconnected for manual-door operation in case of electrical power failure. Notice the electric light socket on the automatic opener unit, which turns on when the door opens to light up the inside of the garage.

Patio Doors

In recent years, a door that has gained increased popularity is the patio door (Fig. 21–16). These doors provide easy access to easy-living areas. Patio doors are available in metal and wood, with most people favoring wood. They are available with double-insulated glass and inserts that give them a mullion effect. Patio doors may be swinging patio doors or sliding patio doors. They are available in metal, wood, or plastic, or a wood frame totally enclosed in vinyl plastic. The wood frame may also be clad on the exterior with aluminum or vinyl, leaving the interior wood bare for painting or staining. Manufacturers furnish instructions on how to install patio doors.

CHAPTER 22

Ceramic Wall and Floor Tile

Ceramic tile is not unlike brick in its origins and general method of production. The ingredients are essentially earth clays that are baked hard, like brick. In particular, however, tile is pressed into much thinner bisques and smaller surface sizes and then baked. The baking and face glazing produce a variety of finishes that make tile suitable for decorative purposes. Unlike brick which is used for structural purposes, tile is used to surface either walls or floors. It has a hard glossy surface that is impervious to moisture and resists soiling. It is easy to keep clean.

Tile generally is used for floors, walls, roof coverings and drain pipe. Some roof coverings are made of baked clays, the same as the drain tile used underground. This chapter discusses ceramic tile used for wall and floor coverings and its installation.

The term ceramic tile distinguishes the earth clay tiles from the metal and plastic tiles. The use of ceramic tile dates back four to six thousand years ago. It was used in the early Egyptian, Roman, and Greek cultures and, because of its durability, is sometimes the only part of structures left to study from the diggings into the ruins of those ancient days (Fig. 22–1).

The introduction of ceramic tile in the United States began about the time of the Philadelphia Centennial Exposition in 1876. Until then, tile was made and used in Europe. Some English manufacturers showed tile products at the exposition, which intrigued American

Fig. 22–1. Reconstructed Egyptian hall showing tile floor.

building material producers. Subsequently a factory was started in Ohio, then another in Indiana.

Tile Applications

The uses of tile range from the obvious to the exotic. The obvious begins with its use in a bathroom (Fig. 22–2), because water does not affect it and glazed tile harmonizes with tubs and other fixtures. Other areas of the home where moisture may affect conventional wall finishes are likely areas for tile. Kitchen-sink tops and backsplashes (Fig. 22–3) are a natural for tile. Fig. 22–4 shows an entire wall in tile, in a breakfast nook. For the same reason, tile is ideally suited to swimming pools. Walls and floors, as well as the pool itself, are covered with ceramic tile in the public pool. You can even find tile on a swimming pool on a deluxe ocean cruiser.

Fig. 22–2. A bathroom is a perfect place for ceramic tile.

Types of Tile

Not so long ago, tile was available in a relatively limited number of colors, shapes, and finishes, such as a few smooth, pastel-colored tiles. But no more. Today, there is a variety available. For walls, there is the

Fig. 22–3. Tile is also used for countertops.

standard 4¼″ × 4¼″ tile, but you can also get tile in a variety of shapes—diamond, curved, and so on—and in a great variety of colors and textures, from sand to smooth.

For the floor there is block tile, a tile that comes in shapes, and mosaic tile with tiles in individual 1-inch squares secured to a mesh background.

Fig. 22–4. Tile is available in block style for floors and walls.

Formerly, tile was installed with portland cement—a so-called mud job—but today the job is usually done, on floors and walls, with adhesive. There is a wide variety of adhesives available, and you should check with a tile dealer to ensure that you select the proper one.

Fig. 22–5. Guidelines are essential for installing tile on walls and floors.

Installing Tile

Dealers also carry detailed instructions—in some cases, color films—that show how to install tile. What follows, however, are the basic steps involved for tiling walls and floors:

Plan the Layout

For walls, one job is the standard tub surround of three walls. Start by drawing horizontal and vertical lines in the center of each wall. On one wall, lay a line of tile along the top of the tub. Look at the tile in relation to the lines. If you will have to install tile that is less than one tile wide, move the lines so that when you install, there will be no tile pieces that are less than one-half a tile wide. In some cases, of course, you will not be able to install tiles of the proper size, but you should try.

Fig. 22–6. Tile being applied to adhesive that has been put on with a toothed (notched) trowel.

Apply Adhesive

Use a notched trowel to apply adhesive; your dealer will tell you the type to use. Hold the trowel at a 45 degree angle and apply only a limited amount of adhesive at a time so that it does not dry out before the tile can be applied.

Place each tile with a twisting motion to embed it well in the cement. Press the tile down, tapping it firmly in place. Keep the joints straight.

Standard tile comes with nibs or spacers on the edges so that the proper joint distance will be maintained. If this is not the case, you will have to be very careful. Use small sticks of the same thickness to maintain joint distance. However you do it, make sure that the tiles are straight.

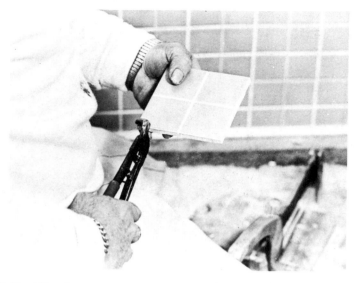

Fig. 22-7. Tile nippers are good for making irregular cuts in the tile.

Fig. 22-8. If tile does not have spacers, you must provide them.

Fig. 22–9. Tap tile carefully using a beater block to be sure it is seated properly.

Fig. 22–10. A rented tile cutter works well for making square cuts.

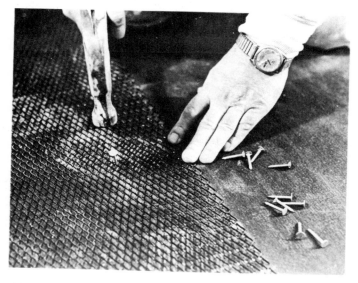

Fig. 22–11. In this installation, mesh is secured to floor. Cement is to be used. It is an old-fashioned "mud job."

Cutting Tile

At some point in the installation it is highly likely that you will have to cut the tile. For this you can use a rented tile cutter—it works like a paper cutter. For tiles that must have small pieces nibbled from them to fit around pipes and the like, you can use nippers—tile dealers usually have these.

When cutting tile, follow these procedures:

1. Cut and place sections one at a time. Cutting all the tile sections you need can lead to errors, because few corners are plumb.
2. Nip off very small pieces of the tile when cutting. Attempting to take a big chunk at one time can crack the tile.

Applying Grout

Grout is the material that fills the joints. It once was available only in white, but today you can get grout in many different colors, and it makes a nice accent material to the tile.

Fig. 22–12. A squeegee or rubber trowel is good for smoothing grout.

The package containing the grout—it comes as a powder you mix with water—will give instructions on its use. It should be applied at least 24 hours after the tile. A squeegee makes a good applicator: lay the grout on and squeegee off the excess. Excess grout can be cleaned off with a damp sponge. To shape the joint lines, you can use the end of a toothbrush. When the grout has dried, spray it with silicone and polish with a cloth.

Doing a Floor

Installing floor tile follows essentially the same procedure: draw lines, move them as needed, apply adhesive, set tiles. If you are tiling a countertop and have both a horizontal and vertical surface, it is easier to install tile on the horizontal area first.

CHAPTER 23

Attic Ventilation

The most nearly universal form of attic ventilation is the gable-end louvered vent, ranging in size from very large triangular units down to a single slit. According to one investigator, "The aesthetic appeal of simple louvers has been a major factor in their continued and popular use." Unfortunately, they have limited control of rain and snow penetration when they are sized for summer ventilation. More complicated, more expensive louvers are equally limited in stopping rain or snow penetration. Canadian building scientist Y. E. Forgues, in his paper, *The Ventilation of Insulated Roofs,* warns,

> *Gable vents have a series of blades sloped at 45 degrees. If not properly designed, they may allow rain or snow to enter the roof space. . . . In regions such as the Prairies, which are subject to fine, wind-driven snow, gable vents may have to be closed in winter to prevent snow from accumulating in attics.*

Many Americans, including this author, have found their attics full of snow after a northeaster because one end of the house's gables faces north. American Vietnam veterans tell of the sand in the barrack attics brought in through the gable-end vents. However, builders are not interested in more expensive vents as long as they believe the simple ones work. They insist that "They've never had any trouble because of gable vents."

The ventilation requirements in the three model building codes came from the FHA Minimum Property Standards, section 403–3.2:

> *Cross ventilation for each separate space with ventilating openings protected against the entrance of rain and snow.*
>
> *Ratio of total net free ventilating area to area of ceiling shall be not less than 1/150, except that ratio may be 1/300 provided:*
>
> *A vapor retarder having a transmission rate not exceeding one perm . . . is installed on the warm side of the ceiling, or*
>
> *At least 50 percent of the required ventilating area is provided by ventilators located in the upper portion of the space to be ventilated (at least 3'-0" above eave or cornice vents) with the balance of the required ventilation provided by eave or cornice vents.*

Moisture Control

For moisture control, the codes require that attic vents in houses with a vapor diffusion retarder in the ceiling shall have a free venting area equal to 1/300 of the attic floor area. NOTE: attic area is the area of the attic floor, not the roof area. The *BOCA* code requires that the vent openings be protected with a corrosion-resistant screening. The 1988 *UBC* requires mesh with $\frac{1}{4}$-inch openings. In order to compensate for the reduction in Net Free Venting Area (NFVA) caused by the 8×8 mesh (64 openings per in^2), the screened opening must be 25 percent larger. If screened and louvered openings are used for the vent, the louver size must be 125 percent larger than the specified area.

Example—A house with an attic floor area of 1000 ft^2 would require a NFVA of $7\frac{1}{2}$ ft^2.

1000 ft$^2 \times \frac{1}{300} = 3\frac{1}{3}$ ft^2 NFVA

3.33 ft$^2 \times 25$ percent plus $= 4\frac{1}{6}$ ft^2 screened area NFVA

3.33 ft$^2 + 4.16$ ft$^2 = 7.49$ ft^2 or $7\frac{1}{2}$ ft^2 louvered area with screens.

This house would require a louvered vent of $7\frac{1}{2}$ ft^2 /2 $= 3\frac{3}{4}$ ft^2 in each gable end of the house. Unfortunately, the 1/300 ratio is less than one-half square inch per square foot of attic area and is inadequate. *BOCA* officials reply to this inadequacy is that the code deals with *minimums*.

Summer Ventilation

The need for attic ventilation is greatest in summertime. But if any large reduction in attic temperatures is to be achieved by gravity air flow, the vent areas will have to be greatly increased. The vent area is actually determined by the stack height, that is, the difference in height from the inlet vent to the outlet vent. If half of the vents are low—in the eaves—and the other half are near the ridge the air flow will be much greater than if all the vents were high. The greater the stack height, the greater the gravity air flow. It is necessary to depend on gravity air flow rather than the greater air flow caused by wind because the **prime** need for ventilation is **not** when the wind is blowing, but rather when *there is no wind to force air into and out of the attic.*

In a house with a 6-foot stack height, the NFVA will have to be 1/50 of the attic area. If, for example, half the intake vent is screened, and the other half of the vent is a louvered screened exhaust vent located high in the attic, a 1000-ft^2 attic will need a screened intake of 12 ft^2 and two louvered exhausts of 11 ft^2 each. Not only is the size of these vents impractical, they could lead to lawsuits because there is no practical way to control rain and snow penetration. Worse yet, if the stack height is less than 6 feet, even larger vents would be required. The sizes of the exhaust and intake areas are related to the total area needed, so any reduction in the exhaust area will require a considerable increase in the intake area in order to get the same gravity air-flow. When exhaust/intake NFVAs are balanced, maximum cooling takes place.

If a gable-end louvered vent is to be used with soffit vents, it is practically impossible to provide enough exhaust vent to equal the soffit vent area. Even if the exhaust vent is one-half the area of the soffit vent, the gable-end louvered vent will have to be extremely large.

Example—A 25′ wide house with a 4′ high ridge will have a stack height of 2′-8″. The gable-end louvered vents would have to be 12′-6″ wide at the base and 2′ high.

Clearly the codes are of no help. Their minimums are better than nothing, but the better-than-nothing penalty is ice-dams, and all the trouble that results. Ice-dams will be found on houses without attic ventilation, but 95 percent of the ice-dams are found on houses with gable-end louvered vents. Using the stack height to size the NFVA does provide vents with greater air flow, which results in better cooling.

But rain and snow penetration could lead to lawsuits. Other means of attic ventilation must be found. But first we must establish criteria which all attic ventilation schemes must meet:

1. There must always be an exhaust vent in a negative pressure area.
2. The separate ventilation devices must operate as a *system*.
3. The *system* must operate independently of the wind.
4. The system must work at the most crucial time: when there is no wind.
5. The system must substantially reduce snow and rain penetration.
6. Airflow must be along the underside of the roof sheathing.
7. The flow of air must be consistent and in the same direction, *regardless* of wind direction.
8. 1½ cubic feet per minute of air, per square foot of attic floor area, must be moved along the underside of the roof sheathing for effective ventilation.

Airflow Patterns

We begin our analysis by looking at the airflow patterns in attics with different types of attic vents, and with the wind blowing directly at them or perpendicular (right angles) to them.

Gable-End Louvered Vents

Fig. 23–1 shows the wind blowing directly at the gable-end louvered vent. When the wind strikes the building it creates an area of high positive pressure, and the vent it strikes becomes an *intake* vent. The wind flows up, around, and over the structure creating areas of negative pressure, or suction, and causes vents located on the negative side to become *exhaust* vents.

As long as the wind blows directly at the vent, there will always be an exhaust vent in a negative pressure area. The wind enters the positive vent, drops and moves over the insulation in the attic floor, moves up and exits through the negative vent. Criteria 1 and 2 are met. But the remaining criteria are not.

But the wind does not blow steadily in one direction; shifts in direction change positive pressure areas to negative and vice versa. Wind perpendicular to the ridge makes each vent become and intake/exhaust

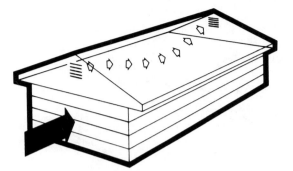

Fig. 23–1. **Air flow pattern with wind parallel to ridge.** *(Courtesy Air Vent, Inc.)*

vent (Fig. 23–2). The louvers now work independently of each other, not as a system. The underside of the sheathing is not ventilated, little moisture is removed. There is no air movement due to thermal effect. None of the criteria are met.

Soffit Vents

Soffit vents provide effective air flow regardless of wind direction, and there is a balance between intake and exhaust areas (Figs. 23–3 and 23–4). The air flow is confined to the attic floor, where it could reduce

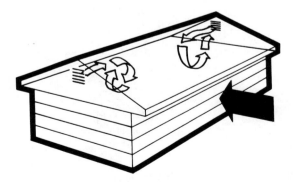

Fig. 23–2. **Air flow pattern with wind is perpendicular to ridge.** *(Courtesy Air Vent, Inc.)*

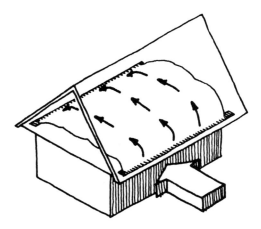

Fig. 23–3. Air flow pattern with soffit vents when wind is parallel to soffits.

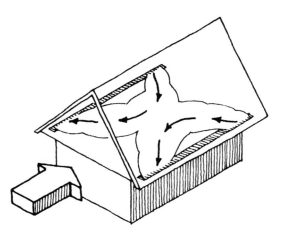

Fig. 23–4. Air flow pattern in soffit vents when wind is perpendicular to soffits.

the effectiveness of the insulation. The underside of the sheathing is not ventilated, and very little moisture is removed. There is no air movement, due to thermal effect, when there is no wind. What little effectiveness these vents have is due to wind.

Gable/Soffit Vent Combination

There is a common but mistaken belief that more is better: combine gable and soffit vents to work as a system. But as the airflow patterns in Figs. 23–5 and 23–6 show, regardless of wind direction, the gable and soffit vents work independently of each other. Soffits vents meet criterion 1, but not the remainder of the criteria.

There are other attic ventilation schemes: roof louvers only (either mushroom rectangular vents or turbine vents), roof louvers with soffit vents, and attic fans. With a single roof louver, air movement is only within a small area surrounding the louver. With two roof louvers only, the air movement is from one louver vent to the other. The louvers wash only the roof area immediately next to them.

Roof louvers with soffits vents are heavily wind-dependent. When the wind is perpendicular to the ridge, only the soffit vents work. When the wind of parallel to the ridge, only the roof louvers work.

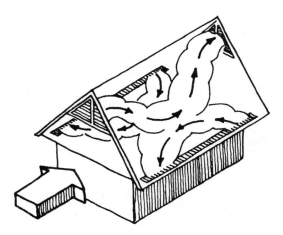

Fig. 23–5.   Air flow patterns with gable and soffit vents when wind is parallel to ridge.

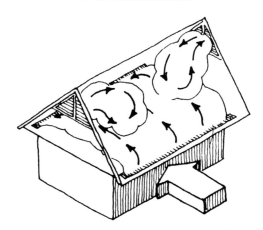

Fig. 23–6 Air flow pattern with gable and soffit vent with wind perpendicular to ridge.

Some manufacturers of modular houses equip attics with fans in the mistaken belief that they *make a difference* by reducing attic temperatures and therefore reduce air conditioning loads. But numerous studies by universities, public utilities, the federal government, and others cast doubt on these claims. Fans may help lower attic temperatures; however, this has little effect on the total cooling load. Energy is not saved. The fans increase the electric bill without reducing the air conditioning bill. If the fan is located on the windward side of the roof, the wind reduces its effectiveness. Locating the fan on the leeward side helps, but no more than 50 percent of the roof sheathing can be washed (ventilated). Most utilities do not recommend fans.

Turbine Vents

Recently, turbine vents have been the subject of heated debate and controversy. As we have seen, a roof vent moves air only in a small area immediately surrounding the vent. Even if soffit vents are combined with the roof vent, the airflow pattern is irregular. Air enters one soffit and exits through the other soffit vent. Little of the air flow finds its way up to the roof vent.

A turbine vent is actually nothing more than a roof vent with a rotating ventilator cap (Fig. 23–7). At low wind speeds when the attic's need for ventilation is greatest, the turbine performs no better or worse than other roof vents. High winds cause excessive spinning which causes excessive wearing of the bearings. Turbine vents often stick and fail to rotate.

Repeated testing, and wind tunnel testing of turbine vents over a number of years, reveals how poorly they work. These tests show that air may enter the turbine rather than be exhausted by it. Many believe that while the turbine is turning, no rain or snow can enter. But just as it can and does enter regular roof louvers, it enters the turning turbine. Hundreds of houses in Oklahoma have several inches of sand over the ceiling insulation brought in by turbine ventilators.

Testing at Texas A&M has shown little difference between the airflow with a turbine ventilator and an open stack, or a gravity-type ventilation system. The same conclusions were reached at Virginia Polytechnic Institute, based on wind tunnel testing of turbine ventilators. One engineer at another university has observed that how well turbine ventilators work depends on who is doing the testing.

Texas A&M concluded that the only attic exhaust airflow through a turbine ventilator under normal wind conditions is the buoyant air resulting from the density difference between the hot attic and the cooler outside air. The turbines fail—as do gable end louvered vents and soffit vents—*when there is no wind.*

Is There a Solution?

Our search has shown that attic ventilation is not a simple straightforward matter of merely installing gable-end louvered vents, or soffit vents, or turbine vents. This simplistic approach ignores the fact that adding attic ventilation is not the same as seeking to reduce attic temperatures and to remove moisture.

Building codes are not alone in recommending, without question, the MPS ratios. Nearly all books on carpentry, energy-efficient construction, architectural construction, and virtually all magazines for homeowners and do-it-yourselfers parrot these ratios. The manufacturers of attic ventilation products are of little help because they are engaged in claims and counterclaims over net free venting areas, whether

Fig. 23–7. Turbine vent.

or not external baffles on ridge vents are necessary, and if ridge and soffit vents must be balanced.

Many writers agree that "the most effective type of attic ventilation is had with continuous ridge and soffit vents." But they do not tell us why they believe this, or what combinations of ridge and soffit vents to use and why. One researcher challenges both the manufacturers and the code writers on using screening for soffit vents, but he provides us with no guidance.

The eight criteria tell us that there must be a *system* of attic ventilation, and what that system must do. The analysis of airflow patterns shows that gable end vents, roof/turbine vents, all with or without soffit vents, and attic fans, fail to meet the criteria. We are left with ridge vents and whether or not they comply with the criteria. Unfortunately the criteria do not help us sort out the maze of conflicting advice:

1. For soffit vents use any of the following: aluminum, vinyl, or masonite panels with holes or slots; circular screened vents of various diameters; screening; single-louvered or double-louvered strips; rectangular screened louvered vents. All but the strips and panels can be placed in each bay, every other bay, or every fifth bay, and anywhere in the soffit.

2. Install any type of continuous ridge vent with any kind of soffit vent.

3. Install only a ridge vent with an external baffle and only double-louvered soffit vents.

4. The soffit vents can be located anywhere in the soffit.

5. Keep soffit vents as far away from outside wall as possible. Install them next to fascia board.

Ridge Vents

The ridge vent, located at the highest point of the roof, provides the 50 percent of the NFVA in a high location. The remaining 50 percent is divided between the low soffit vents: 25 percent in each soffit. This balance provides equal vent areas for equal airflow into and out of the attic. For maximum cooling effect, input and output NFVA should be balanced. An unbalanced system's effective ventilation rate is reduced to the airflow through the smaller NFVA.

Figs. 23–8 and 23–9 show that the airflow pattern with ridge and soffit vents remains the same regardless of wind direction. The airflow pattern is from the low soffit vents up alongside the sheathing and exiting at the ridge vent. Because the airflow is always in the same direction, it creates an inertia of constant air movement. This explains why the ridge vent is so efficient. Some researchers have questioned whether the airflow pattern is as claimed. They argue that according to fluid mechanics when the air enters the small soffit slot and suddenly empties into the large attic space, it will spread throughout the attic. But recent measurements have confirmed that the airstream flows alongside the roof sheathing and within 3 or 4 inches of the roof sheathing.

Baffles

Crucial to the performance of a ridge vent is the external baffle. Unfortunately there is no agreement on this among ridge vent manufacturers. Several companies produce baffleless ridge vents, and claim they operate trouble-free.

Construction Research Laboratory of Miami, Florida is a highly respected testing laboratory specializing in testing of attic vents. Be-

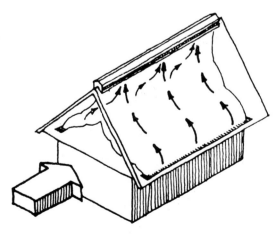

Fig. 23–8. Airflow pattern with ridge and soffit vents with wind parallel to ridge.

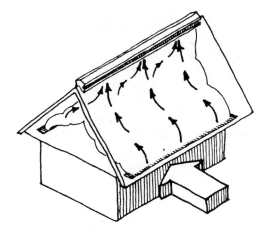

Fig. 23–9. Airflow patterns with ridge and soffit vents with wind perpendicular to ridge.

tween 1983 and 1987 they ran a series of tests of ridge vents with and without baffles, to measure wind-driven rain penetration into the vents. The wind speeds were 50, 75 and 100 mph. The simulated rain rate was 8 inches per hour, the rate used for Dade County, Florida evaluation. The water spray was added to the airstream upwind of the test specimen at a rate equal to an 8 inches per hour of rain.

In one series of tests at 50 mph and lasting 5 minutes, water entered the entire length of the baffleless ridge vent and dropped down into the space below. The total water leakage was 1.5 quarts, or approximately 9 ounces per minute. Under the same test conditions, the ridge vent with the external baffle allowed only 0.2 ounces of water penetration in 5 minutes, 0.6 ounces at 75 mph, and 1.2 ounces at 100 mph in 5 minutes.

Another series of tests on a different brand of ridge vent with external baffle showed zero water penetration at 50, 75, and 100 mph wind speeds. Ridge vents without baffles took on 0.3 gallons of water per hour per foot of vent to as much as 1 gallon with a wind speed of 100 mph.

Photographs of the testing show clearly that the exterior baffle causes the air and rain stream to jump up and over the top of the vent. The baffle is important for a number of reasons: (1) The winds may jump over the entire roof or follow the roof surface depending on roof pitch and wind direction. (See 1989 ASHRAE *Handbook of Funda-mentals,* chapter 14.) The baffle *helps* to kick the wind over the ridge vent. (2) Snow-laden winds of 25 mph or greater can penetrate the smallest of openings. The baffle helps reduce this by causing the wind to jump over the ridge vent. (3) The movement of the wind over the ridge vent creates a venturi action which causes both the windward and leeward sides of the ridge to be in a negative pressure area. That is, it *sucks* air out of the attic. Soffit vents are not necessary for this to happen.

Soffit Vents

Most of the soffit vents made of vinyl, aluminum, and masonite have either circular holes or slots. Typically, the vinyl soffits have .125-inch (⅛) diameter holes spaced 0.5 inch (½) oc. A full square foot of these soffit vents would *initially* provide an NFVA of 6 in^2 to the ridge vent's 9 in^2. There are regional differences, but even in the South, houses that should have 3-foot or wider soffits have less than one foot of overhang. If the average soffit is six inches wide, the installed NFVA of these soffit vents is about 3 in^2. As the soffit NFVA is reduced, a point is reached where the ridge vent reverses: the windward side becomes an intake vent, and the leeward side becomes an exhaust vent. When the ratio of ridge NFVA to soffit NFVA is 9 to 6 there is little danger of ridge vent reversal until the holes in the soffit vent become blocked with dirt. When the ratio is 9 to 4 or 9 to 3, ridge vent reversal is highly probable.

Some ridge vent manufacturers claim they can find no reason why ridge and soffit vents should be balanced. Yet they admit that the ridge vent works better with soffit vents. When the inlet/outlet vents are balanced, maximum flow and cooling takes place. However, although an *absolute* balance is not necessary, the question of how much imbalance and whether more or less soffit venting is needed must be answered.

The exact point at which the imbalance will cause the ridge to reverse, is unknown. The ridge vent is driven by the soffit vents—air from the soffit vents must be fed to the ridge vent. The soffit vents can be unbalanced in the direction of 10 to 15 percent more venting area. However, excessive soffit venting will result in the air moving from one section of the soffit vent to the next rather than to the ridge vent. As the soffit NFVA is reduced, the ridge vent reverses, and snow-laden wind can enter the windward side and drop snow on the attic floor. Air Vent Incorporated installs a fiberglass filter in their ridge vent that prevents snow penetration.

Field experience, and hundreds of thousands of dollars of damage from too little attic ventilation, indicates that when the soffit vent NFVA is about 3 or 4 in^2 the ridge vent will reverse. This information and the following guidelines help in selecting soffit vents.

1. The soffit vent must be continuous, and span the width between the rafters.
2. The soffit vent NFVA should match the ridge vent NFVA.
3. Do not use screening, vinyl, aluminum, or masonite soffit panels with circular holes. The holes will eventually block up from dirt and reduce NFVA. Screening will also block up from dust and dirt. Screening may be 70 percent open or 70 percent closed. If the contractor uses a screening that is 70 percent closed, what is the NFVA of the screening? The masonite will get painted and even if the NFVA is adequate, the painting will reduce it.
4. Locate the soffit vent next to the facia board. The closer the vent is located to the exterior wall, the closer it gets to the area of positive wind pressure. This increases the chances of rain or snow penetration.
5. Avoid single-louvered soffit vents. There are two ways to install them, one of which is wrong. Eliminating them gets rid of the worry of whether they were correctly installed, and also possible lawsuits resulting from water damage.
6. Use only a double-louvered soffit vent. It cannot be installed backwards. The louvers face in opposite directions (Fig. 23–10) which helps creates a turbulence that helps keep snow from entering the attic space.

Remember that a 1-foot wide soffit will allow only a 10-inch wide piece of masonite, vinyl, or aluminum. This reduces the NFVA to 5 in^2, which may be adequate, *as long as the vents are not painted or enough holes eventually blocked by dirt to reduce the NFVA to the danger point.*

These guidelines are not rules-of-thumb. Wind tunnel testing shows snow or rain entering the soffit vents when they are located next to the exterior wall. Gable-end louvered vents or soffit vents, alone or in combination, provide no air circulation in the upper part of the attic for moisture removal. Placing a small circular screened vent in the middle of the soffit does not bathe the complete width of the rafter bay. Pockets of moisture remain, and the ridge reverses because of the imbalance. The ridge works part time, which is better than not at all. But the penalty is water penetration.

In new construction keeping the soffit vent next to the facia board is easier to do than when retrofitting. Air Vent Incorporated manufactures a Model SV202 (Fig. 23–10) which has a vertical flange. The flange is nailed to the rafter or truss ends before the facia board is installed.

When vinyl siding is used, the SV202 is installed at the rafter/truss ends, and a J-channel holds up the other end of the soffit vent and the solid vinyl soffit. Again, this should be discussed with the vinyl siding contractor at the pre-framing conference.

When a customer wants gable-end louvered vents for looks, or because the house is a colonial, but agrees to ridge/soffit vents, the inside of the gable-end vents must be blocked to prevent air from entering.

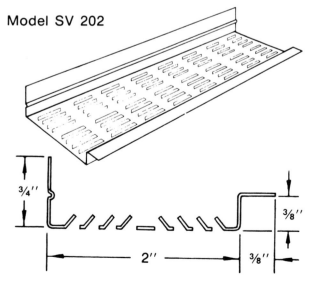

Fig. 23–10.   Double-louvered soffit vent. *(Courtesy of Air Vent, Inc.)*

Drip-Edge Vents

Drip-edge vents (Fig. 23–11) seem to make a lot of sense: in one operation the eave drip-edge and the vent are installed. Unfortunately, a good idea turns out to be a bad one in practice.

One of the giants of the residential construction business regularly used drip-edge vents. But field reports from Toledo (Ohio), Buffalo (New York), Minnesota, and other areas told of snow being driven through the vent six and eight feet into the attic. Construction Research Laboratory was hired to find the reason for the snow penetration.

What the laboratory discovered was that snow will always penetrate the vent unless a baffle is installed to prevent this. No commercially available drip-edge vent with a baffle is available.

Another problem with many of the drip-edge vents is that the NFVA is too little—about 4 in². Some drip-edge vents are easily damaged when a ladder is placed against them. Many builders report that they do not keep the rainwater away from the siding. Ice-dams are a common problem with drip-edge vents.

Shingle-Over Ridge Vents

Although aluminum ridge vents are still available, plastic ridge vents are slowly replacing them. Fig. 23–12 is an example of one such vent. Called SHINGLEVENT II™, it is made from polyethylene and manufactured by Air Vent, Inc. The cap shingles install over the top, which is rounded to prevent cracking of the cap shingles. There have been increasing field reports of cap shingle cracking caused by the sharp point of some plastic ridge vents.

Fig. 23–11. Drip-edge vent.

Fig. 23–12. Plastic shingle-over vent with shingle cap on top. Note exterior baffle. *(Courtesy of Air Vent, Inc.)*

Hip Roofs

Hip roofs, depending on the pitch, often have little ridge, and venting them can be a problem. Install double-louvered soffit vents in all four eaves.

Should ridge vents be installed on the hip rafter? With low pitch roofs it may work; use a ridge vent without an external baffle. The external baffle can provide a channel for water which could enter the ridge and the attic space. This can easily happen on a high-pitch hip roof.

Is Attic Ventilation Necessary?

There are eight different schools of attic ventilation, one of which argues that attic ventilation is not necessary. Another school insists that in superinsulated attics (R-40 and higher), ventilation does not work, and may actually cause trouble. The non-venting of cathedral ceilings has resulted in millions of dollars of roof damage because of leaks in the sheetrock air retarder. Cathedral ceilings and Casablanca fans, eyebrow, and other recessed lights can be a disastrous combination if the attic space is not ventilated.

If the ceiling sheetrock air retarder is continuous and unbroken—no penetrations—venting of the cathedral ceiling is not necessary. Unfortunately, the shingle manufacturers will void the shingle warranty if the attic space is not ventilated.

As a contractor ask yourself:

1. Am I willing to gamble that the shingles will not fail from heat damage—as the manufacturers claim?
2. Who will pay for the supervision of the sealing of the air retarder?
3. Is it possible to seal the air retarder that is punctured with Casablanca fans, recessed lighting fixtures, and skylights?

The sealing of the air retarder must be supervised by someone who is experienced. Do not make the mistake of assigning it to one of the crew. Even if the ceiling is ventilated, the air retarder should be as leakproof as possible. It may not be possible to avoid penetrating the ceiling with the stack vent. However, bathroom fans should never be vented vertically. The stack effect keeps a constant stream of warm air rising, which keeps the backflow valve constantly open. Pre-plan the bathroom exhaust fans; place them in interior walls, and route the ductwork down and out through the joist header. Cold air cannot move up the duct work, the stack effect is eliminated, as is the possibility of moisture damage.

CHAPTER 24

Radon

Radon and Its Sources

Radon is a naturally occurring radioactive gas found in soils and rocks that make up the earth's crust. Radon gas comes from the natural breakdown or decay of radium. But it does not remain a gas for very long. It has a half-life of 3.8 days. If you filled a gallon jug with radon on Monday morning, by late Thursday night only one-half gallon of radon would be left; in another 3.8 days, the one-half gallon would decay down to one-fourth gallon, and so on. When radium decays to become radon, it releases energy. It is the rapid release of energy that causes radon and radon decay products, called *progeny* or *radon daughters,* to be dangerous to health. It is the decay of the short-lived daughters that is dangerous, not the decay of radon.

Radon daughters present in the air we breathe will be inhaled. Because they are not gases, they can stick to the lungs. The energy given off as the daughters decay can strike the lung cells, damage the tissue, and lead to lung cancer. For the sake of simplicity, the term radon will be used throughout this chapter.

How Does Radon Get into a House?

Radon enters a structure through the soil beneath the house, through cracks or holes in the slab floor, the foundation, chimneys, fireplaces, and so on (Fig. 24–1). However, the most common way that radon enters is when the house is depressurized or under negative pressure. Fireplaces, chimneys, furnaces, and gas stoves require combustion air. Unless these appliances are supplied with outside air, they will take air from the house. They are like fans and suck the air out of the house, as do bathroom fans and range hoods vented to the exterior.

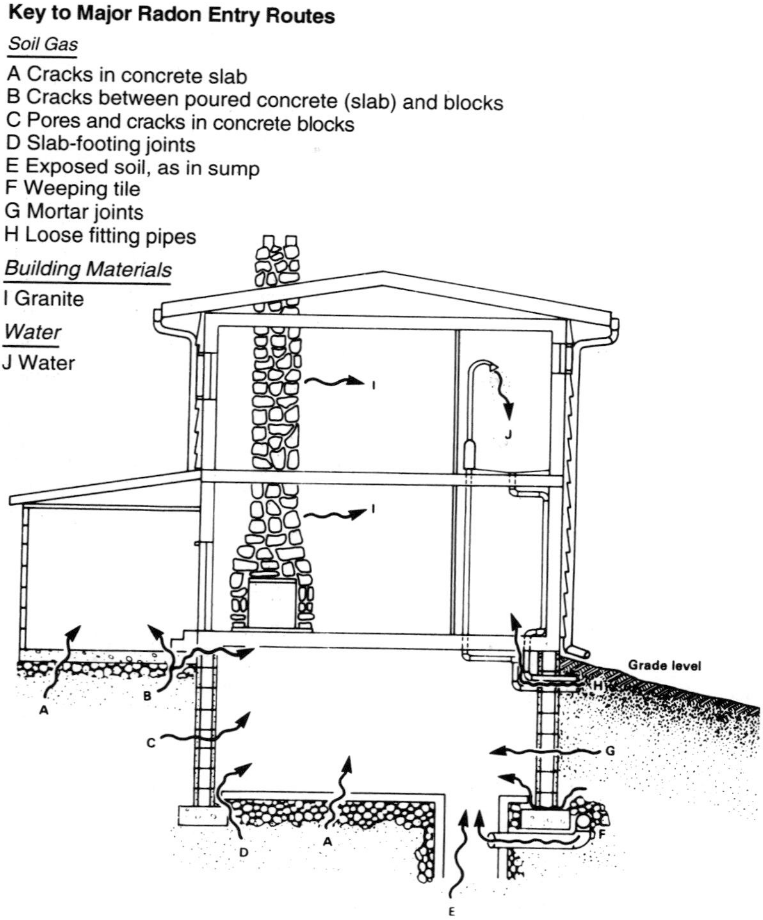

Key to Major Radon Entry Routes

Soil Gas

A Cracks in concrete slab
B Cracks between poured concrete (slab) and blocks
C Pores and cracks in concrete blocks
D Slab-footing joints
E Exposed soil, as in sump
F Weeping tile
G Mortar joints
H Loose fitting pipes

Building Materials

I Granite

Water

J Water

Fig. 24–1.   Major radon entry routes into detached houses.

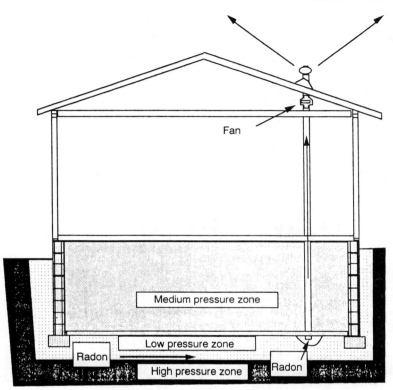

Fig. 24–2.   Active stack sub-slab depressurization system.

This sucking action reduces the amount of air in the house and depressurizes it, putting it under negative pressure. Wind and stack effect also help depressurize the structure.

The stack effect is caused by the tendency of warm buoyant air to rise and leak out of the upper portion of a structure. As the warm air leaves, cold air is drawn into the lower parts of the structure. The rate of infiltration varies considerably from day to day, hour to hour, depending on wind speed and direction, open or closed windows or vents.

A house under negative pressure allows radon gas to move into the holes and cracks in basements and slab-on-grade foundations. The higher air pressure in the soils is drawn into the basement by the lower

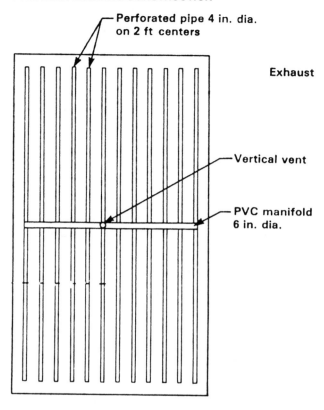

Top view — network laid under slab

Fig. 24–3. Sub-slab pipe network.

indoor air pressure. This is *pressure-driven radon.* Radon can enter buildings even if there is no pressure difference. It will spread from an area of higher concentration to an area of lower concentration by diffusion. This is *diffusion-driven radon.*

If well water is supplied by groundwater in contact with radium-bearing formations, it will bring radon into the house. The health risk from breathing radon gas released from the water, as in a shower, is ten times greater than drinking water containing radon. Building materials

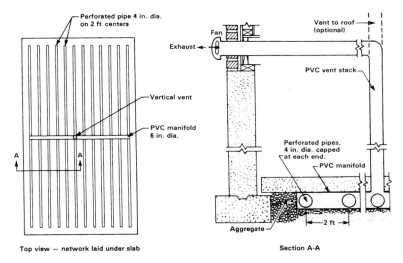

Top view — network laid under slab   Section A-A

Fig. 24–4.   Active system using pipe network.

can be contaminated with radium, but there is little information on this subject.

Radon Reduction Methods

Site Evaluation

High radon levels are found in areas of uranium or phosphate mines; but there are also certain geological formations containing high radon levels. One such formation, for example, is the Reading Formation which extends across Pennsylvania up into New Jersey and New York. Unfortunately, there is no simple reliable method of identifying potential problem sites. Although researchers have made substantial progress in identifying areas of high radon risk, it requires many, many measurements of a particular site to determine the risk. At this point Value Analysis might be used to decide which is more cost effective: building a radon-resistant structure or spending money on site evaluation to possibly avoid the need for a radon-resistant building. Although research continues, the data that has been accumulated is useful only on a community-wide basis, not for individual lots.

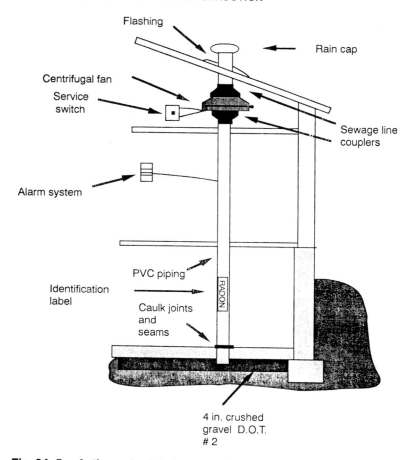

Fig. 24–5. Active sub-slab depressurization system exhausted vertically.

There are a number of ways to prevent radon from entering a house: (a) sealing all holes, cracks, exposed soil, sumps, floor drains; (b) reversing the direction of radon flow so that the air movement is from the house to the soil and outside air; (c) avoiding using water containing radon through the use of aeration or carbon filters; (d) avoiding the use of radium contaminated building materials; (e) providing for a sub-slab depressurization or pressurization system during new construction; (f) installing mechanical barriers to block soil-gas entry; (g) avoiding risky sites; (h) using planned mechanical systems; (i) supplying fresh air to dilute the radon; (j) reversing the airflow by keeping the basement under positive pressure.

Sub-Slab Ventilation

Unfortunately, space restrictions prevent a comprehensive review of all the techniques for dealing with radon. Sub-slab ventilation has worked well and is the system we will discuss.

Soil gas can accumulate in the aggregate under the slab and enter the house through any openings: where the floor slab meets the foundation wall, at gas, electric, or water pipe penetrations. The purpose of the sub-slab ventilation system is to exhaust the soil gas out of the aggregate before it can enter the house. There are two systems: the passive system, which may consist of a single pipe or network of pipes, and the active system, which also consists of a pipe or network of pipes and an exhaust fan.

The passive system depends on the suction created by the stack effect to remove the radon. A non-perforated pipe is installed vertically downward through the floor slab and into the aggregate. The pipe is extended vertically up through the house and exits through the roof (Fig. 24–2). It is important that this passive stack be continuously vertical. A more elaborate version is to have a network of pipes (Fig. 24–3).

The active system uses a fan to create the suction. It may be a single pipe system or a network system Fig. 24–4). The active system may be exhausted vertically (Fig. 24–5) or horizontally as shown in Fig. 24–4.

For more information consult the following Environmental Protection Agency (EPA) publications: *Radon Reduction Techniques for Detached Houses,* and *Radon-resistant Construction Techniques for New Construction.*

Appendix

Energy Savings Worksheet

If you want a more accurate estimate of your energy savings than the ones given in Chapter 17, you may use the Worksheet given in this Appendix. Step-by-step instructions are as follows:

1. Examine air-conditioning unit, determine SEER (for a key to abbreviations, see page 636). Divide SEER by 3.413 to obtain efficiency or COP and enter result in Box A. Typical efficiencies are given in Table X. If SEER is unknown, enter 2.3 in Box A.
2. Examine heating equipment. Determine whether it is a gas furnace, oil furnace, heat pump, electric furnace, or electric baseboard heating. Determine efficiency, and enter in Box B. Typical efficiencies are given in Table X. If efficiency is unknown, enter 0.65 in Box B.
3. Obtain cost of electricity, either by examining your electric bills or by contacting your utility. Multiply the cost in cents per kilowatt-hour by 2.93 and enter result in Box C.
4. Obtain cost of heating fuel, either by examining your fuel bills or by contacting your utility.
   If you heat with gas, multiply the cost in dollars per CCF (or therm) by 10 and enter result in Box D.
   If you heat with oil, multiply the cost in dollars per gallon by 7.15 and enter result in Box D.

If you heat with electricity (including a heat pump), multiply the cost in cents per kilowatt-hour by 2.93 and enter result in Box D.

5. Divide the value in Box C by the value in Box A and enter result in Box E.

6. Divide the value in Box D by the value in Box B and enter result in Box F.

7. Inspect your attic to determine the type and level of conventional attic insulation, the area of the ceiling, and whether or not the cooling ducts run through the attic.

The level of insulation may be estimated with the following chart for insulation thickness (in inches) as a function of insulation type and level:

Type of Insulation	R-11	R-19	R-30	R-38
Fiberglass batts	3.5"	6.25"	9.75"	12.5"
Loose-fill fiberglass	4.75"	8.25"	12.75"	16"
Loose-fill cellulose	3.75"	6.50"	10.50"	13"

The area of the ceiling is determined by estimating the length and width (in feet) of the ceiling and multiplying these two values together. Enter this value in Box 1.

8a. If you plan to install a radiant barrier (RB), go to Table Y1. Locate a city that is near your location and then read off the value for that city for the level of insulation in your attic.

Then multiply this value by one of the following factors depending upon the type of radiant barrier you plan to install, and enter the result in Box 2:

Configuration Factor

For low range of values for dusty attic floor RB	0.16
For high range of values for dusty attic floor RB	0.65
For RB attached to rafter bottoms, and with no ducts in attic	0.78
For RB attached to rafter bottoms, and with ducts in attic,	
and with R-11 conventional attic insulation	0.98
or with R-19 conventional attic insulation	1.07
or with R-30 conventional attic insulation	1.15
or with R-38 conventional attic insulation	1.22
For RB draped over tops of rafters or attached to roof deck,	
and with no ducts in attic	0.68
For RB draped over tops of rafters or attached to roof deck,	
and with ducts in attic, and with R-11 conventional attic insulation	0.86
or with R-19 conventional attic insulation	0.93

or with R-30 conventional attic insulation	1.01
or with R-38 conventional attic insulation	1.07

8b. If you plan to install more insulation, go to Table Y3. Locate a city near your location and read off the value for that city and for the initial and final levels of attic insulation. Note that values in the table may be added in steps. For example, if you start with R-11 insulation and want to go to the R-38 level, add the values for going from R-11 to R-19, for R-19 to R-30, and for R-30 to R-38. Enter the value in Box 2.

9a. If you plan to install a radiant barrier, go to Table Y2. Locate the same city that you used for Step 8a and read off the value for that city for the level of insulation in your attic.

Then multiply this value by one of the following factors depending upon the type of radiant barrier you plan to install, and enter the result in Box 3:

Configuration Factor

For low range of values for dusty attic floor RB	0.24
For high range of values for dusty attic floor RB	0.61
For RB attached to rafter bottoms	0.88
For RB draped over tops of rafters or attached to roof deck	0.82

9b. If you plan to install more insulation, go to Table Y4. Locate the same city that you used for Step 8b and read off the value for that city and for the initial and final levels of attic insulation. Note that values in the table may be added in steps. For example, if you start with R-11 insulation and want to go to to R-38 level, add the values for going from R-11 to R-19, for R-19 to R-30, and for R-30 to R-38. Enter the value in Box 3.

10. Multiply the values in boxes 1, 2, and E together, and divide the result by 1,000,000. Enter the result in Box 4.

11. Multiply the values in Boxes 1, 3, and F together, and divide the result by 1,000,000. Enter the result in Box 5.

12. Add the values in Boxes 4 and 5 together, and enter the result in Box 6. This is the expected savings per year due to adding a radiant barrier or additional attic insulation.

13a. If you plan to install a radiant barrier, determine the estimated cost for installing a radiant barrier in your home. This may be from a quote, or you may estimate the cost by using the values in *Table 1* along with your estimate of the ceiling area. Note that for radiant barriers installed on the rafters or on the roof deck, you will have to estimate that area of the roof and the areas of the gable ends. Enter the estimated cost in Box 7.

13b. If you plan to install additional attic insulation, determine the estimated cost for installing more insulation in your home. This may be from a quote, or you may estimate the cost by using the values in *Table 2* along with your estimate of the ceiling area. Enter the estimated cost in Box 7.

14. Go to Table Z. Locate the census region where you live and read off the value for electricity. Enter this value in Box 8.

15. Go to Table Z. Locate the census region where you live and read off the value for either electricity, oil, or natural gas, depending upon your heating fuel type. Enter this value in Box 9.

16. Multiply the value in Box 4 by the value in Box 8. Enter the result in Box 10.

17. Multiply the value in Box 5 by the value in Box 9. Enter the result in Box 11.

18. Add the value in Box 10 to the value in Box 11 and enter the result in Box 12.

19. Compare the value in Box 12 with the value in Box 7. If the value in Box 12 is greater than or equal to the value in Box 7, then the radiant barrier or additional insulation is an economical investment. If the value in Box 12 is less than the value in Box 7, then the radiant barrier or additional insulation is not an economical investment.

20. A simple payback period may also be determined by dividing the value in Box 7 by the value in Box 6. The result will be the number of years that it takes for the energy savings with the radiant barrier or additional insulation to pay back its initial cost. Note that this procedure is not applicable to the radiant barrier on the attic floor, because the energy savings changes from year to year.

Note: If you are planning to install a radiant barrier on the attic floor on top of the existing attic insulation, you should go through the worksheet twice, using the two factors that are given in Steps 8a and 9a to obtain an estimate of the expected range of energy savings.

Example of Use of Worksheet

I live in Orlando, Florida in a one-level 1800 square foot house. I have a heat pump system that has medium efficiency. My electricity costs 8 cents per kilowatt hour. I have 3.5 inches of fiberglass batt insulation (R-11) in my attic and the air-conditioning ducts are in the attic. A contractor has quoted a price for a radiant barrier installed on the bottoms of my rafters and on the gable ends for $400. Would this be a good investment?

Following the steps outlined in the instructions, the worksheet is filled out. The total present value of energy savings given in Box 12 is $533.14. This value exceeds the quoted cost of the radiant barrier of $400, and thus would be a good investment.

EXAMPLE

ENERGY SAVINGS ESTIMATE FOR RADIANT BARRIERS OR ATTIC INSULATION

WORKSHEET

COST OF ENERGY FOR HEATING AND COOLING

Code: (A)	(B)	(C)	(D)	(E)	(F)
Cooling Equipment Efficiency	Heating Equipment Efficiency	Cooling Fuel Price $/Million BTU 8 × 2.93 =	Heating Fuel Price $/Million BTU 8 × 2.93 =	Cooling Energy Cost $/Million BTU [C÷A]	Heating Energy Cost $/Million BTU [D÷B]
2.6	1.9	23.44	23.44	9.02	12.34
From Table X	From Table X				

For fuel prices:

	Electricity:	$/million BTU = ¢/KWH × 2.93
	Natural Gas:	$/million BTU = ($/therm or $/CCF) × 10
	Fuel Oil:	$/million BTU = $/gal. × 7.15

ESTIMATED ENERGY SAVINGS

Code: (1)	(2)	(3)	(4)	(5)	(6)	(7)
Ceiling Area, Square Feet	Cooling Load Factor 2575 × 0.98 =	Heating Load Factor 275 × 0.88 =	Annual Cooling Savings, $/yr [(1) × (2) × E] ÷ 1,000,000	Annual Heating Savings, $/yr [(1) × (3) × F] ÷ 1,000,000	Total Energy Savings, $/yr [(4) + (5)]	Cost for RB or Insulation, $
1800	2524	242	40.98	5.38	46.36	400
	From Table Y	From Table Y				(Estimated Cost of RBS or Insulation)

ESTIMATED LIFE CYCLE PRESENT VALUE SAVINGS

Code: (8)	(9)	(10)	(11)	(12)
Cooling Discount Factor	Heating Discount Factor	Present Value Cooling Savings, $ [(4) × (8)]	Present Value Heating Savings, $ [(5) × (9)]	Total Present Value Energy Savings, $ [(10) + (11)]
11.50	11.50	471.27	61.87	533.14
From Table Z	From Table Z			

ENERGY SAVINGS ESTIMATE FOR RADIANT BARRIERS OR ATTIC INSULATION

WORKSHEET

COST OF ENERGY FOR HEATING AND COOLING

Code: (A)	(B)	(C)	(D)	(E)	(F)
Cooling Equipment Efficiency	Heating Equipment Efficiency	Cooling Fuel Price $/Million BTU	Heating Fuel Price $/Million BTU	Cooling Energy Cost $/Million BTU [C÷A]	Heating Energy Cost $/Million BTU [D÷B]
From Table X	From Table X				

For fuel prices:
Electricity: $/million BTU = ¢/KWH × 2.93
Natural Gas: $/million BTU = ($/therm or $/CCF) × 10
Fuel Oil: $/million BTU = $/gal. × 7.15

ESTIMATED ENERGY SAVINGS

Code: (1)	(2)	(3)	(4)	(5)	(6)	(7)
Ceiling Area, Square Feet	Cooling Load Factor	Heating Load Factor	Annual Cooling Savings, $/yr [(1) × (2) × E] ÷ 1,000,000	Annual Heating Savings, $/yr [(1) × (3) × F] ÷ 1,000,000	Total Energy Savings, $/yr [(4) + (5)]	Cost for RB or Insulation, $
	From Table Y	From Table Y				(Estimated Cost of RBS or Insulation)

ESTIMATED LIFE CYCLE PRESENT VALUE SAVINGS

Code: (8)	(9)	(10)	(11)	(12)
Cooling Discount Factor	Heating Discount Factor	Present Value Cooling Savings, $ [(4) × (8)]	Present Value Heating Savings, $ [(5) × (9)]	Total Present Value Energy Savings, $ [(10) + (11)]
From Table Z	From Table Z			

Table X. Equipment Efficiencies

	Low	Medium	High	Very High
Gas Furnace (AFUE)	0.50	0.65	0.80	0.90
Oil Furnace (AFUE)	0.50	0.65	0.80	0.90
Heat Pump				
Heating (COP)	1.6	1.9	2.2	2.5
Cooling (COP)	2.1	2.6	3.1	3.4
Air Conditioner	1.8	2.3	2.9	3.5
(COP)				
Electric Furnace	1.0	1.0	1.0	1.0
Electric Baseboard Heating	1.0	1.0	1.0	1.0

Table Y1. Cooling Load Factors for Radiant Barriers (Note: R-11, R-19, R-30, and R-38 refer to the existing level of conventional insulation.)

City	R-11	R-19	R-30	R-38
Albany, NY	876	409	259	211
Albuquerque, NM	1598	851	522	426
Atlanta, GA	1673	832	516	405
Bismarck, ND	706	388	245	191
Chicago, IL	960	475	284	229
Denver, CO	1020	550	357	294
El Toro, CA	1232	636	405	351
Houston, TX	2162	1120	672	521
Knoxville, TN	1597	823	517	411
Las Vegas, NV	2535	1210	703	539
Los Angeles, CA	429	256	168	148
Memphis, TN	1832	907	555	440
Miami, FL	3090	1631	938	727
Minneapolis, MN	769	418	257	204
Orlando, FL	2575	1299	832	662
Phoenix, AZ	3308	1595	942	738
Portland, ME	297	120	82	62
Portland, OR	551	299	178	147
Raleigh, NC	1440	738	460	359
Riverside, CA	1999	931	556	448
Sacramento, CA	1592	849	542	445
Salt Lake City, UT	1286	651	409	332
St. Louis, MO	1466	757	479	369
Seattle, WA	223	119	80	65
Topeka, KS	1523	790	512	397
Waco, TX	2371	1175	713	552
Washington, DC	1221	622	386	301

Figures in table are based on a radiant barrier with an emissivity of 0.05 or less when clean.

Table Y2. Heating Load Factors for Radiant Barriers (Note: R-11, R-19, R-30, and R-38 refer to the existing level of conventional insulation.)

City	R-11	R-19	R-30	R-38
Albany, NY	929	400	193	140
Albuquerque, NM	931	476	299	238
Atlanta, GA	605	282	163	137
Bismarck, ND	1192	513	293	206
Chicago, IL	842	377	210	144
Denver, CO	989	473	277	236
El Toro, CA	792	378	242	197
Houston, TX	387	182	108	80
Knoxville, TN	725	337	206	164
Las Vegas, NV	774	438	277	227
Los Angeles, CA	738	390	227	188
Memphis, TN	630	304	180	164
Miami, FL	99	47	28	26
Minneapolis, MN	1062	447	223	154
Orlando, FL	275	130	77	62
Phoenix, AZ	606	321	191	162
Portland, ME	1112	490	253	194
Portland, OR	937	427	238	186
Raleigh, NC	741	342	219	162
Riverside, CA	892	422	248	189
Sacramento, CA	821	397	236	192
Salt Lake City, UT	906	415	223	187
St. Louis, MO	738	324	169	136
Seattle, WA	904	364	197	133
Topeka, KS	868	379	219	176
Waco, TX	477	225	138	119
Washington, DC	912	386	212	182

Figures in table are based on a radiant barrier with an emissivity of 0.05 or less when clean.

Table Y3. Cooling Load Factors for Additional Insulation (Note: R-11, R-19, R-30, and R-38 refer to the existing and addition levels of conventional insulation.)

City	R-11 to R-19	R-19 to R-30	R-30 to R-38
Albany, NY	1171	258	87
Albuquerque, NM	1100	689	189
Atlanta, GA	1649	508	184
Bismarck, ND	695	226	84
Chicago, IL	1061	293	99
Denver, CO	715	344	117
El Toro, CA	854	384	123
Houston, TX	1310	945	247
Knoxville, TN	1476	527	193
Las Vegas, NV	1960	997	369
Los Angeles, CA	214	122	25
Memphis, TN	1797	584	219
Miami, FL	1694	883	315
Minneapolis, MN	471	259	90
Orlando, FL	1435	691	284
Phoenix, AZ	3175	1334	488
Portland, ME	392	66	27
Portland, OR	368	316	60
Raleigh, NC	1375	434	153
Riverside, CA	1983	713	241
Sacramento, CA	1145	582	194
Salt Lake City, UT	966	462	159
St. Louis, MO	1482	444	186
Seattle, WA	169	73	23
Topeka, KS	991	465	193
Waco, TX	1606	819	317
Washington, DC	1210	392	138

Table Y4. Heating Load Factors for Additional Insulation (Note: R-11, R-19, R-30, and R-38 refer to the existing and addition levels of conventional insulation.)

City	R-11 to R-19	R-19 to R-30	R-30 to R-38
Albany, NY	5358	2751	1030
Albuquerque, NM	3460	1697	626
Atlanta, GA	2660	1332	497
Bismarck, ND	7072	3610	1369
Chicago, IL	4923	2569	952
Denver, CO	4765	2450	872
El Toro, CA	1977	923	336
Houston, TX	1358	632	242
Knoxville, TN	3145	1584	599
Las Vegas, NV	2114	1042	375
Los Angeles, CA	1706	814	295
Memphis, TN	2711	1359	489
Miami, FL	254	121	38
Minneapolis, MN	6399	3323	1239
Orlando, FL	712	390	125
Phoenix, AZ	1444	744	318
Portland, ME	5870	3096	1137
Portland, OR	3980	1992	738
Raleigh, NC	2977	1489	606
Riverside, CA	2302	1121	406
Sacramento, CA	2651	1294	467
Salt Lake City, UT	4623	2321	858
St. Louis, MO	4010	2038	759
Seattle, WA	4328	2295	831
Topeka, KS	4297	2199	802
Waco, TX	1966	968	353
Washington, DC	3999	2014	731

Table Z. Discount Factors Adjusted for Average Fuel Price Escalation (Based on 7 percent discount rate and 25-year life.)

Census Region	Electricity	Fuel Oil	Natural Gas
1	11.68	15.33	13.85
2	11.37	15.56	14.42
3	11.50	15.33	14.36
4	12.12	15.58	14.46
U. S. Average	11.56	15.41	14.33

Region 1: Maine, New Hampshire, Vermont, Massachusetts, Connecticut, Rhode Island, New York, New Jersey, Pennsylvania

Region 2: Ohio, Indiana, Illinois, Michigan, Wisconsin, Minnesota, Iowa, Missouri, North Dakota, South Dakota, Nebraska, Kansas

Region 3: Delaware, Maryland, District of Columbia, Virginia, West Virginia, North Carolina, South Carolina, Georgia, Florida, Kentucky, Tennessee, Alabama, Mississippi, Arkansas, Louisiana, Oklahoma, Texas

Region 4: Montana, Idaho, Wyoming, Colorado, New Mexico, Arizona, Utah, Nevada, Washington, Oregon, California, Alaska, Hawaii

Source: "Energy Prices and Discount Factors for Life-Cycle Cost Analysis 1988," NISTIR 85–3273–3, U. S. Department of Commerce, November 1988.

Key to Abbreviations

AFUE — annual fuel utilization efficiency

ASTM — American Society for Testing and Materials

BTU — British thermal unit

CCF — hundred cubic feet (of natural gas)

COP — coefficient of performance

DOE — U.S. Department of Energy

FSEC — Florida Solar Energy Center

gal. — gallon

KWH — kilowatt (of electricity)

MIMA — Mineral Insulation Manufacturers Association

NAHB — National Association of Home Builders

NFPA — National Fire Protection Association

ORNL — Oak Ridge National Laboratory

RB — radiant barrier

RBS — radiant barrier system

RIMA — Reflective Insulation Manufacturers Association

SEER — seasonal energy efficiency ratio

therm — 100,000 BTU

TVA — Tennessee Valley Authority

UBC — Uniform Building Code

Information Services

U.S. Department of Energy
Office of Scientific and Technical Information
P.O. Box 62
Oak Ridge, TN 37830

National Appropriate Technology Assistance Service
U.S. Department of Energy
P.O. Box 2525
Butte, MT 59702–2525
Telephone: 1–800–428–2525
1–800–428–1718 (In Montana Only)

Conservation and Renewable Energy Inquiry and Referral Service
P.O. Box 8900
Silver Spring, MD 20907
Telephone: 1–800–523–2929
1–800–233–3071

Reflective Insulation Manufacturer Association (RIMA)
661 East Monterey
Pomona, CA 91767
Telephone: 714–620–8011

Florida Solar Energy Center
300 State Road 401
Cape Canaveral, FL 32920–4099

NAHB-National Research Center
400 Prince George's Boulevard
Upper Marlboro, MD 20772–8731

U.S. Department of Commerce
National Technical Information Service
5285 Port Royal Road
Springfield, VA 22161

Electric Power Research Institute
P. O. Box 10412
Palo Alto, CA 94303
Telephone: 415–855–2000

Customer Group
Tennessee Valley Authority
Signal Place, 3B
Chattanooga, TN 37402–2801

Alliance to Save Energy
1725 K Street, NW #914
Washington, DC 20006

Your State Energy Office
To get their phone number, call: 202–639–8749

Glossary

Apron The finish piece that covers the joint between a window stool and the wall finish below.

Admixture A substance other than cement, water, and aggregates included in a concrete mixture for the purpose of altering one or more properties of the concrete.

Aggregate Inert particles such as sand, gravel, crushed stone, or expanded minerals, in a concrete or plaster mixture.

Air-entraining cement A portland cement with an admixture that causes a controlled quantity of stable, microscopic air bubbles to form in the concrete during mixing.

AISC The American Institute of Steel Construction.

Anchor bolt A bolt embedded in concrete for the purpose of fastening a building frame to a concrete or masonry foundation.

Ash dump A door in the underfire of a fireplace that permits ashes from the fire to be swept into a chamber beneath, from which they may be removed at a later time.

Asphalt A tarry brown or black mixture of hydrocarbons.

Asphalt roll roofing A continuous sheet of the same roofing material used in asphalt shingles. *See* **asphalt shingle.**

Asphalt-saturated felt A moisture-resistant sheet material, available in several different thicknesses, usually consisting of a heavy paper that has been impregnated with asphalt.

Asphalt shingle A roofing unit composed of a heavy organic or inorganic felt saturated with asphalt and faced with mineral granules.

ASTM American Society for Testing and Materials, an organization that promulgates standard methods of testing the performance of building materials and components.

Awning window A window that pivots on an axis at or near the top edge of the sash and projects toward the outdoors.

Balloon frame A wooden building frame composed of 2 × 4 studs spaced 16 inches oc which run from the sill to the top plates at the eave.

Baluster A small, vertical member between a handrail and a stair on floor.

Band joist A wooden joist running perpendicular to the joists in a floor and closing off the floor platform at the outside face of the building.

Bar A small rolled steel shape, usually round or rectangular in cross section; a rolled steel shape used for reinforcing concrete.

Baseboard A strip of finish material placed at the junction of a floor and a wall to create a neat intersection and protect the wall against damage from feet, furniture, and floor-cleaning equipment.

Batten A strip of wood or metal used to cover the crack between two adjoining boards or panels.

Bay A rectangular area of a building defined by four adjacent columns; a portion of a building that projects from a facade.

Bead A narrow line of weld metal or sealant; a strip of metal or wood used to hold a sheet of glass in place; a narrow, convex molding profile; a metal edge or corner accessory for plaster.

Beam A straight structural member that acts primarily to resist non-axial loads.

Bearing A point at which one building element rests upon another.

Bearing wall A wall that supports floors or roofs.

Bedrock   A solid layer of rock.

Bending stress   A compressive or tensile stress resulting from the application of a nonaxial force to a structural member.

Bentonite clay   An absorptive, colloidal clay that swells to several times its dry volume when saturated with water.

Bevel   An end or edge cut at an angle other than a right angle.

Bevel siding   Wood cladding boards that taper in cross section.

Billet   A large cylinder or rectangular solid of metal produced from an ingot as an intermediate step in converting it into rolled or extruded metal products.

Bitumen   A tarry mixture of hydrocarbons, such as asphalt or coal tar.

Blind nailing   Attaching boards to a frame or sheathing with toenails driven through the edge of each piece so as to be completely concealed by the adjoining piece.

Blocking   Pieces of wood inserted tightly between joists, studs, or rafters in a building frame to stabilize the structure, inhibit the passage of fire, provide a nailing surface for finish materials, or retain insulation.

Board foot   A unit of lumber volume, a rectangular solid nominally twelve square inches in cross-sectional area and one foot long.

Board siding   Wood cladding made up of boards, as differentiated from shingles or manufactured wood panels.

BOCA   Building Officials and Code Administrators International, Inc., an organization that publishes a model building code.

Bolster   A long chair used to support reinforcing bars in a concrete slab.

Bond   In masonry, the adhesive force between mortar and masonry units, or the pattern in which masonry units are laid to tie two or more wythes together into a structural unit. In reinforced concrete, the adhesion between the surface of a reinforcing bar and the surrounding concrete.

Bottom bars   The reinforcing bars that lie close to the bottom of a beam or slab.

Box beam   A bending member of metal or plywood whose cross section resembles a closed rectangular box.

Bracing   Diagonal members, either temporary or permanent, installed to stabilize a structure against lateral loads.

Brad   A small finish nail.

Bridging   Bracing or blocking installed between steel or wood joists at midspan to stabilize them against buckling and, in some cases, to permit adjacent joists to share loads.

British thermal unit (Btu)   The quantity of heat required to raise one pound of water one degree Fahrenheit.

Broom finish   A skid-resistant texture imparted to an uncured concrete surface by dragging a stiff-bristled broom across it.

Brown coat   The second coat of plaster in a three-coat application.

Btu   *See* **British thermal unit.**

Buckling   Structural failure by gross lateral deflection of a slender element under compressive stress, such as the sideward buckling of a long, slender column or the upper edge of a long, thin floor joist.

Building code   A set of legal restrictions intended to assure a minimum standard of health and safety in buildings.

Built-up roof (BUR)   A roof membrane laminated from layers of asphalt-saturated felt or other fabric, bonded together with bitumen or pitch.

Buoyant uplift   The force of water or liquefied soil that tends to raise a building foundation out of the ground.

BUR   *See* **built-up roof.**

Butt   The thicker end, as the lower edge of a wood shingle or the lower end of a tree trunk; a joint between square-edged pieces; a weld between square-edged pieces of metal that lie in the same plane; a type of door hinge that attaches to the edge of the door.

Butyl rubber   A synthetic rubber compound.

Caisson   A cylindrical sitecast concrete foundation that penetrates through unsatisfactory soil to rest upon an underlying stratum of rock or satisfactory soil; and enclosure that permits excavation work to be carried out underwater.

Calcining   The driving off of the water of hydration from gypsum by the application of heat.

Camber   A slight initial curvature in a beam or slab.

Cambium   The thin layer beneath the bark of a tree that manufactures cells of wood and bark.

Cantilever   A beam, truss, or slab that extends beyond its last point of support.

Cant strip   A strip of material with a sloping face used to ease the transition from a horizontal to a vertical surface at the edge of a membrane roof.

Capillary action   The pulling of water through a small orifice or fibrous material by the adhesive force between the water and the material.

Capillary break   A slot or groove intended to create an opening too large to be bridged by a drop of water, and thereby to eliminate the passage of water by capillary action.

Carbon steel   Low-carbon or mild steel.

Casing   The wood finish pieces surrounding the frame of a window or door; a cylindrical steel tube used to line a drilled or driven hole in foundation work.

Casting   Pouring a liquid material or slurry into a mold whose form it will take as it solidifies.

Cast-in-place   Concrete that is poured in its final location; sitecast.

Caulk   A low-range sealant.

Cellulose   The material of which the structural fibers in wood are composed; a complex polymeric carbohydrate.

Celsius   A temperature scale on which the freezing point of water is established as zero and the boiling point as 100 degrees.

Cement A substance used to adhere materials together; in concrete work, the dry powder that, when it has combined chemically with the water in the mix, cements the particles of aggregate together to form concrete.

Cementitious Having cementing properties; usually used with reference to inorganic substances, such as portland cement and lime.

Ceramic tile Small, flat, thin clay tiles intended for use as wall and floor facings.

Chair A device used to support reinforcing bars.

Chamfer A flattening of a longitudinal edge of a solid member on a plane that lies at an angle of 45 degrees to the adjoining planes.

Chord A member of a truss.

Class A, B, C roofing Roof covering materials classified according to their resistance to fire when tested in accordance with ASTM E108. Class A is the highest, the class C is the lowest.

Clear dimension, clear opening The dimension between opposing inside faces of an opening.

Clinker A fused mass that is an intermediate product of cement manufacture; a brick that is overburned.

CMU *See* **concrete masonry unit**.

Cohesive soil A soil such as clay whose particles are able to adhere to one another by means of cohesive and adhesive forces.

Cold-rolled steel Steel rolled to its final form at a temperature at which it is no longer plastic.

Collar tie A piece of wood nailed across two opposing rafters near the ridge to resist wind uplift.

Column An upright structural member acting primarily in compression.

Column cage An assembly of vertical reinforcing bars and ties for a concrete column.

Compression A squeezing force.

Compression gasket A synthetic rubber strip that seals around a sheet of glass or a wall panel by being squeezed tightly against it.

Compressive strength The ability of a structural material to withstand squeezing forces.

Concrete A structural material produced by mixing predetermined amounts of portland cement, aggregates, and water, and allowing this mixture to cure under controlled conditions.

Concrete block A concrete masonry unit, usually hollow, that is larger than a brick.

Concrete brick A solid concrete masonry unit the same size and proportions as a modular clay brick.

Concrete masonry unit (CMU) A block of hardened concrete, with or without hollow cores, designed to be laid in the same manner as a brick or stone; a concrete block.

Condensate Water formed as a result of condensation.

Condensation The process of changing from a gaseous to a liquid state, especially as applied to water.

Continuous ridge vent A screened, watershielded ventilation opening that runs continuously along the ridge of a gable roof.

Control joint An intentional, linear discontinuity in a structure or component, designed to form a plane of weakness where cracking can occur in response to various forces so as to minimize or eliminate cracking elsewhere in the structure.

Coping saw A handsaw with a thin, very narrow blade, used for cutting detailed shapes in the ends of wood moldings and trim.

Corbel A spanning device in which masonry units in successive courses are cantilevered slightly over one another; a projecting bracket of masonry or concrete.

Corner bead A metal or plastic strip used to form a neat, durable edge at an outside corner of two walls of plaster or gypsum board.

Cornice The exterior detail at the meeting of a wall and a roof overhang; a decorative molding at the intersection of a wall and a ceiling.

Counterflashing   A flashing turned down from above to overlap another flashing turned up from below so as to shed water.

Cove base   A flexible strip of plastic or synthetic rubber used to finish the junction between resilient flooring and a wall.

Creep   A permanent inelastic deformation in a material due to changes in the material caused by the prolonged application of structural stress.

Cripple stud   A wood wall-framing member that is shorter than full-length studs because it is interrupted by a header or sill.

Cross-grain wood   Wood incorporated into a structure in such a way that its direction of grain is perpendicular to the direction of the principal loads on the structure.

Cup   A curl in the cross section of a board or timber caused by unequal shrinkage or expansion between one side of the board and the other.

Curing   The hardening of concrete, plaster, gunnable sealant, or other wet materials. Curing can occur through evaporation of water or a solvent, hydration, polymerization, or chemical reactions of various types, depending on the formulation of the material.

Curing compound   A liquid that, when sprayed on the surface of newly placed concrete, forms a water-resistant layer to prevent premature dehydration of the concrete.

Damper   A flap to control or obstruct the flow of gases; specifically, a metal control flap in the throat of a fireplace, or in an air duct.

Darby   A stiff straightedge of wood or metal used to level the surface of wet plaster.

Daylighting   Illuminating the interior of a building by natural means.

Dead load   The weight of the building itself.

Deadman   A large and/or heavy object buried in the ground as an anchor.

Decking   A material used to span across beams or joists to create a floor or roof surface.

Deformation A change in the shape of a structure or structural element caused by a load or force acting on the structure.

Derrick Any of a number of devices for hoisting building materials on the end of a rope or cable.

Dew point The temperature at which water will begin to condense from a mass of air at a given temperature and moisture content.

Diagonal bracing *See* **bracing.**

Diaphragm action A bracing action that derives from the stiffness of a thin plane of material when it is loaded in a direction parallel to the plane. Diaphragms in buildings are typically floor, wall, or roof surfaces of plywood, reinforced masonry, steel decking, or reinforced concrete.

Dimension lumber Lengths of wood, sawed directly from the log.

Dormer A structure protruding through the plane of a sloping roof, usually with a window and its own smaller roof.

Double glazing Two parallel sheets of glass with an airspace between.

Double-hung window A window with two overlapping sashes that slide vertically in tracks.

Dowel A short rod of wood or steel; a steel reinforcing bar that projects from a foundation to tie it to a column or wall.

Downspout A vertical pipe for conducting water from a roof to a lower level.

Drip A discontinuity formed into the underside of a window sill or wall component to force adhering drops of water to fall free of the face of the building rather than move farther toward the interior.

Drywall *See* **gypsum board.**

Dry well An underground pit filled with stone or other porous material, from which water from a roof drainage system or from footing drains can seep into the surrounding soil.

Duct A hollow conduit, commonly of sheet metal, through which air can be circulated; a tube used to establish the position of a posttensioning tendon in a concrete structure.

Eave The horizontal edge at the low side of a sloping roof.

Edge bead A strip of metal or plastic used to make a neat, durable edge where plaster or gypsum board abuts another material.

Efflorescence A powdery deposit on the face of a structure of masonry or concrete, caused by the leaching of chemical salts by water migrating from within the structure of the surface.

Elastomeric/plastomeric membrane A rubberlike sheet material used as a roof covering.

Elevation A drawing that views a building from any of its sides; a vertical height above a reference point such as sea level.

Elongation Stretching under load; growing longer because of temperature expansion.

End nail A nail driven through one piece of lumber and into the end of another.

EPDM Ethylene propylene diene monomer, a synethic rubber material used in roofing membranes.

Expansion joint A discontinuity extending completely through the foundation, frame, and finishes of a building to allow for gross movement due to thermal stress, material shrinkage, or foundation settlement.

Extrusion The process of squeezing a material through a shaped tube to produce a length of material of desired cross section; an element produced by this process, such as toothpaste.

Face nail A nail driven through the side of one wood member into the side of another.

Fahrenheit A temperature scale on which the boiling point of water is fixed at 212 degrees and the freezing point at 32.

Fascia The exposed vertical face of an eave.

Felt A thin, flexible sheet material made of soft fibers pressed and bonded together. In building practice, a thick paper, or a sheet of glass fibers.

Finger joint   A glued connection between two pieces of wood, using interlocking deeply cut "fingers."

Finish   Exposed to view; material that is exposed to view.

Finish carpentry   The wood exposed to view on the interior of a building, such as window and door casings, baseboards.

Finish floor   The floor material exposed to view, as differentiated from the *subfloor*, which is the loadbearing floor surface beneath.

Firebrick   A brick made from special clays to withstand very high temperatures, as in a fireplace, furnace, or industrial chimney.

Firecut   A sloping end cut on a wood beam or joist where it enters a masonry wall. The purpose of the firecut is to allow the wood member to rotate out of the wall without prying the wall apart, if the floor or roof structure should burn through in a fire.

Fireproofing   Material used around a steel structural element to insulate it against excessive temperatures in case of fire.

Fireresistance rating   The time, in hours or fractions of an hour, that a material or assembly will resist fire exposure as determined by ASTM E119.

Fire resistant   Incombustible, slow to be damaged by fire; forming a barrier to the passage of fire.

Firing   The process of converting dry clay into a ceramic material through the application of intense heat.

First cost   The cost of construction.

Fixed window   Glass that is immovably mounted in a wall.

Flame spread rating   A measure of the rapidity with which fire will spread across the surface of a material as determined by ASTM E84.

Flashing   A thin, continuous sheet of metal, plastic, rubber, or waterproof paper used to prevent the passage of water through a joint in a wall, roof, or chimney.

Float   A trowel with a slightly rough surface used in an intermediate stage of finishing a concrete slab; to use a float for finishing concrete.

Flue A passage for smoke and combustion products from a furnace, stove, water heater, or fireplace.

Flush door A door with smooth, planar faces.

Fly rafter A rafter in a rake overhang.

Foil-backed gypsum board Gypsum board with aluminum foil laminated to its back surface as a vapor retarder and thermal insulator.

Folded plate A roof structure whose strength and stiffness derive from a pleated or folded geometry.

Footing The widened part of a foundation that spreads a load from the building across a broader area of soil.

Form tie A steel rod with fasteners on each end, used to hold together the formwork for a concrete wall.

Formwork Temporary structures of wood, steel, or plastic that serve to give shape to poured concrete, and to support it and keep it moist as it cures.

Foundation The portion of a building that has the sole purpose of transmitting structural loads from the building into the earth.

Framing plan A diagram showing the arrangement and sizes of the structural members in a floor or roof.

French door A symmetrical pair of glazed doors hinged to the jambs of a single frame and meeting at the center of the opening.

Frictional soil A soil such as sand that has little or no attraction between its particles, and derives its strength from geometric interlocking of the particles; a noncohesive soil.

Frost line The depth in the earth to which the soil can be expected to freeze during a severe winter.

Furring strip A length of wood or metal attached to a masonry or concrete wall to permit the attachment of finish materials to the wall using screws or nails.

Gable The triangular wall beneath the end of a gable roof.

Gable roof A roof consisting of two oppositely sloping planes that intersect at a level ridge.

Gable vent A screened, louvered opening in a gable, used for exhausting excess heat and humidity from an attic.

Galvanizing The application of a zinc coating to steel to prevent corrosion.

Gambrel A roof shape consisting of two roof planes at different pitches on each side of a ridge.

Gasket A dry, resilient material used to seal a joint between two rigid assemblies by being compressed between them.

General contractor A contractor who has responsibility for the overall conduct of a construction project.

GFRC *See* **glass-fiber-reinforced concrete.**

Girder A beam that supports other beams; a very large beam, especially one that is built up from smaller elements.

Girt A beam that supports wall cladding between columns.

Glass fiber batt A thick, fluffy, nonwoven insulating blanket of filaments spun from glass.

Glass-fiber-reinforced concrete (GFRC) Concrete with a strengthening admixture of short alkali-resistant glass fibers.

Glazed structural clay tile A hollow clay block with glazed faces, used for constructing interior partitions.

Glazing The act of installing glass; installed glass; an adjective referring to materials used in installing glass.

Glazing compound Any of a number of types of mastic used to bed small lights of glass in a frame.

Glue laminated timber A timber made up of a large number of small strips of wood glued together.

Glulam A glue laminated timber made from single pieces of lumber such as 2×6s.

Grade The surface of the ground; to move earth for the purpose of bringing the surface of the ground to an intended level or profile.

Grade beam A reinforced concrete beam that transmits the load from a bearing wall into spaced foundations such as pile caps or caissons.

Grain In wood, the direction of the longitudinal axes of the wood fibers, or the figure formed by the fibers.

Grout A high-slump mixture of portland cement, aggregates, and water, poured or pumped into cavities in concrete or masonry for the purpose of embedding reinforcing bars and/or increasing the amount of loadbearing material in a wall; a mortar used to fill joints between ceramic tiles or quarry tiles.

Gusset plate A flat steel plate to which the chords are connected at a joint in a truss; a stiffener plate.

Gutter A channel to collect rainwater and snowmelt at the eave of a roof.

Gypsum board An interior facing panel consisting of a gypsum core sandwiched between paper faces. Also called *drywall, plasterboard.*

Gypsum wallboard *See* **gypsum board.**

Hardboard A very dense panel product, usually with at least one smooth face, made of highly compressed wood fibers.

Hawk A metal square with a handle below, used by a plasterer to hold a small quantity of wet plaster and transfer it to a trowel for application to a wall or ceiling.

Header Lintel; band joist; a joist that supports other joists; in steel construction, a beam that spans between girders; a brick or other masonry unit that is laid across two wythes with its end exposed in the face of the wall.

Hearth The incombustible floor area outside a fireplace opening.

Heartwood The dead wood cells nearer the center of a tree trunk.

Heat of hydration The thermal energy given off by concrete or gypsum as it cures.

Heat-strengthened glass Glass that has been strengthened by heat treatment, though not to as great an extent as tempered glass.

High-lift grouting A method of constructing a reinforced masonry wall in which the reinforcing bars are grouted into the wall in story-high increments.

High-range sealant A sealant that is capable of a high degree of elongation without rupture.

High-strength bolt A bolt designed to connect steel members by clamping them together with sufficient force that the load is transferred between them by friction.

Hip The diagonal intersection of planes in a hip roof.

Hip roof A roof consisting a four sloping planes that intersect to form a pyramidal or elongated pyramid shape.

Hollow concrete masonry Concrete masonry units that are manufactured with open cores, such as ordinary concrete blocks.

Hollow-core door A door of two face veneers separated by an airspace, with solid wood spacers around the four edges. The face veneers are connected by a grid of thin spacers running through the airspace.

Hopper window A window whose sash pivots on an axis along the sill, and opens by tilting toward the interior of the building.

Hydrated lime Calcium hydroxide produced by burning calcium carbonate to form calcium oxide (quicklime), then allowing the calcium oxide to combine chemically with water.

Hydration A process of combining chemically with water to form molecules or crystals that include hydroxide radicals or water of crystallization.

Hydrostatic pressure Pressure exerted by standing water.

Hygroscopic Readily taking up and retaining moisture.

Hyperbolic paraboloid shell A concrete roof structure with a saddle shape.

I-beam (obsolete term) An American Standard section of hot-rolled steel.

Ice dam An obstruction along the eave of a roof caused by the refreezing of water emanating from melting snow on the roof surface above.

Insulating glass Double or triple glazing.

Inverted roof A membrane roof assembly in which the thermal insulation lies above the membrane.

Isocyanurate foam A thermosetting plastic foam with thermal insulating properties.

Jack A device for exerting a large force over a short distance, usually by means of screw action or hydraulic pressure.

Jack rafter A shortened rafter that joins a hip or valley rafter.

Jamb The vertical side of a door or window.

Joist One of a group of light, closely spaced beams used to support a floor deck or flat roof.

Joist-band A broad, shallow concrete beam that supports one-way concrete joists whose depths are identical to its own.

Joist girder A light steel truss used to support open-web steel joists.

Key A slot formed into a concrete surface, such as a footing, to lock in the foundation wall.

Kiln A furnace for firing clay or glass products; a heated chamber for seasoning wood.

Knee wall A short wall under the slope of a roof.

Lag screw A large-diameter wood screw with a square or hexagonal head.

Laminate To bond together in layers; produce by bonding together layers of material.

Laminated glass A glazing material consisting of outer layers of glass laminated to an inner layer of transparent plastic.

Landing A platform in or at either end of a stair.

Lap joint   A connection in which two pieces of material are over-lapped before fastening.

Lateral force   A force acting generally in a horizontal direction, such as wind, earthquake, or soil pressure against a foundation wall.

Latex caulk   A low-range sealant based on a synthetic latex.

Leader   A vertical pipe for conducting water from a roof to a lower level.

Let-in bracing   Diagonal bracing nailed into notches cut in the face of the studs so it does not increase the thickness of the wall.

Level cut   A saw cut that produces a level surface at the lower end of a sloping rafter.

Life-cycle cost   A cost that takes into account both the first cost and costs of maintenance, replacement, fuel consumed, monetary inflation, and interest over the life of the object being evaluated.

Light   A sheet of glass.

Lignin   The natural cementing substance that binds together the cellulose in wood.

Linoleum   A resilient floorcovering material composed primarily of ground cork and linseed oil on a burlap or canvas backing.

Lintel   A beam that carries the load of a wall across a window or door opening.

Live load   The weight of snow, people, furnishings, machines, vehicles, and goods in or on a building.

Load   A weight or force acting on a structure.

Loadbearing   Supporting a superimposed weight or force.

Lookout   A short rafter, running at an angle to the other rafters in the roof, which supports a rake overhang.

Louver   A construction of numerous sloping, closely spaced slats used to prevent the entry of rainwater into a ventilating opening.

Low-emissivity coating   A surface coating for glass that permits the passage of most shortwave electromagnetic radiation (light and heat), but reflects most longer-wave radiation (heat).

Low-iron glass Glass formulated with a low iron content so as to have a maximum transparency to solar energy.

Mandrel A stiff steel core placed inside the thin steel shell of a site-cast concrete pile to prevent it from collapsing during driving.

Mansard A roof shape consisting of two superimposed levels of *hip roofs* with the lower level at a steeper pitch.

Mason One who builds with bricks, stones, or concrete masonry units; one who works with concrete.

Masonry Brickwork, blockwork, and stonework.

Masonry cement Portland cement with dry admixtures designed to increase the workability of mortar.

Masonry opening The clear dimension required in a masonry wall for the installation of a specific window or door unit.

Mat foundation A single concrete footing that is essentially equal in area to the area of ground covered by the building.

Meeting rail The wood or metal bar along which one sash of a double-hung or sliding window seals against the other.

Member An element of a structure such as a beam, a girder, a column, a joist, a piece of decking, a stud, or a chord of a truss.

Membrane A sheet material that is impervious to water or water vapor.

Miter A diagonal cut at the end of a piece; the joint produced by joining two diagonally cut pieces at right angles.

Modular Conforming to multiple of a fixed dimension.

Modulus of elasticity An index of the stiffness of a material, derived by measuring the elastic deformation of the material as it is placed under stress, and then dividing the stress by the deformation.

Moisture retarder A membrane used to prevent the migration of liquid water through a floor, wall, or ceiling.

Molding A strip of wood, plastic, or plaster with an ornamental profile.

Mortar A substance used to join masonry units, consisting of cementitious materials, fine aggregate, and water.

Mullion A vertical or horizontal bar between adjacent window or door units.

Muntin A small vertical or horizontal bar between small lights of glass in a sash.

Nail-base sheathing A sheathing material, such as wood boards or plywood, to which siding can be attached by nailing, as differentiated from one, such as cane fiber board or plastic foam board, that is too soft to hold nails.

Nail popping The loosening of nails holding gypsum board to a wall, caused by drying shrinkage of the studs.

Neoprene A synthetic rubber.

Nominal dimension An approximate dimension assigned to a piece of material as a convenience in referring to the piece.

Nonbearing Not carrying a load.

Nosing The projecting forward edge of a stair tread.

Organic soil Soil containing decayed vegetable and/or animal matter; topsoil.

Oriented strand board (OSB) A building panel composed of long shreds of wood fiber oriented in specific directions and bonded together under pressure.

Oxidation Corrosion; rusting; rust.

Panel A broad, thin piece of wood; a sheet of building material such as plywood or particle board; a prefabricated building component that is broad and thin, such as a curtain wall panel; an area within a truss consisting of an opening and the members that surround it.

Panel door A wood door in which two or more thin panels are held by stiles and rails.

Parging Portland cement plaster applied over masonry to make it less permeable to water.

Particle board A building panel composed of small particles of wood bonded together under pressure.

Partition An interior nonloadbearing wall.

Perlite Expanded volcanic rock, used as a lightweight aggregate in concrete and plaster, and as an insulating fill.

Perm A unit of vapor permeability.

Pier A caisson foundation unit.

Pile A long, slender piece of material driven into the ground to act as a foundation.

Pile cap A thick slab of reinforced concrete poured across the top of a pile cluster to cause the cluster to act as a unit in supporting a column or grade beam.

Piledriver A machine for driving piles.

Pitch The slope of a roof or other plane, often expressed as inches of rise per foot of run; a dark, viscous hydrocarbon distilled from coal tar; a viscous resin found in wood.

Pitched roof A sloping roof.

Plainsawing Sawing a log into dimension lumber without regard to the direction of the annual rings.

Plaster A cementitious material, usually based on gypsum or portland cement, applied to lath or masonry in paste form, to harden into a finish surface.

Plasterboard *See* **gypsum board.**

Platform frame A wooden building frame composed of closely spaced members nominally 2 inches (51 mm) in thickness, in which the wall members do not run past the floor framing members.

Plumb Vertical.

Plumb cut A saw cut that produces a vertical (plumb) surface at the lower end of a sloping rafter.

Ply A layer, as in a layer of felt in a builtup roof membrane or a layer of veneer in plywood.

Plywood A wood panel composed of an odd number of layers of wood veneers bonded together under pressure.

Polyethylene A thermoplastic widely used in sheet form for vapor retarders, moisture barriers, and temporary construction coverings.

Polymer A compound consisting of repeated chemical units.

Polymercaptans Compounds used in high-range gunnable sealants.

Polystyrene foam A thermoplastic foam with thermal insulating properties.

Polyurethane Any of a large group of resins and synthetic rubber compounds used in sealants, varnishes, insulating foams, and roof membranes.

Polyurethane foam (PUR) A thermosetting foam with thermal insulating properties.

Polyvinyl chloride (PVC) A thermoplastic material widely used in construction products, including plumbing pipes, floor tiles, wallcoverings, and roofing membranes.

Portland cement The gray powder used as the binder in concrete, mortar, and stucco.

Posttensioning The compressing of the concrete in a structural member by means of tensioning high-strength steel tendons against it after the concrete has cured.

Pour To cast concrete; an increment of concrete casting carried out without interruption.

Precast concrete Concrete cast and cured in a position other than its final position in the structure.

Prehung door A door that is hinged to its frame in a factory or shop.

Pressure-treated lumber Lumber that has been impregnated with chemicals under pressure, for the purpose of retarding either decay or fire.

Prestressed concrete Concrete that has been pretensioned or posttensioned.

Prestressing Applying an initial compressive stress to a concrete structural member, either by pretensioning or posttensioning.

Pretensioning The compressing of the concrete in a structural member by pouring the concrete for the member around stretched high-strength steel strands, curing the concrete, and releasing the external tensioning force on the strands.

Priming Covering a surface with a coating that prepares it to accept another type of coating or sealant.

Quarry tile A large clay floor tile, usually unglazed.

Quartersawn Lumber sawn in such a way that the annual rings run roughly perpendicular to the face of each piece.

Raft A mat footing.

Rafter A framing member that runs up and down the slope of a pitched roof.

Rail A horizontal framing piece in a panel door; a handrail.

Rainscreen principle Wall cladding is made watertight by providing wind-pressurized air chambers behind joints to eliminate air pressure differentials between the outside and inside that might transport water through the joints.

Rake The sloping edge of a pitched roof.

Ray A tubular cell that runs radially in a tree trunk.

Reflective coated glass Glass onto which a thin layer of metal or metal oxide has been deposited to reflect light and/or heat.

Reinforced concrete Concrete work into which steel bars have been embedded to impart tensile strength to the construction.

Relative humidity A decimal fraction representing the ratio of the amount of water vapor contained in a mass of air to the amount of water it could contain under the existing conditions of temperature and pressure.

Resilient flooring A manufactured sheet or tile flooring of asphalt, polyvinyl chloride, linoleum, rubber, or other resilient material.

Resin A natural or synthetic, solid or semisolid organic material of high molecular weight, used in the manufacture of paints, varnishes, and plastics.

Retaining wall A wall that resists horizontal soil pressures at an abrupt change in ground elevation.

Ridge board The board against which the tips of rafters are fastened.

Rise A difference in elevation, such as the rise of a stair from one floor to the next, or a rise per foot of run in a sloping roof.

Riser A single vertical increment of a stair; the vertical face between two treads in a stair; a vertical run of plumbing, wiring, or ductwork.

Rock wool An insulating material manufactured by forming fibers from molten rock.

Rotary-sliced veneer A thin sheet of wood cut by rotating a log against a long, sharp knife blade in a lathe.

Rough carpentry Framing carpentry, as distinguished from finish carpentry.

Roughing in The installation of mechanical, electrical, and plumbing components that will not be exposed to view in the finished building.

Rough opening The clear dimensions of the opening that must be provided in a wall frame to accept a given door or window unit.

Run Horizontal dimension in a stair or sloping roof.

R-value A numerical measure of resistance to the flow of heat.

Sandwich panel A panel consisting of two outer faces of wood, metal, or concrete bonded to a core of insulating foam.

Sapwood The living wood in the outer region of a tree trunk or branch.

Sash A frame that holds glass.

Scratch coat The first coat in a three-coat application of plaster.

Screed A strip of wood, metal, or plaster that establishes the level to which concrete or plastic will be placed.

Seismic Relating to earthquakes.

Seismic load A load on a structure caused by movement of the earth relative to the structure during an earthquake.

Set To cure; to install; to recess the heads of nails; a punch for recessing the heads of nails.

Shading coefficient The ratio of total solar heat passing through a given sheet of glass to that passing through a sheet of clear double-strength glass.

Shaft wall A wall surrounding a shaft.

Shake A shingle split from a block of wood.

Shale A rock formed from the consolidation of clay or silt.

Shear A deformation in which planes of material slide with respect to one another.

Sheathing The rough covering applied to the outside of the roof, wall, or floor framing of a light frame structure.

Shed A building or dormer with a single-pitched roof.

Shim A thin piece of material placed between two components of a building to adjust their relative positions as they are assembled; to insert shims.

Shingle Water-resistant material nailed in overlapping fashion with many other such units to a sloping roof watertight; to apply shingles.

Shiplap A board with edges rabbeted so as to overlap flush from one board to the next.

Sidelight A tall, narrow window alongside a door.

Siding The exterior wall finish material applied to a light frame wood structure.

Siding nail A nail with a small head, used to fasten siding to a building.

Silicone A polymer used for high-range sealants, roof membranes, and masonry water repellants.

Sill The strip of wood that lies immediately on top of a concrete or masonry foundation in wood frame construction; the horizontal bottom portion of a window or door; the exterior surface, usually sloping to shed water, below the bottom of a window or door.

Sill sealer A resilient, fibrous material placed between a foundation and a sill to reduce air infiltration between the outdoors and indoors.

Single-hung window A window with two overlapping sashes, the lower of which can slide vertically in tracks, and the upper of which is fixed.

Slab on grade A concrete surface lying upon, and supported directly by, the ground beneath.

Slag The mineral waste that rises to the top of molten iron or steel, or to the top of a weld.

Slaked lime Calcium hydroxide.

Slump test A test in which wet concrete or plaster is placed in a cone-shaped metal mold of specified dimensions and allowed to slump under its own weight after the cone is removed. The vertical distance between the height of the mold and the height of the slumped mixture is an index of its working consistency.

Slurry A watery mixture of insoluble materials.

Smoke shelf The horizontal area behind the damper of a fireplace.

Soffit The undersurface of a horizontal element of a building, especially the underside of a stair or a roof overhang.

Soffit vent An opening under the eave of a roof used to allow air to flow into the attic or the space below the roof sheathing.

Sole plate The horizontal piece of dimension lumber at the bottom of the studs in a wall in a light frame building.

Solid-core door A flush door with no internal cavities.

Span The distance between supports for a beam, girder, truss, vault, arch, or other horizontal structural device; to carry a load between supports.

Span rating The number stamped on a sheet of plywood or other wood building panel to indicate how far in inches it may be spaced.

Splash block A small precast block of concrete or plastic used to divert water at the bottom of a downspout.

Split jamb A door frame fabricated in two interlocking halves, to be installed from the opposite sides of an opening.

Stile A vertical framing member in a panel door.

Stool The interior horizontal plane at the sill of a window.

Storm window A sash added to the outside of a window in winter to increase its thermal resistance and decrease air infiltration.

Story pole A strip of wood marked with the exact course heights of masonry for a particular building, used to make sure that all the leads are identical in height and coursing.

Straightedge To strike off the surface of a concrete slab using screeds and a straight piece of lumber or metal.

Strain Deformation under stress.

Stress Force per unit area.

Stressed-skin panel A panel consisting of two face sheets of wood, metal, or concrete bonded to perpendicular spacer strips.

Stringer The sloping wood or steel member supporting the treads of a stair.

Structural tubing Hollow steel cylindrical or rectangular shapes.

Subfloor The loadbearing surface beneath a finish floor.

Substructure The occupied, below-ground portion of a building.

Superplasticizer A concrete admixture that makes wet concrete extremely fluid without additional water.

Superstructure The above-ground portion of a building.

Supporting stud A wall framing member that extends from the sole plate to the underside of a header and that support the header.

Tensile stress A stress caused by stretching of a material.

Thermal break A section of material with a low thermal conductivity, installed between metal components to retard the passage of heat through a wall or window assembly.

Thermal bridge A material of higher thermal conductivity that conducts heat more rapidly through an insulated building assembly, such as a steel stud in an insulated stud wall.

Thermal conductivity The rate at which a material conducts heat.

Thermal insulation A material that greatly retards the passage of heat.

Thermal resistance The resistance of a material or assembly to the conduction of heat.

Transit-mixed concrete Concrete mixed in a drum on the back of a truck as it is transported to the building site.

Truss A triangular arrangement of structural members that reduces nonaxial forces on the truss to a set of axial forces in the members.

Type X gypsum board A fiber-reinforced gypsum board used where greater fire resistance is required.

Urea formaldehyde A water-based foam used as thermal insulation.

Vermiculite Expanded mica, used as an insulating fill or a lightweight aggregate.

Waferboard A building panel made by bonding together large, flat flakes of wood.

Water-cement ratio A numerical index of the relative proportions of water and cement in a concrete mixture.

Water-resistant gypsum board A gypsum board designed for use in locations where it may be exposed to occasional dampness.

Water table The level below which the ground is saturated with water; the level to which ground water will fill an excavation.

Water vapor   Water in its gaseous phase.

Wind brace   A diagonal structural member whose function is to stabilize a frame against lateral forces.

Winder   A stair tread that is wider at one end than at the other.

Wind load   A load on a building caused by wind pressure and/or suction.

Wind uplift   Upward forces on a structure caused by negative aerodynamic pressures that result from certain wind conditions.

Zero-slump concrete   A concrete mixed with so little water that it has a slump of zero.

Zoning ordinance   A law that specifies in detail how land may be used in a municipality.

Index

**Over a Century of Excellence
for the Professional
and
Vocational Trades and the Crafts**

Order now from your local bookstore
or use the convenient order form
at the back of this book.

AUDEL

These fully illustrated, up-to-date guides and manuals mean a better job done for mechanics, engineers, electricians, plumbers, carpenters, and all skilled workers.

CONTENTS

ELECTRICAL

House Wiring (Seventh Edition)

ROLAND E. PALMQUIST;
revised by PAUL ROSENBERG

*5 1/2 × 8 1/4 Hardcover 248 pp. 150 Illus.
ISBN: 0-02-594692-7 $22.95*

Rules and regulations of the current 1990 National Electrical Code for residential wiring fully explained and illustrated.

Practical Electricity
(Fifth Edition)

ROBERT G. MIDDLETON;
revised by L. DONALD MEYERS

*5 1/2 × 8 1/4 Hardcover 512 pp. 335 Illus.
ISBN: 0-02-584561-6 $19.95*

The fundamentals of electricity for electrical workers, apprentices, and others requiring concise information about electric principles and their practical applications.

Guide to the 1993 National Electrical Code

ROLAND E. PALMQUIST;
revised by PAUL ROSENBERG

*5 1/2 × 8 1/4 Paperback 608 pp.
100 line drawings
ISBN: 0-02-077761-2 $25.00*

The guide to the most recent revision of the electrical codes—how to read them, under-
stand them, and use them. Here is the most authoritative reference available, making clear the changes in the code and explaining these changes in a way that is easy to understand.

Installation Requirements of the 1993 National Electrical Code

PAUL ROSENBERG

*5 1/2 × 8 1/4 Paperback 261 pp.
100 line drawings
ISBN: 0-02-077760-4 $22.00*

A handy guide for electricians, contractors, and architects who need a reference on location. Arranging all the pertinent requirements (and only the pertinent requirements) of the 1993 NEC, it has an easy-to-follow format. Concise and updated, it's a perfect working companion for Apprentices, Journeymen, or for Master electricians.

Mathematics for Electricians and Electronics Technicians

REX MILLER

*5 1/2 × 8 1/4 Hardcover 312 pp. 115 Illus.
ISBN: 0-8161-1700-4 $14.95*

Mathematical concepts, formulas, and problem-solving techniques utilized on-the-job by electricians and those in electronics and related fields.

Fractional-Horsepower Electric Motors

REX MILLER and
MARK RICHARD MILLER

5 1/2 × 8 1/4 Ha...
ISBN: 0-672-2...
The i... ...tenance, re-
pa... ...e small-to-moder-
ate... ...otors that power home
appl... ...d industrial equipment.

Electric Motors (Fifth Edition)

EDWIN P. ANDERSON
and REX MILLER

5 1/2 × 8 1/4 Hardcover 696 pp.
Photos/line art
ISBN: 0-02-501920-1 $35.00
Complete reference guide for electricians, industrial maintenance personnel, and installers. Contains both theoretical and practical descriptions.

Home Appliance Servicing
(Fourth Edition)

EDWIN P. ANDERSON;
revised by REX MILLER

5 1/2 × 8 1/4 Hardcover 640 pp. 345 Illus.
ISBN: 0-672-23379-7 $22.50
The essentials of testing, maintaining, and repairing all types of home appliances.

Television Service Manual
(Fifth Edition)

ROBERT G. MIDDLETON;
revised by JOSEPH G. BARRILE

5 1/2 × 8 1/4 Hardcover 512 pp. 395 Illus.
ISBN: 0-672-23395-9 $16.95
A guide to all aspects of television transmission and reception, including the operating principles of black and white and color receivers. Step-by-step maintenance and repair procedures.

Electrical Course for Apprentices and Journeymen
(Third Edition)

ROLAND E. PALMQUIST

5 1/2 × 8 1/4 Hardcover 478 pp. 290 Illus.
ISBN: 0-02-594550-5 $19.95
This practical course in electricity for those in formal training programs or learning on their own provides a thorough understanding of operational theory and its applications on the job.

Questions and Answers for Electricians Examinations
(1993 NEC Rulings Included)

revised by PAUL ROSENBERG

5 1/2 × 8 1/4 Paperback 270 pp.
100 line drawings
ISBN: 0-02-077762-0 $20.00
An Audel classic, considered the most thorough work on the subject in coverage and content. This fully revised edition is based on the 1993 National Electrical Code®, and is written for anyone preparing for the various electricians' examinations—Apprentice, Journeyman, or Master. It provides the license applicant with an understanding of theory as well as all definitions, specifications, and regulations included in the new NEC.

MACHINE SHOP AND MECHANICAL TRADES

Machinists Library
(Fourth Edition, 3 Vols.)

REX MILLER

5 1/2 × 8 1/4 Hardcover 1,352 pp. 1120 Illus.
ISBN: 0-672-23380-0 $52.95
An indispensable three-volume reference set for machinists, tool and die makers, machine operators, metal workers, and those with home workshops. The principles and methods of the entire field are covered in an up-to-date text, photographs, diagrams, and tables.

Volume I: Basic Machine Shop

REX MILLER

5 1/2 × 8 1/4 Hardcover 392 pp. 375 Illus.
ISBN: 0-672-23381-9 $17.95

Volume II: Machine Shop

REX MILLER

5 1/2 × 8 1/4 Hardcover 528 pp. 445 Illus.
ISBN: 0-672-23382-7 $19.95

Volume III: Toolmakers Handy Book

REX MILLER

5 1/2 × 8 1/4 Hardcover 432 pp. 300 Illus.
ISBN: 0-672-23383-5 $14.95

Mathematics for Mechanical Technicians and Technologists
JOHN D. BIES

5 1/2 × 8 1/4 Hardcover 342 pp. 190 Illus.
ISBN: 0-02-510620-1 $17.95

The mathematical concepts, formulas, and problem-solving techniques utilized on the job by engineers, technicians, and other workers in industrial and mechanical technology and related fields.

Millwrights and Mechanics Guide (Fourth Edition)
CARL A. NELSON

5 1/2 × 8 1/4 Hardcover 1,040 pp. 880 Illus.
ISBN: 0-02-588591-x $29.95

The most comprehensive and authoritative guide available for millwrights, mechanics, maintenance workers, riggers, shop workers, foremen, inspectors, and superintendents on plant installation, operation, and maintenance.

Welders Guide (Third Edition)
JAMES E. BRUMBAUGH

5 1/2 × 8 1/4 Hardcover 960 pp. 615 Illus.
ISBN: 0-672-23374-6 $23.95

The theory, operation, and maintenance of all welding machines. Covers gas welding equipment, supplies, and process; arc welding equipment, supplies, and process; TIG and MIG welding; and much more.

Welders/Fitters Guide
HARRY L. STEWART

8 1/2 × 11 Paperback 160 pp. 195 Illus.
ISBN: 0-672-23325-8 $7.95

Step-by-step instruction for those training to become welders/fitters who have some knowledge of welding and the ability to read blueprints.

Sheet Metal Work
JOHN D. BIES

5 1/2 × 8 1/4 Hardcover 456 pp. 215 Illus.
ISBN: 0-8161-1706-3 $19.95

An on-the-job guide for workers in the manufacturing and construction industries and for those with home workshops. All facets of sheet metal work detailed and illustrated by drawings, photographs, and tables.

Power Plant Engineers Guide (Third Edition)
FRANK D. GRAHAM;
revised by CHARLIE BUFFINGTON

5 1/2 × 8 1/4 Hardcover 960 pp. 530 Illus.
ISBN: 0-672-23329-0 $27.50

This all-inclusive, one-volume guide is perfect for engineers, firemen, water tenders, oilers, operators of steam and diesel-power engines, and those applying for engineer's and firemen's licenses.

Mechanical Trades Pocket Manual (Third Edition)
CARL A. NELSON

4 × 6 Paperback 364 pp. 255 Illus.
ISBN: 0-02-588665-7 $14.95

A handbook for workers in the industrial and mechanical trades on methods, tools, equipment, and procedures. Pocket-sized for easy reference and fully illustrated.

PLUMBING

Plumbers and Pipe Fitters Library (Fourth Edition, 3 Vols.)
CHARLES N. McCONNELL

5 1/2 × 8 1/4 Hardcover 952 pp. 560 Illus.
ISBN: 0-02-582914-9 $68.45

This comprehensive three-volume set contains the most up-to-date information available for master plumbers, journeymen, apprentices, engineers, and those in the building trades. A detailed text and clear diagrams, photographs, and charts and tables treat all aspects of the plumbing, heating, and air conditioning trades.

Volume I: Materials, Tools, Roughing-In
CHARLES N. McCONNELL;
revised by TOM PHILBIN

5 1/2 × 8 1/4 Hardcover 304 pp. 240 Illus.
ISBN: 0-02-582911-4 $20.95

Volume II: Welding, Heating, Air Conditioning
CHARLES N. McCONNELL;
revised by TOM PHILBIN

5 1/2 × 8 1/4 Hardcover 384 pp. 220 Illus.
ISBN: 0-02-582912-2 $22.95

Volume III: Water Supply, Drainage, Calculations
CHARLES N. McCONNELL;
revised by TOM PHILBIN

5 1/2 × 8 1/4 Hardcover 264 pp. 100 Illus.
ISBN: 0-02-582913-0 $20.95

The Home Plumbing Handbook
(Fourth Edition)
CHARLES N. McCONNELL
8 1/2 × 11   Paperback   224 pp.   210 Illus.
ISBN: 0-02-079651-X $17.00
This handy, thorough volume, a longtime standard in the field with the professional, has been updated to appeal to the do-it-yourself plumber. Aided by the book's many illustrations and manufacturers' instructions, the home plumber is guided through most basic plumbing procedures. All techniques and products conform to the latest changes in codes and regulations.

The Plumbers Handbook
(Eighth Edition)
JOSEPH P. ALMOND, SR.;
revised by REX MILLER
4 × 6   Paperback   368 pp.   170 Illus.
ISBN: 0-02-501570-2 $19.95
Comprehensive and handy guide for plumbers and pipefitters—fits in the toolbox or pocket. For apprentices, journeymen, or experts.

Questions and Answers for Plumbers' Examinations
(Third Edition)
JULES ORAVETZ;
revised by REX MILLER
5 1/2 × 8 1/4 Paperback   288 pp.   145 Illus.
ISBN: 0-02-593510-0 $14.95
Complete guide to preparation for the plumbers' exams given by local licensing authorities. Includes requirements of the National Bureau of Standards.

HVAC

Air Conditioning: Home and Commercial (Fourth Edition)
EDWIN P. ANDERSON;
revised by REX MILLER
5 1/2 × 8 1/4   Hardcover   528 pp.   180 Illus.
ISBN: 0-02-584885-2 $29.95
A guide to the construction, installation, operation, maintenance, and repair of home, commercial, and industrial air conditioning systems.

Heating, Ventilating, and Air Conditioning Library
(Second Edition, 3 Vols.)
JAMES E. BRUMBAUGH
5 1/2 × 8 1/4 Hardcover 1,840 pp. 1,275 Illus.
ISBN: 0-672-23388-6 $53.85
An authoritative three-volume reference library for those who install, operate, maintain, and repair HVAC equipment commercially, industrially, or at home.

Volume I: Heating Fundamentals, Furnaces, Boilers, Boiler Conversions
JAMES E. BRUMBAUGH
5 1/2 × 8 1/4   Hardcover   656 pp.   405 Illus.
ISBN: 0-672-23389-4 $17.95

Volume II: Oil, Gas and Coal Burners, Controls, Ducts, Piping, Valves
JAMES E. BRUMBAUGH
5 1/2 × 8 1/4   Hardcover   592 pp.   455 Illus.
ISBN: 0-672-23390-8 $17.95

Volume III: Radiant Heating, Water Heaters, Ventilation, Air Conditioning, Heat Pumps, Air Cleaners
JAMES E. BRUMBAUGH
5 1/2 × 8 1/4   Hardcover   592 pp.   415 Illus.
ISBN: 0-672-23391-6 $17.95

Oil Burners (Fifth Edition)
EDWIN M. FIELD
5 1/2 × 8 1/4   Hardcover   360 pp.   170 Illus.
ISBN: 0-02-537745-0 $29.95
An up-to-date sourcebook on the construction, installation, operation, testing, servicing, and repair of all types of oil burners, both industrial and domestic.

Refrigeration: Home and Commercial (Fourth Edition)
EDWIN P. ANDERSON;
revised by REX MILLER
5 1/2 × 8 1/4   Hardcover   768 pp.   285 Illus.
ISBN: 0-02-584875-5 $34.95
A reference for technicians, plant engineers, and the homeowner on the installation, operation, servicing, and repair of everything from single refrigeration units to commercial and industrial systems.

PNEUMATICS AND HYDRAULICS

Hydraulics for Off-the-Road Equipment (Second Edition)
HARRY L. STEWART;
revised by TOM PHILBIN

5 1/2 × 8 1/4 Hardcover 256 pp. 175 Illus.
ISBN: 0-8161-1701-2 $13.95

This complete reference manual on heavy equipment covers hydraulic pumps, accumulators, and motors; force components; hydraulic control components; filters and filtration, lines and fittings, and fluids; hydrostatic transmissions; maintenance; and troubleshooting.

Pneumatics and Hydraulics (Fourth Edition)
HARRY L. STEWART;
revised by TOM STEWART

5 1/2 × 8 1/4 Hardcover 512 pp. 315 Illus.
ISBN: 0-672-23412-2 $19.95

The principles and applications of fluid power. Covers pressure, work, and power; general features of machines; hydraulic and pneumatic symbols; pressure boosters; air compressors and accessories; and much more.

Pumps (Fifth Edition)
HARRY L. STEWART;
revised by REX MILLER

5 1/2 × 8 1/4 Hardcover 552 pp. 360 Illus.
ISBN: 0-02-614725-4 $35.00

The practical guide to operating principles of pumps, controls, and hydraulics. Covers installation and day-to-day service.

CARPENTRY AND CONSTRUCTION

Carpenters and Builders Library (Sixth Edition, 4 Vols.)
JOHN E. BALL;
revised by JOHN LEEKE

5 1/2 × 8 1/4 Hardcover 1,300 pp. 988 Illus.
ISBN: 0-02-506455-4 $89.95

This comprehensive four-volume library has set the professional standard for decades for carpenters, joiners, and woodworkers.

Volume 1: Tools, Steel Square, Joinery
JOHN E. BALL;
revised by JOHN LEEKE

5 1/2 × 8 1/4 Hardcover 377 pp. 340 Illus.
ISBN: 0-02-506451-7 $21.95

Volume 2: Builders Math, Plans, Specifications
JOHN E. BALL;
revised by JOHN LEEKE

5 1/2 × 8 1/4 Hardcover 319 pp. 200 Illus.
ISBN: 0-02-506452-5 $21.95

Volume 3: Layouts, Foundation, Framing
JOHN E. BALL;
revised by JOHN LEEKE

5 1/2 × 8 1/4 Hardcover 269 pp. 204 Illus.
ISBN: 0-02-506453-3 $21.95

Volume 4: Millwork, Power Tools, Painting
JOHN E. BALL;
revised by JOHN LEEKE

5 1/2 × 8 1/4 Hardcover 335 pp. 244 Illus.
ISBN: 0-02-506454-1 $21.95

Complete Building Construction (Second Edition)
JOHN PHELPS;
revised by TOM PHILBIN

5 1/2 × 8 1/4 Hardcover 744 pp. 645 Illus.
ISBN: 0-672-23377-0 $22.50

Constructing a frame or brick building from the footings to the ridge. Whether the building project is a tool shed, garage, or a complete home, this single fully illustrated volume provides all the necessary information.

Complete Roofing Handbook (Second Edition)
JAMES E. BRUMBAUGH
revised by JOHN LEEKE

5 1/2 × 8 1/4 Hardcover 536 pp. 510 Illus.
ISBN: 0-02-517851-2 $30.00

Covers types of roofs; roofing and reroofing; roof and attic insulation and ventilation; skylights and roof openings; dormer construction; roof flashing details; and much more. Contains new information on code requirements, underlaying, and attic ventilation.

Complete Siding Handbook (Second Edition)
JAMES E. BRUMBAUGH
revised by JOHN LEEKE

5 1/2 × 8 1/4 Hardcover 440 pp. 320 Illus.
ISBN: 0-02-517881-4 $30.00

This companion volume to the *Complete Roofing Handbook* has been updated to reflect current emphasis on compliance with building codes. Contains new sections on spunbound olefin, building papers, and insulation materials other than fiberglass.

Masons and Builders Library
(Second Edition, 2 Vols.)
LOUIS M. DEZETTEL;
revised by TOM PHILBIN

5 1/2 × 8 1/4 Hardcover 688 pp. 500 Illus.
ISBN: 0-672-23401-7 $27.95

This two-volume set provides practical instruction in bricklaying and masonry. Covers brick; mortar; tools; bonding; corners, openings, and arches; chimneys and fireplaces; structural clay tile and glass block; brick walls; and much more.

Volume 1: Concrete, Block, Tile, Terrazzo
LOUIS M. DEZETTEL;
revised by TOM PHILBIN

5 1/2 × 8 1/4 Hardcover 304 pp. 190 Illus.
ISBN: 0-672-23402-5 $14.95

Volume 2: Bricklaying, Plastering, Rock Masonry, Clay Tile
LOUIS M. DEZETTEL;
revised by TOM PHILBIN

5 1/2 × 8 1/4 Hardcover 384 pp. 310 Illus.
ISBN: 0-672-23403-3 $14.95

WOODWORKING

Wood Furniture: Finishing, Refinishing, Repairing
(Third Edition)
JAMES E. BRUMBAUGH
revised by JOHN LEEKE

5 1/2 × 8 1/4 Hardcover 384 pp. 190 Illus.
ISBN: 0-02-517871-7 $25.00

A fully illustrated guide to repairing furniture and finishing and refinishing wood surfaces. Covers tools and supplies; types of wood; veneering; inlaying; repairing, restoring and stripping; wood preparation; and much more. Contains a new color insert on stains.

Woodworking and Cabinetmaking
F. RICHARD BOLLER

5 1/2 × 8 1/4 Hardcover 360 pp. 455 Illus.
ISBN: 0-02-512800-0 $18.95

Essential information on all aspects of working with wood. Step-by-step procedures for woodworking projects are accompanied by detailed drawings and photographs.

MAINTENANCE AND REPAIR

Building Maintenance
(Second Edition)
JULES ORAVETZ

5 1/2 × 8 1/4 Paperback 384 pp. 210 Illus.
ISBN: 0-672-23278-2 $11.95

Professional maintenance procedures used in office, educational, and commercial buildings. Covers painting and decorating; plumbing and pipe fitting; concrete and masonry; and much more.

Gardening, Landscaping and Grounds Maintenance
(Third Edition)
JULES ORAVETZ

5 1/2 × 8 1/4 Hardcover 424 pp. 340 Illus.
ISBN: 0-672-23417-3 $15.95

Maintaining lawns and gardens as well as industrial, municipal, and estate grounds.

Home Maintenance and Repair: Walls, Ceilings and Floors
GARY D. BRANSON

8 1/2 × 11 Paperback 80 pp. 80 Illus.
ISBN: 0-672-23281-2 $6.95

The do-it-yourselfer's guide to interior remodeling with professional results.

Painting and Decorating
REX MILLER and GLEN E. BAKER

5 1/2 × 8 1/4 Hardcover 464 pp. 325 Illus.
ISBN: 0-672-23405-x $18.95

A practical guide for painters, decorators, and homeowners to the most up-to-date materials and techniques in the field.

Tree Care (Second Edition)
JOHN M. HALLER

8 1/2 × 11 Paperback 224 pp. 305 Illus.
ISBN: 0-02-062870-6 $16.95

The standard in the field. A comprehensive guide for growers, nursery owners, foresters, landscapers, and homeowners to planting, nurturing, and protecting trees.

Upholstering
(Third Edition)
JAMES E. BRUMBAUGH

5 1/2 × 8 1/4 Hardcover 416 pp. 318 Illus.
ISBN: 0-02-517862-8 $25.00

The esentials of upholstering are fully explained and illustrated for the professional, the apprentice, and the hobbyist. Features a new color insert illustrating fabrics, a new chapter on embroidery, and an expanded cleaning section.

AUTOMOTIVE AND ENGINES

Diesel Engine Manual
(Fourth Edition)
PERRY O. BLACK;
revised by WILLIAM E. SCAHILL

5 1/2 × 8 1/4 Hardcover 512 pp. 255 Illus.
ISBN: 0-672-23371-1 $15.95

The principles, design, operation, and maintenance of today's diesel engines. All aspects of typical two- and four-cycle engines are thoroughly explained and illustrated by photographs, line drawings, and charts and tables.

Gas Engine Manual
(Third Edition)
EDWIN P. ANDERSON;
revised by CHARLES G. FACKLAM

5 1/2 × 8 1/4 Hardcover 424 pp. 225 Illus.
ISBN: 0-8161-1707-1 $12.95

How to operate, maintain, and repair gas engines of all types and sizes. All engine parts and step-by-step procedures are illustrated by photographs, diagrams, and troubleshooting charts.

Small Gasoline Engines
REX MILLER and
MARK RICHARD MILLER

5 1/2 × 8 1/4 Hardcover 640
ISBN: 0-672-23414-9 $1
Practical inform
maintain
en
e ─cycle
ers, edgers,
owers, emergency
rs, outboard motors, and
ot ent with engines of up to ten
ho epower.

NEW EDITION FOR 1993

Truck Guide Library (3 Vols.)
JAMES E. BRUMBAUGH

5 1/2 × 8 1/4 Hardcover 2,144 pp. 1,715 Illus.
ISBN: 0-672-23392-4 $50.95

This three-volume set provides the most comprehensive, profusely illustrated collection of information available on truck operation and maintenance.

Volume 1: Engines
JAMES E. BRUMBAUGH

5 1/2 × 8 1/4 Hardcover 416 pp. 290 Illus.
ISBN: 0-672-23356-8 $16.95

Volume 2: Engine Auxiliary Systems
JAMES E. BRUMBAUGH

5 1/2 × 8 1/4 Hardcover 704 pp. 520 Illus.
ISBN: 0-672-23357-6 $16.95

Volume 3: Transmissions, Steering, and Brakes
JAMES E. BRUMBAUGH

5 1/2 × 8 1/4 Hardcover 1,024 pp. 905 Illus.
ISBN: 0-672-23406-8 $16.95

DRAFTING

Industrial Drafting
JOHN D. BIES

5 1/2 × 8 1/4 Hardcover 544 pp. Illus.
ISBN: 0-02-510610-4 $24.95

Professional-level introductory guide for practicing drafters, engineers, managers, and technical workers in all industries who use or prepare working drawings.

Answers on Blueprint Reading
(Fourth Edition)
ROLAND PALMQUIST;
revised by THOMAS J. MORRISEY

5 1/2 × 8 1/4 Hardcover 320 pp. 275 Illus.
ISBN: 0-8161-1704-7 $12.95

Understanding blueprints of machines and tools, electrical systems, and architecture. Question and answer format.

HOBBIES

Complete Course in Stained Glass
PEPE MENDEZ

8 1/2 × 11 Paperback 80 pp. 50 Illus.
ISBN: 0-672-23287-1 $8.95

The tools, materials, and techniques of the art of working with stained glass.